America by Design

America by Design

Science, Technology, and the Rise of Corporate Capitalism

David F. Noble

OXFORD UNIVERSITY PRESS
Oxford   New York   Toronto   Melbourne

Oxford University Press
Oxford   London   Glasgow
New York   Toronto   Melbourne   Wellington
Nairobi   Dar es Salaam   Cape Town
Kuala Lumpur   Singapore   Jakarta   Hong Kong   Tokyo
Delhi   Bombay   Calcutta   Madras   Karachi

Library of Congress Cataloguing in Publication Data

Noble, David F.
America by design.

Includes bibliographical references and index.
1. Technology—Social aspects—United States.
2. United States—History.   3. Production (Economic
theory).   4. Capitalism.   5. Science and industry—
United States.   I. Title
T14.5.N6   301.24'3   76-47928
ISBN 0-19-502618-7 pbk.

This reprint, 1980.

For Ched

The gains in technics are never registered automatically in society; they require equally adroit inventions and adaptations in politics; and the careless habit of attributing to mechanical improvements a direct role as instruments of culture and civilization puts a demand upon the machine to which it cannot respond. . . .

No matter how completely technics relies upon the objective procedures of the sciences, it does not form an independent system, like the universe; it exists as an element in human culture and it promises well or ill as the social groups that exploit it promise well or ill. The machine itself makes no demands and holds out no promises: it is the human spirit that makes demands and keeps promises.

Lewis Mumford,
Technics and Civilization

Contents

Foreword

The notion of technological determinism has dominated popular understanding of the industrial revolution. Changes in technology are assumed to have been the principal cause of industrialization, and the whole process is seen purely as a technological revolution. Yet new inventions, new processes, and new applications of scientific discoveries do not in themselves dictate changes in production. Unless accompanied by changes in social relations, especially in the organization of labor, technological changes tend to be absorbed into existing social structures; far from revolutionizing society, they merely reinforce the existing distribution of power and privilege.

In the Middle Ages, technical improvements such as the horse collar, the windmill, and the sawmill did not revolutionize production or weaken the domination of the feudal nobility. In the eighteenth and nineteenth centuries, on the other hand, technological changes did have a revolutionary effect because they were part of a political and social revolution—the overthrow of arbitrary authority, the destruction of mercantilist restraints on trade, the emergence of a landless proletariat, and the separation of the worker from his tools. These upheavals laid the basis of the factory system, rendered handicraft production obsolete, and surrounded the new mode of manufacture with a political climate highly conducive to its development.

The second phase of industrial development—the subject of the present study by David F. Noble—began to unfold when the capitalist, having expropriated the worker's property, gradually expropriated his technical knowledge as well, asserting his own mastery over production. Before the fruits of modern science could be applied to industrial production, the work process had to be split up into hundreds of separate operations performed by workers who no longer understood the relation of one operation to another, and who therefore exercised no control over the process. Social innovation thus laid the basis for technical innovation, and the pioneers of industrial technology played a central role in social engineering as well.

It is for these reasons that Noble insists that technology has to be seen as "social production," and the professional engineer as an expert not only in applied science but in the management of social relations. Noble shows how the engineering profession emerged from the workshop and the school, how the "school culture" wing of the profession succeeded in making academic credentials the prerequisite of admission, and how academically trained engineers replaced "rule-of-thumb" methods with esoteric knowledge over which they themselves had established a monopoly. The professionalization of engineering and the establishment of engineering education as a recognized branch of higher learning forged a link between the corporation and the university that remains unbroken to this day. The corporation thus shifted to the university, an institution partly or wholly financed by the state, such secondary costs of production as personnel training and basic research. Early experiments with in-house research and personnel training, especially in the electrical industry, came under rising criticism and gradually gave way to a "cooperative" movement in which, as at Antioch, academic and industrial training were combined—this being itself merely an intermediate step toward the full-fledged assumption of industrial responsibilities by the modern "multiversity."

The expropriation of the workers' technical knowledge had as a logical consequence the growth of modern management, in which technical knowledge now came to be concentrated. As the scientific management movement split up production into its component procedures, reducing the worker to an appendage of the machine, a great expansion of technical and supervisory personnel took place in order to oversee the productive process as a whole. The old-style entrepreneur, who carried most of his firm's operations in his head, gave way to the university-trained industrial manager, often an engineer. Noble's analysis of the movement of engineers into management undercuts any temptation to see engineers and managers as distinct strata. The engineers failed to develop a point of view of their own, as Veblen had hoped when he predicted that the managers' quest for profits would increasingly conflict with the engineers' devotion to technical efficiency. Instead, the engineers placed their expertise at the service of a productive system efficient in its details but supremely wasteful and irrational in its general tendency.

The industrial system's dependence on military spending and war provides the clearest indication of its overriding irrationality. It is significant that, as Noble shows, engineers regarded the First World War not as a disaster for civilization but as a "unique opportunity" to put their ideas into practice—to bring the corporations, the universities, and the state into closer partnership. Without moralizing, Noble paints a vivid picture of "the unreality of the war which these men fought and the strange lighthearted spirit in which they perceived its horrors"; the development of this cult of war is perhaps the most telling indictment of the productive system over which they presided.

On this last point, however, Noble leaves it to the reader to make his own judgments. His book is deeply critical of industrial society without offending the reader with left-wing pieties, so often associated with radical scholarship. The book reflects a new maturity among scholars on the left, a new rigor and detachment that are quietly transforming historical scholarship. Books like Carol Gruber's *Mars and Minerva,* Stuart Ewen's *Captains of Consciousness,* Harry Braverman's *Labor and Monopoly Capital,* and now David Noble's *America by Design,* have immeasurably enriched historical understanding of the "second industrial revolution" of the twentieth century. They have superseded Marxian and non-Marxian accounts alike.

The impact of this new work has yet to make itself fully felt. When it does, the crudity of earlier interpretations will become evident, as well as the futility of drawing the conventional distinctions between economic, social, and political history. It is the underlying unity of these different lines of development that recent interpretations have exposed: the social transformation implicit in technological change, the transformation of American culture by advertising, the mutual dependence of industry and education. The very complexity of modern society at times almost paralyzes the will to understand it—at no time more effectively than during the supposedly radical decade of the sixties. The revival of the will to master the industrial order—to penetrate its disguises and mystifications—is in many ways the most encouraging development of the seventies.

CHRISTOPHER LASCH

Acknowledgments

Like history itself, the writing of history is a social affair. In a study of this scope, the author must invariably depend not only on the work of countless scholars but also on the advice and labors of many other people. Among those who provided invaluable assistance during the research phase of this enterprise were: E. N. Hartley, Professor of History and Institute Archivist, M.I.T.; Karl Wildes, Professor Emeritus, Department of Electrical Engineering, M.I.T.; Ruth Helmuth, Archivist, Case Western Reserve University; Shonnie Finnegan, Archivist, State University of New York at Buffalo; Ruth T. Wallace, Archivist, University of Pittsburgh; Paul McClure, Archivist, National Academy of Sciences; Harry John, Archivist, National Records Center of the National Archives; Brad Smith, Librarian, University of Rochester; W. Leighton Collins and the late Leslie B. Williams, Director Emeritus and former Director of the American Society for Engineering Education; and Andrey A. Potter, Dean Emeritus, Purdue University School of Engineering. The directors of the libraries at the University of Rochester and Purdue University, Ben Bowman and Joseph Dagnese, generously provided me with office space and thereby rendered my task easier. The initial year of research was funded through a grant from the Manpower Administration of the U.S. Department of Labor. Final revisions of the manuscript were made during my first year of tenure as Mellon Fellow in Humanities and Engineering at M.I.T.

Throughout the five-year period during which this study took shape, I was fortunate to have received advice and encouragement from a number of people. Loren Baritz early counseled me to take my time with the study, sound and unusual advice in this era of tight markets and fast hustles. Among those friendly critics who read at least one version of the entire manuscript and forced me to rethink some of the key themes were: Herbert G. Gutman, Christopher Lasch, Gerard D. McCauley, David Montgomery, Nathan N. Neel, Douglas D. Noble, Martin J. Sklar, Katherine Stone, Charles D. Wendell, William A. Williams, and Harold D. Woodman. Thanks go as well to Eugene D. Genovese, Thomas Parke Hughes, and Russell Jacoby, all of whom read an earlier, extended version of the Introduction; to Wayne Urban, who read the material on educational reform; to Ashbel Green of Knopf, for his editorial artistry; and to Cheryl Noble, for her consistently astute criticism and attention to the more profound questions.

D. F. N.

Introduction

Modern Americans confront a world in which everything changes, yet nothing moves. The perpetual rush to novelty that characterizes the modern marketplace, with its escalating promise of technological transcendence, is matched by the persistence of pre-formed patterns of life which promise merely more of the same. Each major scientific advance, while appearing to presage an entirely new society, attests rather to the vigor and resilience of the old order that produced it. Every new, seemingly bold departure ends by following an already familiar path. This book is an attempt to begin to explain this strange state of affairs: a remarkably dynamic society that goes nowhere.

How has it happened that one product of nineteenth-century industrial capitalism, the private corporation, has been able at once to stimulate and contain another, seemingly more powerful than itself, scientific technology? How has a relatively static, possibly outmoded capitalist social structure managed to endure—indeed, to be reinforced by—a revolution in social production that otherwise has swept away all vestiges of the past, all memories? In search of clues, this study traces the interwoven history of the twin forces which together gave shape to modern America—scientific technology and corporate capitalism—by focusing upon their common medium, modern engineering. At the same time, it poses anew some larger questions that have long preoccupied social theorists: to what extent is social development attributable to advances in technics, and in precisely what manner are the two linked?

For most historians of the modern period, a more or less crude determinism has become a staple of explanation. Whether they have been trying to account for the rise of large-scale industrial units, the development of educational institutions, or changes in ethics, historians commonly have alluded to the demands of modern technology as a causative factor. Rarely, however, have they gone on to examine critically those demands, and, as a result, modern technology has remained a phantom, a conveniently vague device for explaining historical developments by explaining them away. For example, a major characteristic that distinguishes the present century from the one which preceded it is less the actual presence of modern technology—which has marked both centuries—than the degree to which this special activity has come to define the activity of the society as a whole. How is this transformation to be explained? How did a nineteenth-century revolution in technics, whereby science became wedded to the useful arts, give rise to the technological society of the twentieth century? Did the revolution in technics spawn a social revolution?

It has become fashionable to account for the myriad social changes attendant upon the extension of technological activity tautologically, by simple reference to the supposedly essential nature of that activity: it expands. Thus, a stock device of recent social analysis is to view modern technology as though it had a life of its own, an internal dynamic which feeds upon the society that has unleashed it. Propelled according to its own immanent logic and operating through witting and unwitting human agency, it ultimately outstrips the conscious activities which gave birth to it, creating a society in which people are but functional parts of the mechanism. There is a core of truth in the view —a common theme in modern mythology—that human creations tend to assume an existence independent of their creator's will. Problems arise, however, when, in more or less subtle ways, the metaphor is substituted for history, when the rich complexity of the social process is reduced to the inexorable logic of a formalistic technology. Artificially abstracted from the world in which people actually live, such a conception distorts both technology itself and the society which gives it meaning.

Unfortunately, such facile explanations of history do enjoy wide currency, and are daily reinforced by the common conceptual habit of distinguishing between "technology," on the one hand, and "society" (or "culture"), on the other, as if the two were made of altogether different stuff. Within such a framework, it follows that, since society contains all that is human, technology must be something other than human, a disembodied historical force impinging upon the affairs of men.* Not all social theorists, of course, have gone so far.

The classical Marxian view of the role of technology in capitalist society is more subtle and compelling. Here the fundamental relationship between society (social relations) and technology (forces of production) is a dialectical one, and thus, in essence, an identity, with the two being but different aspects of the single process of social production. Insofar as the emerging capitalist social relations between classes make possible the creation of a social surplus, they make possible as well the development of more sophisticated productive forces which both reflect and reinforce those social relations. The internal evolution of these same productive forces, however, entails a transformation of the actual social activity of production—material changes which steadily lay the foundation for a more humane social order, and thus pose a challenge to those social relations. The historical process of social production embraces this reciprocal interrelationship between productive forces and social relations; in producing for itself, society is also producing and reproducing itself, its work habits, institutions, relations between people, and dominant perceptions of reality. The crucial factor in Marx's theory of technological-social change is the twofold significance of the development of the productive forces: they both reinforce the existing social order and undermine it. Social relations and forces of production are thus at the same time in correspondence and in contradiction.

In Marx's view of industrial capitalist society, the pace of social production is quickened, owing to the competitive struggle for capital accumulation and the demand for ever more potent and profitable means of production. The particular social relations of capitalism which make this accumulation possible, relations grounded in the capitalist's ownership of the means of production (including his workers' labor time) and his private appropriation of the surplus social product, make possible as well the rapid development of sophisticated forces of production: scientific, labor-saving, socially integrated. And these new

---

*Such habits of thinking are important to historians, not as explanations of history in themselves but rather as examples of a recurring aspect of history: the mystification of history. Studying them as such does more than merely fill out the record with contemporary perceptions; it guards against their being replicated by the historian himself. For it is the primary task of the historian to demystify history, to render it intelligible in human rather than in super-human or non-human terms, to show that history is a realm of human freedom as well as necessity. By describing how people have shaped history in the past, the historian reminds us that people continue to shape history in the present.

forces of production in turn hold out the promise of a new social order, one marked not by the capitalist trauma of overproduction, economic crises, wasteful competition, routine exploitation of the many by the few, mindless detail labor, and physical drudgery, but rather by collective ownership of the means of production by society as a whole, cooperative enterprise, rational allocation of resources to meet social needs, and the fuller human development of the social individual. Thus, in the light of these new forces of production, the relations of capitalism tend increasingly to appear anachronistic—stark and oppressive vestiges of a more primitive past, fetters upon further social development. Given the requisite revolutionary consciousness on the part of the exploited class—itself a recognition of the disparity between the actuality and the potential of social production—the contradiction erupts into revolution and the old capitalist integument is sloughed off. In Marx's words, "Forces of production and social relations—two different sides of the development of the social individual—appear to capital as mere means, and are merely means for it to produce on its limited foundation. In fact, however, they are the material conditions to blow this foundation sky-high."[1]

Marx's essentially liberatory view of modern technology was echoed by the technocratic Veblen. For him the critical contradiction of industrial capitalism was that between business and industry, the price system and the engineer, between the irrational dictates of the market and the supremely rational imperatives of science (tied to the primitive and noble instinct of workmanship). Like many of his contemporaries who were also critical of the existing order, Veblen was cautiously confident that the superiority of scientific industry would lead ultimately to a "social overturn" of the market society. General Electric's chief engineer, the immigrant German Social Democrat Charles Steinmetz, subscribed to a not uncommon variant of this theme, in which the rise of the giant corporate firm was construed to represent the rational reorganization and centralization of the means of production, and thus a necessary prelude to socialism.[2]

Other twentieth-century observers have been far less sanguine about the emancipatory potential of modern technology. Max Weber and, more recently, Jacques Ellul, while agreeing that capitalism would perhaps be eclipsed or at least be diminished in significance by technological progress, have warned that such progress itself—the inexorable advance of *technique* and functional bureaucracy—would generate a quiet, more subtle transformation, weaving a paralyzing web of instrumentality. Not alone capitalism but the modern technical and administrative apparatus of production forged by capitalism would enslave modern man. In a similar vein, Lewis Mumford has argued persuasively that throughout the history of Western civilization, undue emphasis upon the development of technology (and its correlate, warfare) at the expense of other, equally important activities of human society has resulted in the atrophy of many human capacities and an overwhelming preoccupation with, and obeisance to, the myth of the machine.[3]

Still other critics have tried to reconcile this skeptical view of technological progress with the classical Marxian theory. Georg Lukács, for example, attempted to explain why sufficient revolutionary consciousness never emerged in Europe, despite the requisite advances in productive forces, by examining the mystifying and numbing effect modern capitalist production had on the mind of the worker, the twofold process of alienation and reification described earlier by Marx. Max Horkheimer and Theodor Adorno sought to locate the seeds of twentieth-century authoritarianism not merely in the history of capitalism but in the scientific revolution of the Enlightenment as well. Herbert Marcuse, perhaps more clearly than the others, confronted the dilemma head on. While criticizing Weber for his ahistorical use of the concept of rationalization, and arguing—as if to convince himself—that the apparent demands of modern technology were in reality but highly refined forms of capitalist domination, Marcuse posed the critical question: How has technology become solely a vehicle of class hegemony and not also a vehicle of liberation? "What is at stake," he wrote elsewhere, "is the compatibility of technical progress with the institutions in which industrialization developed." What has become of the fundamental contradiction between social relations and forces of production? Why has not the tension between them become inescapably apparent? How has it happened instead that "with technical progress as its instrument, . . . unfreedom is perpetuated and intensified," the only "novel feature" being the "overwhelming rationality in this irrational enterprise"? This, in essence, is the question that concerns us here.[4]

This study examines the concurrent emergence of modern technology and the rise of corporate capitalism as two sides of the same process of social production in America. It assumes no rigid separation between the forces of production and social relations, but rather takes them, as Marx did, to be fundamentally interrelated: to study the one is to study the other. For technology is not simply a driving force in human history, it is something in itself human; it is not merely man-made, but made of men. Although it may aptly be described as a composite of the accumulated scientific knowledge, technical skills, implements, logical habits, and material products of people, technology is always more than this, more than information, logic, things. It is people themselves, undertaking their various activities in particular social and historical contexts, with particular interests and aims. As Marcuse observed, "We do not ask for the influence or effect of technology on the human individuals. For they are themselves an integral part and factor of technology, not only as the men who invent or attend to machinery, but also as the social groups which direct its application and utilization. Technology, as a mode of production, as the totality of instruments, devices, and contrivances which characterize the machine age, is thus at the same time a mode of organizing and perpetuating (or changing) social relationships, a manifestation of prevalent thought and behavior patterns, an instrument for control and domination."[5] An essentially human phenomenon, technology is thus a social process; it does not simply stimulate social development from outside but, rather, constitutes fundamental social development in itself: the preparation, mobilization, and habituation of people for new types of productive activity, the reorientation of the pattern of social investment, the restructuring of social institutions, and, potentially, the redefinition of social relationships.

This social process called technology, moreover, does not exist simply for itself, in a world of its own making. It is, rather, but one important aspect of the development of society as a whole. Since those who comprise society are at the same time the human material of which technology is composed, technology must inescapably reflect the contours of that particular social order which has produced and sustained it. And like any human enterprise, it does not simply proceed automatically, but rather contains a subjective element which drives it, and assumes the particular forms given it by the most powerful and forceful people in society, in struggle with others. The development of technology, and thus the social development it implies, is as much determined by the breadth of vision that informs it, and the particular notions of social order to which it is bound, as by the mechanical relations between things and the physical laws of nature. Like all others, this historical enterprise always contains a range of possibilities as well as necessities, possibilities seized upon by particular people, for particular purposes, according to particular conceptions of social destiny.

The primary thesis of this book is that the history of modern technology in America is of a piece with that of the rise of corporate capitalism. Both contributed to a transformation of the *modus operandi* of industrial capitalism—the one providing the wherewithal for unlimited productive growth by implicating science in the production process, the other offsetting the destructive tendencies inherent in an unchecked competitive market economy by making possible the regulation of production, distribution, and prices—but neither overcame the basic social relations of capitalist society, the relations between the capitalist, who owns and controls the means of production, and the worker, who must sell his labor for a wage in order to survive. Indeed, if anything, these relations and the ideology fostered by them were intensified by the transformation, capitalism having been strengthened, not weakened, by the change. Merely a "technological veil," as Marcuse phrased it, was drawn over "the reproduction of inequality and enslavement."[6]

The present study tries to account for this change without change. It suggests that one partial explanation for the survival of capitalist social relations, despite the most dramatic advances in productive forces, has to do with the nature of modern engineering, the source of those technological advances.* For the professional engineers who emerged during the second half of the last century in America as the foremost agents of modern technology became as well the agents of corporate capital. Thus, from the outset, they hardly proceeded accord-

---

*The claim here is a modest one. Because they embodied the union of business with science, engineers naturally sought, in their technical work, to resolve the tension between the dictates of the capitalist system and the social potentials implicit in technological development. For them, technology was exclusively a means of strengthening capitalism, rather than something which pointed beyond it. And since they pretty much defined what form technological advances would take, technology tended to evolve in close conformity with capitalist requirements. This is not to say, of course, that they were altogether successful in this regard, that technological progress no longer held out the promise of a new society. Indeed, it was partially for this reason, because it did, that the engineers steadily extended the range of their professional activities to include the deliberate fostering and strengthening of the social relations of corporate capitalism, to encourage working people to work within the system, rather than against it, for a better life. It should perhaps also be emphasized that the efforts of the engineers cannot alone account for the survival of capitalism in America. Other factors played an important part: the remarkable economic growth made possible by America's increasingly dominant military position in the world and thus her access to global markets and industrial resources; the politicization of the economy and the importance of the state as a stabilizing force; and the deliberate creation of a consumer culture, through advertising, to absorb and defuse potentially revolutionary energies. It is, of course, impossible to conduct a "controlled experiment" to assess the relative significance of these various factors. All that is being suggested here is that the role of the engineers in the creation of modern corporate industry must be taken into account before any satisfactory explanation can be attempted.

ing to the dictates of some logically consistent "technical reason," blindly advancing the frontiers of human enterprise, but rather informed their work with the historical imperatives of corporate growth, stability, and control: as their technology progressed, so too did the science-based industrial corporations which they served.

The engineers, moreover, went a step further to ensure that their technical work meshed with the imperatives of corporate social relations; rather than restricting their attention to technical matters, they consciously undertook to structure the labor force and foster the social habits demanded by corporate capitalism.* In short, they ventured to design a new (yet old) social order, one dominated by the private corporation and grounded upon the regulated progress of scientific technology. Forces of production and social relations, industry and business, engineering and the price system—the two poles in the dialectic of social production—collapsed together in the consciousness of corporate engineering, under the name of management.

The first part of this book examines the process whereby, in Marx's words "modern industry . . . makes science a productive force distinct from labor and presses it into the service of capital." It traces this capitalist wedding of science to the useful arts along three intersecting paths—the rise of science-based industry, the development of technical education, and the emergence of the professional engineers—showing how each of these nineteenth-century developments reflected and contributed to the social process of technology as corporate social production. Through consolidation, patent monopoly, and merger, the science-based electrical and chemical industries came, by the turn of the century, to be dominated by a handful of giant firms. Through control over the educational process and licensing, the professional engineers gradually gained a monopoly over the practice of scientific technology. And through the massive employment of technically trained people, the industrial corporations secured a monopoly over the professional engineers (who, "in following their own reason," as Marcuse observed, "followed those who put their reason to profitable use"). In this way,

---

*The engineers thus worked to meet the interwoven imperatives of scientific technology and corporate capitalism in two ways: in their technical work proper, as designers of the machinery and processes of production; and in their broader activities, as managers, educators, and social reformers. This book focuses upon the latter. Another study, in progress, looks more closely at the technical work itself (in particular, the development of automated machine tools) in order to determine precisely how extra-technical considerations, such as managerial control over the workforce, inform machine design and deployment.

modern technology became a class-bound phenomenon, the racing heart of corporate capitalism.[7]

Some engineers, moreover, went beyond this merely objective participation in class control. As educators in the technical schools, leading members of the professional societies, and, most importantly, managers and executives in the new science-based industrial corporations, they ultimately came to view the various activities in which they participated as comprising a unified whole. In their minds, the nineteenth-century process of social production that had produced the corporations, the technical schools, and the professional engineers had now become conscious, a world-view of corporate reform. Part Two describes their subsequent efforts to design America in accordance with this new perception, to create the apparatus for the corporate-sponsored progress of modern technology and simultaneously to establish the patterns of twentieth-century routine which would ensure the stability and growth of corporate capitalism.* Their activities covered a wide range, including industrial and scientific standardization, patent reform, the organization of industrial and university research, and the transformation of both public-school and higher education.

Like all human undertakings, the work of the corporate reformers did not proceed automatically, according to plan. It took the form of an intense struggle (one which has not altogether ended) which was fought out in the factories, the halls of government, the courts, the schools, and the streets. But these engineers held some important advantages which their opposition lacked. First, they enjoyed great social power and prestige, based upon their exalted positions within the new corporations and a web of political, economic, social, and family ties among the country's propertied elite. Second, and of equal importance, they marched under a banner which confounded their opposition: the banner of science. The corporate engineers proceeded in the realm of social reform much as they had in the shops and laboratories—systematically and, to their minds, scientifically. Whatever their particular purpose, they sought their guide in the authority of science and defended their actions in the name of science.

Thus, predictably, as they became conscious of the process of social production that was modern technology, the engineers at once undertook to fashion for themselves a technology *of* social production: mod-

*It should perhaps be stressed at the outset that—as William Appleman Williams reminds us—consciousness of purpose, as it is described in this volume, is not the same as conspiracy.

ern "scientific" management, the focus of the final chapter. In modern management, these engineers who had been trained in science and weaned upon large-scale corporate enterprise fused the imperatives of corporate capitalism and scientific technology into a formal system. Representing a shift in engineering focus from the natural to the social realm, from productive forces to social relations, modern management constituted a deliberate attempt to ease the tension between the two, making both fit within the confines of corporate society. Moreover, as these engineers became managers in industry, private capital itself began to assume the appearance of modern technology, the management experts lending to the power of capital the sanction of objective science. Not alone the actual machinery of production but the entire bureaucratic operation of corporate enterprise took on the guise of an efficient, well-oiled mechanism—the very embodiment of technical reason—against which individual opposition could not but appear "irrational."

Most importantly, therefore, this book is an attempt to lift the technological veil. For as technology has increasingly placed the world at people's fingertips, those people have become less able to put their finger on precisely what technology is. A general mystification evolved just as modern technology was becoming a dominant aspect of social life. Paradoxically perhaps, the popular perception of technology as an autonomous prime mover in history, independent of human will, became widespread at precisely the time when the social process that is modern technology in fact began to come under the conscious control of human authority, in the specific form of private corporate capital.

The imperatives of the automatic market, professional specialization, and rationalized management, coupled with the corporate monopolization of technological intelligence, have contributed to the appearance of technology as an autonomous force in history. Insofar as social analysis merely replicates such a perception of reality without penetrating beneath the apparent technological necessity, it further contributes to the general mystification and reinforces the particular social relations which are thereby obscured. This study is an attempt to reintegrate the mystified conception of technology with the actual activities from which it has been abstracted, to reveal it as the human enterprise that it is, and can be, by substituting history for metaphor.

Part One

Technology as Social Production

Industry, Education, and Engineers

I

The Wedding of Science
to the Useful Arts – I

The Rise of Science-
Based Industry

Science and capitalism press forward by nature. When the two are combined, the pace of social production quickens into a sustained drive. With scientific investigation and discovery as the engine of competitive innovation, capitalism becomes revolutionary at the core and competitors are compelled routinely to anticipate the future in order to survive. Those who are able to harness science itself, therefore, and direct it for their own ends, have gained a considerable advantage. For them, the competitive task of anticipating the future has become easier since they now have the means for determining that future themselves. So it was with the alchemists of the late nineteenth century, and their successors of the twentieth, who undertook to transform science into gold and, in the process, gave rise to modern science-based industry.

Benjamin Franklin, at once a prominent scientist and a successful man of practical affairs, proclaimed toward the end of his life that science would serve the coming century as "handmaiden to the arts." A generation later, in a series of lectures at Harvard, the physician Jacob Bigelow introduced into general usage the word which came to signify that union: "technology." "There has probably never been an age," he observed, "in which the practical applications of science have employed so large a portion of talent and enterprise of the community, as in the present. To embody ... the various topics which belong to such an undertaking, I have adopted the general name of Technology, a word sufficiently expressive, which is found in some of the older dictionaries, and is beginning to be revived in the literature of practical men at the present day. Under this title is attempted to include an account ... of the principles, processes, and nomenclatures of the more conspicuous arts, particularly those which involve applications of science, and which may be considered useful, by promoting the benefit of society, together with the emolument of those who pursue them."[1]

Both Franklin and Bigelow were ahead of their times. The small world in which science and the useful arts were first brought together, a world inhabited primarily by gentlemen like themselves who cultivated an interest in both, remained restricted throughout much of the nineteenth century. For the most part, the scientists who occupied the universities pursued what they called natural philosophy and aimed in their investigations toward the discovery of the loftier metaphysical truths about the nature of the universe and the role of the mortal within it; in general, they had nothing but disdain for the practical application of their work and its correlate, money-making. The practical men who inhabited the shop and the countinghouse, on the other hand, had little use for theories and speculation; in their all-consuming search for utility and profit, they relied exclusively upon traditional lore and the hard-won lessons of experience.

Thus, the weight of centuries of separate, isolated traditions impeded the union of science to the useful arts foreseen by Bigelow and Franklin. When this union did occur on a scale significant enough to alter the nature of both science and industry, it was only because the arts had become the spur to science rather than the other way around. The scientific community overcame its Platonic prejudices, at first haltingly and then with a rush, only after the men in the workshops and offices of industry had begun to appropriate the haphazard discoveries of science for their own practical and pecuniary ends. The locus of modern technology in America was thus the domain of "manufactures" rather than the realm of science, and modern science-informed technology was characterized from the outset by the overriding imperative of manufactures: profitable utility. From the start, modern technology was nothing more nor less than the transformation of science into a means of capital accumulation, through the application of discoveries in physics and chemistry to the processes of commodity production.

Science was introduced into the world of production through the combined efforts of innovative craftsmen in the shops who had read about the discoveries of university-based science in the popular scientific publications of the day, and prescient capitalists who had learned from the same literature, from inventive employees, or from university-based associates. Where the former saw in new scientific knowledge about the nature of the material world the basis for new or refined methods of production, the latter recognized the potential for enhancing profitability. Of course, it was often true that these two visions took shape as one in a single mind: the capitalist was frequently an inventor of sorts, while the inventive craftsman shared not a little of the entrepreneurial spirit of the capitalist. The distinction is drawn merely to emphasize the fact that, in a capitalist society, the development of modern technology presupposed both. The efforts of the inventor remained stillborn without access to or collaboration with a source of capital.[2]

Modern science-based industry—that is, industrial enterprise in which ongoing scientific investigation and the systematic application of scientific knowledge to the process of commodity production have become routine parts of the operation—emerged very late in the nineteenth century. It was the product of significant advances in chemistry and physics and also of the growing willingness of the capitalist to embark upon the costly, time-consuming, and uncertain path of research and development. This willingness reflected both the intensifying demand to outproduce competitors at home and abroad and the unprecedented accumulation of sufficient surplus capital—the product of traditional manufactures, financial speculation, and industrial consolidation—with which to underwrite a revolution in social production. As Marx observed, at a time when such development had barely begun to unfold, science-based industry sees the light of day "only once heavy industry has reached an advanced stage, when machinery itself has yielded very considerable resources. Invention then becomes a branch of business, and the application of science to immediate production [aims] at determining the inventions at the same time it solicits them."[3]

Of the new industries which emerged between 1880 and 1920 and transformed the nature of social production in America, only two grew out of the soil of scientific rather than traditional craft knowledge: the electrical and chemical industries. The creation of both presupposed and stimulated advances in physics and chemistry and would have been unthinkable without some basic knowledge about the behavior of atoms, molecules, gases, light, magnetism, and electricity. Of the two, the electrical industry arose first, in the 1880s; by the turn of the century it had become a major force in the world of production, dominated by a handful of large, powerful, and dynamic corporations. Although a healthy heavy-chemical and powder industry existed throughout the second half of the nineteenth century in the United States, it was not until World War I, with the seizure of German-owned patents and the establishment of a protective tariff, that a domestic dyestuff industry, and thus a chemical industry proper, was established on a par with the electrical industry.

As the first science-based industries in the country, the electrical and chemical industries set the pattern of production and management for modern industry as a whole. Moreover, they produced the people—industry-minded physicists and chemists and, especially, the electrical and chemical engineers—who would carry the scientific revolution into the older and the new industries: extractive, petroleum, steel, rubber, and, most important of all in terms of American economic development, automotive. In all of these industries the systematic introduction of science as a means of production presupposed, and in turn reinforced, industrial monopoly. This monopoly meant control not simply of markets and productive plant and equipment but of science itself as well. Initially the monopoly over science took the form of patent control—that is, control over the *products* of scientific technology. It then became control over the *process* of scientific production itself, by means of organized and regulated industrial research. Finally it came to include command over the social prerequisites of this process: the development of the institutions necessary for the production of both scientific knowledge and knowledgeable people, and the integration of these institutions within the corporate system of science-based industry. "The scientific-technical revolution," as Harry Braverman has explained, "cannot be understood in terms of specific innovations. . . ." Rather, it "must be understood in its totality as a mode of production into which science and exhaustive engineering have been integrated as part of ordinary functioning. The key innovation is not to be found in chemistry, electronics, automatic machinery . . . or any of the products of these science-technologies, but rather in the transformation of science itself into capital."[4]

In the period between 1880 and 1920 the first and second generations of men who created and ran the modern electrical industry formed the vanguard of science-based industrial development in the United States. These were the people who first successfully combined the discoveries of physical science with the mechanical know-how of the workshop to produce the much-heralded electrical revolution in power generation, lighting, transportation, and communication; who forged the great companies which manufactured that revolution and the countless electric utilities, electric railways, and telephone companies which carried it across the nation. At the same time, it was they who introduced the now familiar features characteristic of modern science-based industry: systematic patent procedures, organized industrial-research laboratories, and extensive technical-training programs.

The source of the electrical revolution was the electrical manufacturing industry, the producer of the complex machinery and equipment which made it possible. By the turn of the century three companies had come to dominate this industry. General Electric and Westinghouse had grown out of the interrelated development of electric lighting, power, and traction. The American Telephone and Telegraph Company had arisen as both product and producer of the telephone.

Although the arc light was demonstrated as early as 1808 by Sir Humphry Davy, commercial development had to await the necessary improvements in constant-current dynamos. The early pioneers in arc lighting were therefore also pioneers in the development of better and cheaper sources of current, primarily through improvements upon Faraday's dynamo of 1831. The major breakthroughs, technically speaking, came in the 1870s. The dynamo developed by Charles F. Brush fared the best in a comparative test held by the Franklin Institute in 1877, and Brush went on to make improvements in the arc light, setting up the California Electric Light Company and the Brush Electric Company, in 1879 and 1881, respectively.

Neither the arc lamp nor the dynamo proved patentable in court tests, however, and, as a result, the manufacture of arc-lighting systems became fiercely competitive. The most successful challenger to Brush was Elihu Thomson, a Philadelphia high-school "professor" and electrical wizard who had teamed up with a colleague, E. J. Houston, to form the American Electric Company in 1880. A few years later the New Britain, Connecticut, businessmen who had provided the initial financial support for the venture sold the company to a group from Lynn, Massachusetts, headed by Charles A. Coffin, and its name was changed to the Thomson-Houston Electric Company. Primarily because of the inventive genius of Thomson and his skill at managing his small band of researchers, the company soon developed the best arc-lighting system on the market. At the same time it patented all the improvements for which a patent could be granted—current regulator, air-blast commutator, and lightning arrester—in an effort to monopolize the infant industry. In addition to securing patents on its own inventions, the Thomson-Houston company, under the strong leadership of Coffin and with the solid financial backing of Boston financiers, initiated a policy of either buying out or merging with competitors, with the purpose of both eliminating rivals and securing control over essential patents. According to the company's annual report of February 1890, "From 1882 until 1888 active competition was experienced from numerous manufacturing companies of more or less strength, which was so severe as to seriously affect the profits of the business. . . . Since 1887, however, alliances have been made with most of the manufacturing companies controlling important patent systems."[5] By 1890, then, through the success of its own lighting system and the aggressive merger policy made possible by ample financial support, Thomson-Houston had become the dominant firm in the arc-lighting industry.

The development of incandescent lighting also dates back to the pioneering experiments of Sir Humphry Davy, but it was only with the appearance of Herman Sprengel's mercury vacuum pump in 1865 and the improvements in the dynamo of the 1870s that incandescent lighting became commercially viable. The central figure in this work, of course, was Thomas Edison. After achieving considerable financial success in the manufacture of stock tickers for Western Union, Edison had set himself up in Menlo Park, New Jersey, with a well-equipped laboratory and a permanent staff "to spend full-time in making new inventions." By deliberately transforming invention from a haphazard phenomenon into a routine, businesslike enterprise, he expected to make "a minor invention every ten days, and a big one every six months or so." Edison's investigations and experiments were inextricably informed by economic considerations; as in all engineering work, the profit motive did not lie behind the inventive activity but was bound up with it.* In 1877 he turned his attention to the problem of incandescent lighting. As one historian of modern industry has observed, Edison "chose electric lighting over other possible areas such as the telegraph, the telephone, and the phonograph because he felt there were bigger rewards in lighting. . . . He decided that he would have to set the price for light equivalent to the price of gas light. With the price thus fixed, the path to greatest profits was to reduce costs as much as possible. To lower costs, he made a series of inventions which were designed to economize the materials used in his incandescent lighting system."[6]

To realize his plans for a commercially viable system, Edison required financial support beyond his own resources and formed a company, the Edison Electric Light Company, toward that end. His backers included several bankers, the president and general counsel of

---

*Edison did not fit the mythical image of the humble tinkerer. As Thomas P. Hughes has made clear, Edison was familiar with the latest scientific work and conceived his projects "systematically," consciously incorporating economic factors within technical designs.

Western Union, and a J. P. Morgan partner, Egisto P. Fabbri, who served as director and treasurer of the company until 1883. Edison's ultimate success depended heavily upon the Morgan support; although not an official of the Edison company himself, J. P. Morgan took an active interest in Edison's work, having visited him at Menlo Park in 1879. During the 1880s the Morgan interests invested heavily in Edison lighting companies in Boston and New York, and Morgan was one of the first New York residents to have his home supplied with its own electric-lighting plant. In 1892 Morgan played a leading role in the formation of the General Electric Company and served on that corporation's executive board until his death in 1913.[7]

By October 21, 1879, Edison had produced the first successful incandescent electric light and a scant three years later, thanks to the ample financial support he enjoyed, had begun operation of the Pearl Street central power station in New York, under the auspices of the Edison Electric Illuminating Company. In 1889 the various Edison manufacturing companies, which had been established to produce the lamps, dynamos, and street mains, were merged, together with the Sprague Electric Railway and Motor Company (a pioneer in electric traction), to form the Edison General Electric Company. At this point Edison himself withdrew from active participation in the electrical manufacturing business.

The most important impetus behind the formation of the General Electric Company, according to Thomson-Houston lawyer Frederick P. Fish, was the patent situation. (Fish was one of the country's leading patent attorneys and he was soon to become president of AT&T, largely as a result of his prowess in handling that company's patent cases.) Edison General Electric, through its acquisition of the Sprague company, controlled some vital electric railway patents in addition to the important Edison patents. Thomson-Houston, on the other hand, held some key patents needed by Edison General Electric, particularly in lighting. Because each company owned patents required by the other, neither could develop lighting, railway, or power equipment without fear of infringement suits and injunctions. To get beyond this impasse, the two companies merged in 1892, to form the General Electric Company.[8]

The major competitor of General Electric in the electrical manufacturing field was Westinghouse. George Westinghouse had invented the air brake in 1869 and founded the Westinghouse Air Brake Company in Pittsburgh the same year. In 1881 he set up the Union Switch and Signal Company to attack the problems of automatic switching and signaling devices, and these led him into the electrical manufacturing field. He entered the incandescent-lighting market after he had secured the services of an electrical engineer, William Stanley, who held some lamp and dynamo patents. In 1885 Westinghouse and Stanley revolutionized the electrical industry with the introduction of their alternating-current system, a new method of transmitting electricity. Having purchased the key patents for the system from Nikola Tesla, Westinghouse dominated the alternating-current field for a few years, until Thomson-Houston began successfully producing its own such system.[9]

By the 1890s two large companies, General Electric and Westinghouse, monopolized a substantial part of the American electrical manufacturing industry, and their success and expansion had been in large measure the result of patent control. In addition to the original patents upon which the companies were established, they had gained control over others, and hence over large parts of the electrical manufacturing market, through acquisition of the patent rights of individual inventors, acquisition of competing firms, mergers with competitors, and the systematic and strategic development of their own patentable inventions. As these two giants steadily eliminated smaller competitors, the competition between them intensified. By 1896 over three hundred patent suits between them were pending and the cost of litigation for each company was enormous. In that year, therefore, after a few years of preliminary negotiation, the two companies entered into a patent agreement under a joint Board of Patent Control. The immediate motivations behind the agreement were recent decisions in favor of General Electric's claims on certain railway apparatus and in favor of Westinghouse's claims on Tesla's patents on the rotating field. Now the companies agreed to pool their patents, with General Electric handling 62.5 percent of their combined business. Thus yet another means of patent and market control had been developed: corporate patent-pooling agreements. Designed to minimize the expense and uncertainties of conflict between the giants, they greatly reinforced the position of each vis-à-vis lesser competitors and new entrants into the field.[10]

The emergence of the electrical communications industry during the same period similarly involved the expansion of some companies, primarily through strategies of patent monopoly, at the expense of others. The electromagnetic telegraph, patented by Samuel F. B. Morse in 1837, had become commonplace by the time Alexander Graham Bell and his assistant Thomas Watson began investigating the possibilities of sending several messages over the same wire by perfecting the multiple telegraph. Bell's experiments convinced him that he could transmit the human voice by wire, and in 1876, at the Philadelphia Centennial Exposition, he was able to demonstrate his new "talking machine" to an enthusiastic and dumbfounded audience. In the two years that followed, Bell obtained his two patents for the telephone.

Even before his inventions had reached the stage of patent application in Washington, however, Bell had started to exploit them with the creation of the Bell Patent Association, the nucleus of the first Bell industrial organization. Between 1876 and 1894 Bell and his Boston investors, Thomas Sanders and Gardiner G. Hubbard, "applied themselves to the exploitation of a desirable monopoly" based upon the Bell patents. Hubbard, Bell's father-in-law, brought to the venture valuable experience as well as money. As an attorney with McKay Shoe Machinery Company, which became the cornerstone of the United Shoe Machinery Company in 1899, Hubbard had learned the techniques of patent control and the value of leasing machines and collecting royalty on them, rather than selling them outright. The sale of service rather than patented machines and the vigilant surveillance and domination of the field through patent monopoly thus underlay the policies of the Bell system from the outset. Although the Bell people were able to overcome initial competition from Western Union in 1878, they were soon confronted with rival telephone manufacturing and service companies which sprang up everywhere between 1879 and 1893. While the validity of the Bell patents was finally sustained in 1888 by the U.S. Supreme Court, the Bell organization was nevertheless compelled to institute over six hundred infringement suits in order to protect them.[11]

The life of a patent is seventeen years, after which time the invention becomes public domain. No one understood this fact better than Theodore N. Vail, who became general manager of the National Bell Telephone Company in 1879. From the outset he undertook to "occupy the field," as he termed it, so as to ensure that the Bell company would outlive its original patents. In the Western Union suit hearings he explained how his company proceeded to surround itself "with everything that would protect the business, that is the knowledge of the business, all the auxiliary apparatus; a thousand and one little patents and inventions with which to do the business which was necessary, that is what we wanted to control and get possession of." In order to do this, Vail elaborated, the company very early established an "experimental" or "engineering" department "whose business it was to study the patents, study the development and study these devices that either were originated by our own people or came in to us from the outside. Then early in 1879 we started our patent department, whose business was entirely to study the question of patents that came out with a view to acquiring them, because . . . we recognized that if we did not control these devices, somebody else would." Frederick P. Fish put the matter succinctly in an address to the American Institute of Electrical Engineers: "The businessmen of the organization knew that . . . every added invention would strengthen their position not only during the seventeen years of the main patent but during the seventeen years' term of each and every one of the patents taken out on subsidiary methods and devices invented during the progress of commercial development. [Therefore] one of the first steps taken was to organize a corps of inventive engineers to perfect and improve the telephone system in all directions, . . . that by securing accessory inventions, possession of the field might be retained as far as possible and for as long a time as possible."[12]

In the Bell system, as in General Electric and Westinghouse, such efforts were standard operating procedure, institutionalized as the very core of the business. And, like the other manufacturing companies, the Bell organization enlarged its patent holdings through purchase of patents and competing companies. The corporate structure of the Bell system was changed a number of times between 1877 and 1900. The American Telephone and Telegraph Company, established in 1885 as a subsidiary of American Bell, acquired the assets of the parent company, while the Western Electric Company, organized in 1881, became the manufacturing subsidiary of the system. After 1894, with the expiration of the original Bell patents, competition within the telephone industry intensified, and by 1907 the independent companies had three million telephones in service compared with AT&T's 3.1 million.[13] At this critical juncture J. P. Morgan took over the reins of the company, as he had done at General Electric, and, with Vail at the helm, instituted a successful policy of merger with independents.

Although the industry's own historians are fond of tracing the lineage of the American chemical industry back into the seventeenth century, it is more realistically a product of the rapid expansion of manufactures in the second half of the nineteenth century. And although chemical companies, like steel, textile, paper, railroad, and petroleum concerns, early established the practice of calling upon trained chemists for expert advice, the industry did not really become science-based, with organized research and development, until the establishment of the domestic dyestuff industry during World War I.

Likewise, it was not until World War I that consolidations within the industry gave rise to enterprises with positions as commanding as those of the firms which dominated the electrical industry.

The chemical industry emerged primarily to supply manufacturers with the chemicals basic to their trades: acids, alkalies, and inorganic salts.[14] Before 1850, along with small producers of medicines and gunpowder, the makers of sulfuric acid dominated the industry. By the 1880s a domestic alkali industry had emerged to compete with imports, and at that time also a small dyestuff and organic-chemicals industry began to grow in the face of German monopoly. In the first half of the century the industry supplied makers of textiles, paper, leather, glass, soap, and paints. By the end of the century the demand had expanded dramatically in these industries and now included also the producers of petroleum products, rubber, electrical equipment, bicycles, fertilizers, and insecticides. Within a decade were added synthetic textiles, plastics, and, most important of all, automotive products. With the expanding demand for new products came new chemical methods: catalytic and electrochemical processes, organic synthesis, liquefaction of air, hydrogenation of oils.

In the heyday of such expansion, industry leaders tried to form pools and gentlemen's agreements to control prices and divide markets, but most of these efforts failed to stem the tide of competition and falling prices. The first trade group, the Manufacturing Chemists Association, established in 1872, operated primarily as a social club and a small tariff lobby. On the whole, the rapidly expanding industry was highly competitive, with many small family-owned plants producing for local markets. Nevertheless, a few larger companies, often the earliest in the field, enjoyed healthy shares of the market.

The sulfuric-acid field was dominated by a handful of relatively large concerns, among them two of the earliest producers, Grasselli and Kalbfleisch. W. H. Nichols entered the field in 1870 with the establishment of the Nichols Chemical Co. When the contact catalytic process for producing sulfuric acid was introduced in Germany in the late 1870s, American manufacturers were forced to adopt it in order to compete with imports. The new process was first put into operation at the New Jersey Zinc Company in 1899, and that same year Nichols used it to cajole and club competitors into the first chemical merger. The General Chemical Company brought together the resources of twelve companies and nineteen plants, and by the end of the century Americans were outproducing European competitors in sulfuric acid.

There were no American producers of alkalies before 1880, and manufacturers relied upon imports for their supply of soda ash and caustic soda. In 1880 William Cogswell and Rowland Hazard secured a license for the Solvay ammonia-soda alkali process and three years later launched the first soda-ash plant in the United States, the Solvay Process Company of Syracuse, New York. In 1902 Hazard's son, Frederick Hazard, started the Semet-Solvay Company to make and install coke ovens, a line of activity that led into coke-byproduct manufacture and coal-tar chemicals. For eight years, until 1892, the Solvay Process Company had no major competitors in the alkali field. But then the courts ruled that supplementary patents on improvements of the original Solvay patents on the ammonia soda-ash and Frasch hot-water sulphur processes, both of which had expired, could not be used to prevent others from employing similar methods. In that year, as a result, two new companies appeared, the Michigan Alkali Company and the Mathieson Alkali Company. The former was established by J. B. Ford, founder of the Pittsburgh Plate Glass Company, with the intention of supplying the glass plants with soda ash. Mathieson, of Saltville, Virginia, was the first alkali plant in the South, and the first American producer of bleaching powder. A third concern, Columbia Chemical Company, was formed soon thereafter by the Pittsburgh Plate Glass Company after the departure of Ford, and with the backing of Henry Frick, Andrew Mellon, and other leading Pittsburgh industrialists.

The biggest boon to the alkali industry came in the late 1880s with the introduction of electrolytic processes for producing salts and caustic soda. The process designed by MIT graduate Ernest LeSeuer was the basis of the LeSeuer Electrochemical Company; Elon Hooker took advantage of the cell developed by Clinton Townsend and Elmer Sperry to form the Hooker Electrochemical Company; Mathieson used the Castner cell to liberate chlorine from salt; and Herbert H. Dow, an ambitious and inventive graduate of Case Institute, developed his own process of producing bromine and then chlorine from brine, thereby giving birth to the Dow Chemical Company, of Midland, Michigan. Dow later pioneered in the organic-chemical field, becoming the largest producer of phenols and the first maker of carbon tetrachloride and dichlorobenzene, all of which were key ingredients of the chemistry-based warfare of World War I.

The alkali business exemplified a new feature of the chemical industry: vertical integration. Not only did the alkali producers secure their own extractive plants to supply them with raw materials but the consumers of alkalies undertook to gain control over alkali production. Thus, three of the first five ammonia-soda plants were financed by glass interests, and the very first electrolytic plant was bought by a papermaker, the Brown Paper Company.

The American chemical industry in the late nineteenth century was dominated by the producers of heavy chemicals for industrial consumers; the latter, in turn, demanded even greater volumes of chemicals, purer chemicals, and new chemicals. The expansion of American manufactures thus presupposed and stimulated the expansion of the chemical industry, as well as associated extractive industries (mining of phosphates, copper, sulfur, potash, nitrates). Through continuous innovation, growth, and diversification, the U.S. chemical industry by 1910 had developed the plant and know-how to produce any important industrial chemical. In addition, several companies had emerged to compete with German producers of fine chemicals and medicinals: Mallinckrodt in 1867, Parke, Davis and Eli Lilly in the 1870s, and Upjohn, G. D. Searle, and Abbott Labs in the 1880s. By the turn of the century, Dow, Monsanto, and Merck also manufactured American fine chemicals, although consumers continued to rely most heavily upon German imports. In the explosives field, Du Pont, established in 1802, was the dominant company, producing, in addition to black powder, nitroglycerine and dynamite.

Despite the impressive industrial applications of scientific discoveries that accompanied the rise of the American heavy-chemical industry—notably the use of electrochemical processes—the core of science-based chemical industry, and the basis of German industrial superiority, remained the coal-tar dyestuff industry. Locus of synthetic organic chemistry, it was also the primary realm of organized, systematic scientific research. Within a decade after the synthesis of the first aniline dye in 1856, there was already a small-scale dyestuff industry in the United States. The Albany Aniline and Chemical Works was established in 1868, and the Schoellkopf Aniline Co., the cornerstone of the National Aniline and Chemical Company merger of 1917, was formed in 1879. Growth of the American dyestuff business, however, was prevented by a number of factors. The powerful textile and paper industries, which depended heavily upon natural and synthetic dyes, prevented the erection of a high protective tariff for the fledgling industry so that they could take advantage of cheap German imports. In addition, the German companies profited greatly from the unique American patent system; they were able to secure "product patents" on new chemical combinations of matter, thus preventing others from making the substance even if by a different process, and they were not required to "work" their patents in the United States in return for monopoly protection. By 1912, according to a study done by Bernard Herstein of the U.S. Tariff Board, ninety-eight percent of applications for patents in the chemical field had been assigned to German firms and were never worked in the United States.[15] Thus by the outbreak of World War I, although seven U.S. firms produced synthetic dyes in this country, they were forced to rely upon imported German intermediates, having never been able to develop the means of scientific research vital to an autonomous organic synthetic chemical industry. Only one American firm, the Benzol Products Company, established in 1910 under the General Chemical Company, produced intermediates.

The war, with its unprecedented need for organic-based explosives, and thus a domestic industry independent of Germany, changed this situation dramatically. The U. S. government, through Alien Property Custodian A. Mitchell Palmer, seized all German-owned patents. Palmer began selling the German patents to American companies for the highest bids, but smaller companies vigorously and successfully opposed this practice, arguing that the government was replacing the German monopoly with an American monopoly. Grasselli alone had received 1200 key patents. In the face of such opposition, Palmer yielded to his associate Francis P. Garvan's suggestion that a private foundation be established to hold the patents in trust and issue licenses to American companies on a non-exclusive basis. Thus the Chemical Foundation was formed to promote the American chemical industry. Between 1917 and 1926, 735 of the seized patents were issued to American companies and, not surprisingly, the concerns which had already gained the strongest position in the industry were strengthened further. Du Pont received licenses on three hundred patents, and the National Aniline and Chemical Company obtained one hundred. Other companies which profited immensely from the foundation's patent policy included Eastman Kodak, Union Carbide, the Newport Company (naval stores and wood-distillers), General Chemical, Bausch and Lomb (optical glass), and Bakelite (plastics). Immediately after the Armistice a tariff was established to protect the new American coal-tar products manufacturers. Science-based chemical industry in the United States had become a reality.[16]

The lesson of the war was that large-scale continuous operation and extensive organized research and development were the essentials of financial success in the chemical industry, and that these demanded big companies, corporate organization, and stable markets. By 1920 the days of family-owned, privately financed chemical companies had vanished; the era of corporatization, diversification, and consolidation had arrived. During the 1890s, when huge trusts were formed in other industries—sugar, steel, oil—no financial house "ever dreamed of a chemical combine."[17]

Before World War I, excepting the fertilizer field, there had been only two major consolidations in the chemical industry, and neither had involved outside financing. The successful development of the German (Badische) contact sulfuric-acid process spurred acid producers in this country to unite into the General Chemical Company to develop that process for themselves. The Mutual Chemical Company, formed in 1906, was created to foster the domestic commercial application of the Schultz chrome-tanning process. The consolidations in the postwar period—there were some five hundred mergers in the industry during the 1920s—were often similarly prompted by technical considerations. In addition, however, they involved large-scale diversification and consolidation through stock ownership, holding companies, and Wall Street orchestration. Three huge companies emerged to dominate the American chemical industry: Union Carbide and Carbon, Du Pont, and Allied Chemical and Dye. At the same time, other large firms, like American Cyanamid, Monsanto, Dow, Kodak, and Merck expanded to fill in the gaps.

The formation of Union Carbide and Carbon Corporation in 1917 followed logically from the technical interdependence of five companies. The oldest, National Carbon, processed coke to make carbon products of importance to the growing electrical industry: carbon electrodes for arc lamps, brushes for electric motors and dynamos, and batteries. In addition, it produced carbon electrodes for electric furnaces which were used in the production of metal alloys. Union Carbide employed electric furnaces to produce calcium carbide, and Electro Metallurgical (Electromet) needed them to manufacture metal alloys. Prest-O-Lite produced acetylene, which was generated in a calcium carbide reaction, and Linde extracted oxygen from air, which, together with acetylene, made possible oxyacetylene welding and cutting equipment. During the 1920s Union Carbide and Carbon developed and dominated the field of aliphatic chemicals, liquid gas, ferroalloys, and carbon products. In 1939 the Bakelite Corporation became a unit of Union Carbide, which added greatly to its already strong position in the field of synthetic thermoplastic resins.

Until 1880 Du Pont's activities were restricted to the explosives field. Its preeminence was guaranteed by the tremendous volume of war orders from the Allies and the U.S. government. With surplus profits from explosives, Du Pont undertook extensive research activities and also expanded into the heavy-chemical field with the purchase of Harrison Brothers (manufacturers of nitric and sulfuric acids) and into plastics and paints with the acquisition of the Arlington Company. In addition, partly under the stimulus of the seized German patents, Du Pont established new departments in coal-tar chemicals, synthetic fibers, nitrocellulose lacquer, and synthetic ammonia, and it undertook as well a joint effort with General Motors, Standard Oil, and Dow to develop tetraethyl lead for no-knock gasoline.

Subsequently, in 1926 Du Pont acquired the National Ammonia Company; in 1927 substantial shares in both General Motors and United States Steel; in 1928 the Grasselli Chemical Company; and in 1929 the Krebs Pigment and Chemical Company. By 1929 the expansion of the company initiated by Pierre S. Du Pont had carried it into every branch of the chemical industry except alkalies, fertilizers, and pharmaceuticals.

The Allied Chemical and Dye Corporation was formed in December 1920, at the initiative of W. H. Nichols of the General Chemical Company, as a holding company for the consolidation of five large concerns: Barrett Company (coal-tar products), General Chemical Company (acids), National Aniline and Chemical Company (dyestuffs), Semet-Solvay Company (coke and its byproducts), and the Solvay Products Company (alkalies and nitrogen products). Ranging across the entire chemical industry, Allied was created for both financial and technological purposes. A letter sent to the original stockholders emphasized the scientific basis for the consolidation. "Intensive progressive research is ... an especially important feature of the chemical manufacturing business. In the opinion of the Committee, the promotion of such research, through combination of the material and practical resources of the consolidating companies, is alone a compelling reason for the proposed consolidation."[18] Here, as elsewhere, monopoly and science went hand in hand, the one reinforcing the other.[19]

The electrical and chemical industries formed the vanguard of modern technology in America. In addition to their own rapid development, they fostered the gradual electrification and chemicalization of the older craft-based industries, which, with the steady infusion of personnel from the electrical and chemical fields, increasingly adopted the scientific habit.[20] Among the earliest of these were the so-called chemical-process industries: petroleum refining, wood distillation, extractive and metallurgical, sugar refining, rubber, canning, paper and pulp, photography, cement, lime and plasters, fertilizers. Later these were followed by steel, ceramics and glass, paints and varnishes, soap, leather, textiles, and vegetable oils.

The most important of these industries, because of their effect on the whole economy and the rapidity with which they adopted the scientific approach, were those linked to the development of the automobile: petroleum refining (lubricants, gasoline), rubber (natural and synthetic), and the automotive industry itself. The corporations which dominated these industries thus joined the electrical and chemical giants, during the 1920s, in the promotion of science-based industry. These included Standard of New Jersey, Standard of Indiana, Gulf Refining, Atlantic Refining, Texaco, Shell, Cities Service, Universal Oil Products, Goodyear, Firestone, B. F. Goodrich, General Tire, U.S. Rubber, General Motors, Ford, International Harvester, Chrysler, and Hudson.

These large corporations had been created to control intensifying competition and falling prices, to stabilize expanding markets and guarantee steady supplies of raw materials, to provide profits for Wall Street financiers, speculators, and investors, and, especially in the electrical and chemical industries, to protect and exploit new inventions, techniques, and processes. At the same time, their consolidation of resources provided the material means for science-based industry: the control and purchase of patents, the employment and training of scientific personnel, and the operation of large-scale systematic industrial research and development. Products in whole or in part of the haphazard progress of modern technology during the nineteenth century, these corporations undertook in the twentieth to transform that very technological progress into a well-ordered product of their own. A large majority of the human agents of scientific revolution in industry—notably the electrical and chemical engineers—became employees of such enterprises. Moreover, as this new breed worked up into managerial and executive positions within the science-based corporations, they came to identify the advance of modern technology with the advance of these corporations. To their minds, the scientific transformation of America and the corporate transformation of America had become one and the same.

2

The Wedding of Science
to the Useful Arts – II

The Development of
Technical Education

In the early nineteenth century the colleges were firmly in the hands of the classicists and the clerics, and there was considerable academic disdain for the study of experimental science and even more for the teaching of the "useful arts." Technical education in the United States, therefore, developed in struggle with the classical colleges, both inside and outside of them. One form of this development was the gradual growth of technological studies within the classical colleges, resulting from the reorientation of natural philosophy toward the empirical, experimental, scientific search for truth and from the pressures of some scientists and powerful industrialists for practical instruction; the other was the rise of technical colleges and institutes outside of the traditional colleges in response to the demands of internal improvement projects like canal-building, railroads, manufactures, and, eventually, science-based industry.

Again, it was Benjamin Franklin who early recognized the importance of technical education for the development of the country. As early as 1749, in a pamphlet entitled *Proposals for the Education of Youth in Pennsylvania,* Franklin called for higher educational instruction in mathematics, natural philosophy, the "mechanic powers," hydrostatics, pneumatics, surveying, navigation, architecture, optics, the chemistry of agriculture, and trade and commerce. With a few other prominent Philadelphians, he established in 1756 the Public Academy in the City of Philadelphia. The effort had meager success, however, against the forces of traditional education. In 1779 the Pennsylvania General Assembly abrogated its charter—it had fallen "under the taint of Episcopacy"—and by 1811 the practical orientation of the instruction had been diluted: science was confined to the senior year, consisting only of astronomy, natural philosophy, and the principles of chemistry and electricity; surveying and navigation were dropped altogether. Not long after this the projected King's College in New York constituted a similar attempt at practically oriented higher education. The original proposal for the college listed courses "in the Arts of numbering and measuring; of Surveying and Navigation; the knowledge of Nature . . . and everything useful for the comfort, the convenience and elegance of life, in the Manufactures relating to any of these things. . . ." But this effort also was stillborn, and Columbia College emerged along conventional lines.[1] In 1815 Count Rumford willed Harvard $1000 a year for a course of lectures to teach "the utility of the physical and mathematical sciences for the improvement of the useful arts, and for the extension of the industry, prosperity, happiness and well-being of society." The first holder of this chair was the physician Jacob Bigelow.[2]

Since the classical colleges, with their traditional elitist and religious orientation, obstructed the development of technical education in America, that education took root outside of them, at the initiative of men of affairs and in response to a popular movement for democratic schooling. Up until 1816 the number of engineers, or of men who called themselves engineers, never averaged more than two per state; the early internal improvements in the country and the planning of the national capital were directed by European engineers. But the surge of canal-building following the success of the Erie Canal created a demand for technically skilled workmen to oversee the operations, and the development of the railroad and machine industries intensified that demand. At the same time, the swelling ranks of mechanics and other skilled craftsmen who manned the industrial machine shops and railroad yards required greater access to education and science in order to develop their skills. One writer in *The Inventor* characteristically urged that mechanics be sent to the study and students to the workshop, with the theme "educate labor and set knowledge to work." In the same vein Professor J. B. Turner declared that the book learning offered by the classical colleges produced only "laborious thinkers," while what industry required was "thinking laborers."[3]

The call of mechanic and manufacturer was heeded. "I have established a school in Troy," Amos Eaton boasted in 1824, "for the purpose of instructing persons . . . in the application of science to the common purposes of life."[4] Eaton, an applied scientist with a Baconian spirit who studied chemistry with Benjamin Silliman at Yale, joined with

Stephen van Rensselaer, a wealthy landowner and capitalist, to start the Rensselaer School in 1823. In 1849 the school was reorganized by B. Franklin Green, after a careful study of technical education in Europe, along the lines of the Ecole Centrale des Arts et Manufactures. Renamed the Rensselaer Polytechnic Institute, it signaled the ascendancy of professional training for engineers in America. At the same time, the U.S. Military Academy at West Point, which had incorporated applied-science instruction in the curriculum under the direction of Superintendent Sylvanus Thayer, had begun to produce civil engineers with training in chemistry, physics, and higher mathematics as well as practical engineering.

The success of these pioneering efforts outside of the established colleges had its effect on them, at first with the inclusion of experimental science and later with the incorporation of practical studies in the curriculum. This effect was reflected in the fact that such outstanding physical scientists of the nineteenth century as Joseph Henry, Alexander Bache, the founder of the Franklin Institute, Henry Rowland, and J. Willard Gibbs had all been trained originally as engineers. Similarly, it was an RPI graduate, the chemist Eben Horsford, who persuaded the New England mill owner Abbott Lawrence to give Harvard funds for scientific studies. With a grant of $50,000 to underwrite the operation of the new Lawrence Scientific School, Lawrence specified that he wanted an institution in which scientific education was to be applied to engineering, mining, metallurgy, and the invention and manufacturing of machinery. "Where can we send those who intend to devote themselves to the practical applications of science?" Lawrence wrote the treasurer of Harvard in 1847. "How educate our engineers, our miners, machinists and mechanics? We need a school for young men . . . who intend to enter upon an active life as engineers or chemists or as men of science, applying their attainments to practical purposes." In addition to the purely technical aspects of such training, Lawrence foresaw the need for management training. "Hard hands are ready to work upon materials," he wrote, "and where shall sagacious heads be taught to direct those hands?"[5] Despite the explicit intention of Lawrence, however, the directors of Harvard took only a half-step toward practical instruction; Louis Agassiz, the noted geologist and zoologist, received most of the funds. Not until 1854 did the first engineer graduate from Harvard, and by 1892 there had been only 155. Harvard's reluctance to realize Lawrence's intentions was an important factor contributing to the establishment of the Massachusetts Institute of Technology in 1861.

Founded by the geologist William Barton Rogers and like-minded scientific and civic leaders of Boston, MIT embarked upon a broad range of scientific and technical instruction. Its purpose was its motto: *Mens et Manus*. Interestingly, the name "Technology" was proposed by Jacob Bigelow, one of the new school's supporters, to indicate that the study of science at MIT, rather than being a form of polite learning, would be directed toward practical ends. The MIT school of industrial science opened in 1865, and within a decade individual laboratory instruction in the physical sciences had been introduced as the focus of engineering education; science, as Benjamin Franklin had foreseen, had indeed become handmaiden to the arts.

At Yale the efforts to establish applied scientific study were more fruitful than at Harvard. In 1846 the Yale corporation reluctantly allowed two professors, John P. Norton and Benjamin Silliman, to establish extension courses to teach agricultural chemistry and other practical subjects, largely in response to the pressure of Norton's father, a powerful alumnus of the college. Until 1860 the enterprise was housed in the chapel attic and relied exclusively upon fees for support. In that year Joseph E. Sheffield endowed the venture, thereby creating the Sheffield Scientific School. It was in this school that the first Ph.D. in chemistry in the United States was granted and the first course in mechanical engineering was begun.[6]

While Yale and Harvard attracted the largest contributions for their scientific and technical work, other colleges had also begun to offer such courses. Union College established its civil-engineering course in 1845, followed by Brown two years later, and Dartmouth formed the Chandler Scientific School in 1851. The University of Michigan commenced instruction in engineering in 1852, under the direction of Brown graduate DeVolson Wood, and Cornell set up the Sibley College of Engineering in 1868. Within a decade Robert Thurston had moved from Stevens Institute to Cornell and had introduced laboratory instruction in physical science there. By the late 1870s, after scientific research in chemistry had begun in earnest at Johns Hopkins, courses in "industrial chemistry" or "chemical technology" were undertaken at various colleges.[7] At Columbia, Charles F. Chandler introduced applied-chemistry courses and founded the School of Mines; similar instruction was introduced by Samuel P. Sadtler at the University of Pennsylvania; by Edward Hart at Lafayette College; by William McMurtie and S. W. Parr at the University of Illinois; by A. B. Prescott and E. D. Campbell at the University of Michigan; and by Willis Whitney and A. A. Noyes at MIT.

Without doubt, the big leap forward in technical education in America came in 1862 when Congress passed the Morrill Act granting federal aid to the states for the support of colleges of agriculture and the mechanic arts. State legislatures that had been deaf to all appeals for technical instruction now quickly accepted the federal grants and voted to create the new type of school, while established colleges caught the spirit and added departments of engineering. In the first decade following the passage of the Morrill Act, the number of engineering schools jumped from six to seventy. By 1880 there were eighty-five, and by 1917 there were 126 engineering schools of college grade in the United States. Between 1870 and the outbreak of the First World War, the annual number of graduates from engineering colleges grew from 100 to 4300; the relative number of engineers in the whole population had multiplied fifteenfold.[8]

With the wedding of science to the useful arts, the former became more empirical and then practical while the latter became more scientific. Empiricism was introduced into scientific study as a means of understanding metaphysical truths, a guide to reflection. In the late nineteenth century, however, this process underwent a subtle inversion whereby practical experience, the handmaiden of science in the search for truth, made science its own handmaiden. Mansfield Merriman, president of the infant Society for the Promotion of Engineering Education, described in his presidential address of 1896 how this came about. "First, the principles of science were regarded as principles of truth whose study was ennobling because it attempted to solve the mystery of the universe; and second, the laws of the forces of nature were recognized as important to be understood in order to advance the prosperity and happiness of man. The former point of view led to the introduction of experimental work, it being recognized that the truth of nature's laws could be verified by experience alone; the latter point of view led to the application of these laws in industrial and technical experimentation."[9]

The first view of science led to a progressive relaxation of opposition to science in the classical colleges, on the grounds that empirical investigations were an aid to metaphysical speculation, but opposition to the second view, that science had practical implications, continued unabated. Thus, when Ira Remsen went to teach chemistry at Williams College and requested funds for laboratory facilities like those in German universities, he was met with a telling response: "You will please keep in mind," the school officials admonished him, "that this is a college and not a technical school. The students who come here are not to be trained as chemists or geologists or physicists. They are to be taught the great fundamental truths of all sciences. The object aimed at is culture, not practical knowledge."[10]

But science had gained a foothold. By 1895 Palmer Ricketts, director of RPI, could reflect in hindsight that "the youngest of us here remembers how many of the academic schools were unwillingly forced to add scientific departments in compliance with public demand," and DeVolson Wood of Michigan could allude to the "favorable auspices" for the annual meeting of the Society for the Promotion of Engineering Education in 1894, "because the antagonisms of the past between classical education and scientific education have passed away." Indeed, by the turn of the century a growing number of American colleges were awarding Ph.D.'s in the sciences and could already claim a number of distinguished physical scientists, men such as A. A. Michelson, Simon Newcomb, Henry Rowland, T. W. Richards, J. Willard Gibbs, George Ellery Hale, R. W. Wood, S. P. Langley, and Ira Remsen. At the same time, however, those who pursued a more practical approach to scientific study were not so readily welcomed in academe, a fact which preoccupied these engineering educators. The applied scientists in the engineering departments continued to occupy a second-class status within the academy, and, ironically, the scientists with their newly won respectability often enough lined up with the classicists against the technical educators across the campus.[11]

Their apparent inferior position galled and intimidated the engineers. Samuel Warren of RPI labeled the classicists a "crew of disreputable divinities," and President Francis Amasa Walker of MIT railed against them as he insisted upon the superiority of technical training. "Too long have we submitted to be considered as furnishing something which is, indeed, more immediately and practically useful than the so-called liberal education, but which is, after all, less noble and fine. Too long have our schools of applied science and technology been regarded as affording an inferior substitute for classical colleges. Too long have the graduates of such schools been spoken of as though they had acquired the arts of livelihood at some sacrifice of mental development, intellectual culture, and grace of life. . . . I believe that in the schools of applied science and technology is to be found the perfection of education for young men."[12]

For engineering educators within the liberal arts colleges, men who had to deal with their inferior status on a daily basis, such strength of conviction was hard to come by. More often, the condescension of the humanists and scientists led to self-recrimination and doubt. E. A. Fuertes, civil-engineering professor at Cornell, lamented that "the reason why our profession suffers in the way of which we complain is because it is not like the French body of engineers, which is composed of men who are *ipso facto* cultured gentlemen of great social power. The reason why we have not yet such powers is because we do not deserve it; there cannot be any other reason; it is the only reason that could exist."[13] For the engineering educators, second-class status within the academy was unbearable, and their preoccupation with it made it a topic of discussion at every meeting of the SPEE. For many, however, the solution to the problem was obvious: they must either increase the scientific content of their courses, in order to capitalize on the growing respectability of science, or increase their offerings in "culture studies." They did both.

From about 1870 on, the engineering curricula became distinctly more scientific and the focus of scientific study was shifted from laws of nature to principles of design. This tendency reflected both the engineering educators' quest for academic respectability and the increasing complexity of engineering problems, which defied the traditional cut-and-try approach. Most early instruction in the engineering departments and technical schools had placed great emphasis upon practicality, the *raison d'être* of engineering, in opposition to the "useless" cultural fare of the classical colleges. Such instruction was marked by liberal amounts of shop work, especially at such schools as Worcester and Cornell, and exercises in the drafting room and the field, with only occasional demonstrations by science instructors. After 1870, however, there was a slow but decided trend toward the use of laboratory methods of scientific investigation and experimentation in the solution of engineering problems, and engineering schools gradually undertook construction of laboratory facilities. In civil engineering, for example, there was a new emphasis upon mathematics and physical theory in design, and construction and mining engineers came to rely upon the fundamentals of chemistry. In mechanical engineering also, under the influence of science-minded engineers like Robert Thurston, instruction came increasingly to be based upon the scientific principles of hydraulics, thermodynamics, and the strength of materials, subjects which were often classed as branches of physics.[14]

This trend toward scientific engineering was most pronounced, of course, in the newer branches of engineering which accompanied the rise of science-based industry: electrical and chemical engineering. Instruction in these fields was as much the product of the departments of physics and chemistry as it was an offshoot of mechanical engineering, and the first teachers were men who had been trained as chemists and physicists. The first electrical-engineering courses were offered in the 1880s at universities like Wisconsin, Cornell, and MIT; "industrial chemists" were trained in the late 1880s and 1890s at Wisconsin, Michigan, and MIT, but instruction in chemical engineering proper, based upon the concept of unit operations, did not begin until MIT launched its School of Chemical Engineering Practice in 1917. In both of these fields the evolution was from the scientific toward the technical, rather than the reverse as in the other branches of engineering, and so instruction was grounded upon scientific theory from the outset. Moreover, as William Wickenden later observed, "this increasing scientific emphasis was greatly advanced by the close bond between the schools and the newer industries which had developed directly out of scientific research and technique, notably the electrical industry. An understanding arose, almost from the beginnings of the industry, under which the employers assumed practically the entire responsibility for the practical training of the student and the college was left to devote itself to the scientific foundations."[15]

This trend toward science and mathematics in the engineering curricula raised as many problems as it solved, however. At the turn of the century the great majority of practicing engineers were still those who had received their training in the "school of experience" rather than in the colleges of engineering. There were in fact, as Monte Calvert has described, a distinct traditional "shop culture" in mechanical engineering and a "field culture" in civil engineering which were very much in conflict with the newer "school culture" of the younger engineers. The increasingly scientific nature of the college training added to this tension. Engineers of the "rule-of-thumb," "cut-and-try" method resented the pretensions of the younger, scientifically oriented, "hypothetical" engineers, who invariably worked under them before rising into managerial positions. "Time was," remarked Ashbel Welch before an 1876 joint meeting of the new American Society of Civil Engineers and the American Institute of Mining and Metallurgical Engineers, "there was some truth in the saying, that the stability of a structure was inversely as the science of the builder." Such sentiments had hardly died out among practicing engineers; indeed, many of them continued to hold this view well into the twentieth century even as the school culture began to dominate the profession. William Burr, for example, reported in 1894 that "many men engaged in practical duties of an engineering nature frequently, and perhaps usually, complain that ... young engineers almost invariably have failed to possess" the capacity to deal effectively with practical problems.[16]

The experience-trained engineers were in a very powerful position vis-à-vis the schools of engineering, since it was they, and not the college-trained engineers, who dominated the industries which employed the graduates. Finding the graduates ill-equipped to handle the routine problems of industry, these practicing engineers put considerable pressure upon the schools to adapt their curricula to meet the demands of the "real world." By the close of the century their impact was reflected in the frantic attempts by the schools to provide shop training for their students, alongside the courses of scientific principles and whatever "cultural" courses there were.[17]

The conflicting demands of engineering educators for academic respectability, of a growing profession for esoteric knowledge, and of employers for practically trained men placed a great strain upon the educators, many of whom were also practicing engineers and regarded themselves as professionals. Not surprisingly, their overriding concern during the last decade of the century was that of trying to meet all the requirements of their calling within the standard four-year curriculum. As the embodiment of the wedding of science to the useful arts, the engineer reflected also the tensions inherent in that marriage. William Burr of Columbia College tried to reconcile the work of the shop with that of the laboratory, by emphasizing the scientific importance of the former and the practical importance of the latter. The demands for scientific training, he assured the practical men, "are not the opinions of theorists nor the erratic and irresponsible utterances of impractical men. . . ." The "ideal engineering education," he argued, "consists of a most thorough training in mathematics and physics, but adapted in its entire matter and method to the subsequent engineering practice." However, while "the mechanical engineering student must [therefore] take a comparative large amount of workshop practice, it must be for the training of a mechanical engineer and not for the purpose of attaining the skill of a mechanic."[18]

The efforts of the engineering educators to meet the demands of their profession, academic status, and employers of engineers were never wholly satisfactory. Indeed, the problems have remained the focus of their concern throughout the twentieth century. By the end of the nineteenth century the schools lagged seriously behind the rapidly changing needs of the science-based industries. Even in the larger institutions like Yale, Michigan, Wisconsin, MIT, and Purdue, where adequate shop facilities were available for training in "real world" techniques, the equipment quickly became obsolescent with advances in industry. To fill the gap between formal education and the requirements of employment, therefore, major industrial concerns established in-house training programs, the more or less elaborate "corporation schools." The R. Hoe Publishing Company was the first to do so, in 1875, and the electrical industry followed suit on a large scale shortly thereafter. By the turn of the century such training programs were provided in many large companies. The electrical industry had perhaps the largest of these, and American electrical-engineering graduates normally went to the "test course" or "special apprentice courses" at Lynn, Schenectady, Pittsburgh, or Chicago to complete their education before embarking upon professional careers. At GE, as at the other companies, "the test course was a path between college and business." The corporation-school idea spread considerably between 1890 and 1915; by the second decade of the century, however, new far-reaching approaches toward "education-industry cooperation" were being formulated to bridge the gap between the classroom and the workplace, between scientific theory and engineering practice.[19]

The problems presented by the strictly scientific and technical aspects of engineering education were compounded, in the closing decades of the nineteenth century, by others involving the proper role of the "cultural courses" in the education of engineers. Engineering educators sought to determine the proper relationship between such courses as history, literature, rhetoric, political economy, moral philosophy, and languages—which they ambiguously labeled "humanities" or "culture studies"—and the purely technical instruction, a relationship which would meet the requirements of professionalism, academic status, and engineering practice. Since the first technical schools arose in opposition to the classical colleges, they initially refused to offer any "culture studies" at all. Amos Eaton of RPI had nothing but scorn for the established colleges; the RPI brochure of 1826 boasted that the school "promises nothing but experimental science. . . . Its object is single and unique; and nothing is taught at the school but those branches which have a direct application to the 'business of living.' "[20] However, while the technical institutes, such as Worcester, tended to follow this path, the new schools of engineering adopted a more relaxed posture toward their "adversaries" in the classical colleges. MIT and Cornell, for example, pioneered in establishing "humanities" courses for engineering students which ran concurrently with the technical curriculum. When the classical colleges eventually established their

AMERICA BY DESIGN / *30*

own engineering schools, these adopted a similar pattern of concurrent instruction, as did the majority of engineering schools thereafter; in time even RPI followed suit. The Morrill Act also adopted this approach to technical education; it specified that the new land-grant schools "shall be, without excluding other scientific and classical studies . . . to teach such branches of learning as are related to agriculture and the mechanic arts . . . in order to promote the liberal and practical education of the industrial classes in the several pursuits and professions of life."[21]

The concurrent curriculum, however, though widely adopted, did not altogether eliminate the antagonism between the "business of living" and "culture" or the debates over the training required for each. While some engineering educators, seeking to enhance their professional prestige, tried to emulate the refined airs of the liberal-arts professors, others attacked such pretensions as elitist and vacuous. "[A] broad cultivation," one professor argued, "is the only effectual corrective for the narrow and malformed excellence in some special direction, which falls lamentably short of the vigorous and well-rounded product of the ideal education in engineering. . . . The first and fundamental requisite in the ideal education of young engineers," he added forcefully, "[must be] a broad, liberal education in philosophy and the arts." An anecdote offered by Professor Robert Thurston, on the other hand, suggested a contrary view. Scorning the inclusion of culture studies in the engineering curriculum, he reflected that "the most singular mixtures of literature, history, and other non-professional studies with engineering were often prescribed, as where, in one now famous institution of learning, 'biblical exegesis' constituted a portion of the regular course in civil engineering, or whereas, in the early days of Cornell University, Roman history was similarly imbedded in a course nominally . . . in civil engineering 'like a flyspeck on a white wall,' as the finally emancipated head of the department was heard to say."[22]

There was thus considerable debate in the 1890s among engineering educators about the future of the concurrent curriculum. Henry Eddy concluded that as "the two kinds of study interfere with each other . . . it seems clear that the culture studies must soon disappear from our engineering courses." President Walker of MIT, an economist, strongly disagreed. He predicted that "doubtless more of the economic, historical, and philosophical studies will be introduced, to supplement, by their liberalizing tendencies, the work of the sciences in making the pupils exact and strong. Possibly, some ultimate form for institutions of higher learning may yet be developed, which shall embody much of both the modern school of technology and the old-fashioned college."[23]

This debate sustained within the engineering schools the same antagonism as had existed originally between the technical schools and the classical colleges. Engineering educators wanted to enhance their professional and academic prestige by means of a broadened "liberal" education for the engineer, and at the same time to maintain their role as "real world" revolutionaries within the academy. Their debates reflected this ambivalence. By the turn of the century, however, a number of farsighted engineering educators began to recognize distinctions within that realm which they loosely called "culture studies" or "humanities," distinctions which prompted a reevaluation of their usefulness. While Walker, for example, lumped together "economic, historical, and philosophical studies," a former SPEE president told engineering educators that he had recently become "impressed with the great importance of some other subjects . . . which are not taught in our schools of engineering but which are taught in other departments of our large universities, subjects which in general are under the term humanities. . . . Some of these subjects are becoming real sciences," he pointed out, "and not simply formulated theories and opinions, and the engineering students ought to know something about them."[24]

The emergence of the social sciences—economics, political science, psychology, and sociology—led some engineering educators to redefine the role of "culture studies" in the engineering curriculum. They began to see that some non-engineering courses could have practical value in addition to mere status value—that, rather than simply making the students more refined and "cultured" gentlemen, they could make them more effective engineers. Stressing the importance of knowledge in the social sciences at a time when increasing competition, expanding plants and markets, and intensifying labor conflict were raising difficult problems for the managers of industry, Mansfield Merriman suggested that "the main line of improvement to secure better results will be . . . in partially abandoning the idea of culture and placing the instruction on a more utilitarian basis." William Burr similarly voiced the growing sentiment of practicing engineers in industry when he argued that "engineering education must enable engineers to meet men as well as matter." As more and more engineers worked their way up into the managerial positions of corporations, particularly within the electrical and chemical industries, they began to find fault with the training they had received, training which had not prepared them for the challenges of corporate management. Roughly two-thirds of all engineering graduates were becoming managers in industry within fifteen years after graduation, and they increasingly put pressure on the engineering schools to provide training for all aspects of an engineering career rather than the merely technical. As a result, while the more traditional offerings of the "cultural studies" curriculum—languages, Bible study, ancient history, moral philosophy, and rhetoric—were gradually phased out of engineering programs, new subjects labeled "humanities" and later the "humanistic-social stem"—political science, economics, psychology, sociology—were introduced. While some engineering educators continued to try to broaden the curriculum in order to enhance their academic and professional prestige, others, at the behest of the managers in the new science-based corporations, began to call for a "liberal" education for other reasons. As one engineer phrased it, "A liberal education gives power over men."[25]

3

The Wedding of Science
to the Useful Arts – III

The Emergence of the
Professional Engineer

In the eighteenth century, and to a lesser extent up through the middle of the nineteenth, contact between the separate realms of science and manufactures was limited to the personal association of upper-class gentlemen with feet in both worlds. By mid-century, however, with the gradual reorientation and increasing popularization of science, such contact, while remaining haphazard, became more common outside the elite circles, and the chance adaptation of scientific methods and discoveries to the practical ends of commercial enterprise eventually gave rise to both science-based industry and schools of scientific technology.[1] In addition, a new social type emerged to personify this union: the engineer. "Science and invention were joining hands; their lusty issue was Engineering."[2] The engineer in America, the legitimate child of the epochal wedding of science to the useful arts, was the human medium through which it would work its profound social transformation. He was, as engineers themselves tirelessly boasted, a new breed of man, the link between "the monastery of science and the secular world of business," whose calling, engineering, bridged "the gulf between the impersonal exact sciences and the more human and personal affairs of economics and sociology."[3] By the end of the nineteenth century the province of engineering—modern technology—had already attracted "the brightest and most gifted men."[4] Many of those ambitious and fortunate young men who would formerly have devoted themselves to religion, politics, philosophy, or art turned instead to explore, map out, and lay claim to this vast new realm of human enterprise.

Modern technology, as the mode of production specific to advanced industrial capitalism, was both a product and a medium of capitalist development. So too, therefore, was the engineer who personified modern technology. In his work he was guided as much by the imperatives that propelled the economic system as by the logic and laws of science. The capitalist, in order to survive, had to accumulate capital at a rate equal to or greater than that of competitors. And since his capital was derived ultimately from the surplus product of human labor, he was compelled to assume complete command over the production process in order to maximize productivity and efficiently extract this product from those who labored for him. It was for this reason that mechanical devices and scientific methods were introduced into the workshop. It was for this reason also that the modern engineer came into being. From the outset, therefore, the engineer was at the service of capital, and, not surprisingly, its laws were to him as natural as the laws of science. If some political economists drew a distinction between technology and capitalism, that distinction collapsed in the person of the engineer and in his work, engineering.

Even in his strictly technical work the engineer brought to his task the spirit of the capitalist. His design of machinery, for example, was guided as much by the capitalist need to minimize both the cost and the autonomy of skilled labor as by the desire to harness most efficiently the potentials of matter and energy. The technical and capitalist aspects of the engineer's work were reverse sides of the same coin, modern technology. As such, they were rarely if ever distinguishable: technical demands defined the capitalist possibilities only insofar as capitalist demands defined the technical possibilities. The technical work of the engineer was little more than the scientific extension of capitalist enterprise; it was through his efforts that science was transformed into capital. "The symbol for our monetary unit, the dollar," Henry Towne wrote in 1886, "is almost as frequently conjoined to the figures of an engineer's calculations as are the symbols indicating feet, minutes, pounds, or gallons." "The dollar," he later told Purdue engineering students, "is the final term in every engineering equation." Nearly a century later A. A. Potter, the dean of American deans of engineering, summed up the spirit of his vocation in a similar vein: "Whatever the numerator is in an engineering equation, the denominator is always a dollar mark." The economic inspiration inherent in technical work, of course, did not altogether rule out the possibility of conflict between the demands of technological superiority and of market expediency. When such conflict did arise, however, there was never any doubt about the outcome. The president of the Stevens Institute Alumni Association, for example, was unequivocal on this point, and his address to students in 1896 had no trace of ambiguity: "The financial side of engineering is always the most important; the sooner the young engineer recedes from the idea that simply because he is a professional man, the position is paramount, the better it will be for him. He must always be subservient to those who represent the money invested in the enterprise."[5]

The first American professional engineers were civil engineers, graduates of West Point, RPI, and the canal and early railroad projects. In contrast to the treatment they received at the hands of the genteel academicians, these engineers of the first half of the nineteenth century commonly enjoyed the high status of gentlemen, and often had "esquire" affixed to their name. In a democratic age in which politicians championed the common man who worked with his hands, the engineer was popularly scorned for being aristocratic and for representing the owners of large corporations against the interests of small contractors. The criticism was appropriate. "The skillful engineer," Benjamin Wright reminded his colleagues in 1832, "will determine what is feasible at an expense adapted to the object, and what is otherwise, having in view the present and future remuneration of the stockholders."[6]

Civil engineers initially held positions of broad responsibility in the early canal companies. Their work involved proprietary and managerial functions in addition to the strictly technical, and they often used the experience gained in such capacities to leave the company and set up their own. As soon as the canal and later railroad projects became large enough to provide employment for the growing number of engineers, however, the entrepreneurial route was closed off and the engineers became strictly organization men within large bureaucracies. Although some prominent engineers were able to establish themselves as independent consultants, the majority, especially those coming out of the new technical schools, were absorbed as salaried employees, technical advisors, and project supervisors in large corporations. They were held responsible for the success of company operations, but were without real decision-making authority over the work to be done.[7]

With the founding of the American Society of Civil Engineers in 1852, the civil engineers became the first professional body of engineers in America. Almost immediately they began to grapple with the contradictions inherent in engineering professionalism, struggling to attain professional autonomy and define standards of ethics and social responsibility within a context of professional practice that demanded subservience to corporate authority. During the last three decades of the century they were joined by the mining and mechanical engineers, who attended the rise of modern industry, and by the electrical and chemical engineers, who ushered in the new science-based industries. The American Institute of Mining and Metallurgical Engineers was founded in 1871, the American Society of Mechanical Engineers in 1880, the American Institute of Electrical Engineers in 1884, and the American Institute of Chemical Engineers in 1908. By 1900 college enrollment in mechanical engineering already outnumbered that in civil engineering three to two, and electrical engineering was running a close third.[8]

The mining engineers, always a small minority within the profession, did little to promote professionalism. In the heyday of booming mining operations and global exploration for the raw materials demanded by expanding industry, they functioned much like the early civil engineers, as promoters, entrepreneurs, and company officials. Such was not the case with the mechanical engineers or their offspring, the electrical and chemical engineers.[9]

The mechanical engineers did not split off from civil engineering as did the mining engineers, but emerged from another source entirely: the machine shop. During the first half of the nineteenth century the vital work of metal-working machine shops, railroad-yard engine shops, steamship works, and manufacturing towns was performed by mechanics—skilled craftsmen who had not acquired the status or label of engineer. Their work, however, was the *sine qua non* of modern industry—what Marx called "machinofacture." Their machine shops, as the repositories of mechanical knowledge, produced the machinery upon which industry was built.[10]

The early mechanical engineers were skilled mechanics who had worked their way up to become shop managers, or owners of their own shops. Mechanical engineering was thus, like civil and mining engineering, as much a business as a profession, and as the business aspects of engineering changed with the expansion of industrial enterprises, so too did the nature of engineering professionalism. The new schools of technology provided a steady outpouring of scientifically trained engineers while industrial consolidation and bureaucratization increasingly circumscribed entrepreneurial opportunities and restricted access to top managerial positions within the existing companies. At the same time, the rapidly expanding field of mechanical engineering, reflecting industrial diversification and the growth of scientific and technical knowledge, fostered a strong complementary tendency toward specialization. As a result, the large majority of younger engineers was forced into subordinate corporate employment. While responsible for performing and supervising technical work, they enjoyed neither the prerogatives of ownership nor any ultimate authority over company operations.

The mechanical-engineering profession was thus never a homogeneous body of professionals but rather a loosely defined assemblage of primarily shop-trained industrialists, corporation executives, and shop managers, on the one hand, and school-trained corporate employees, on the other. The men who had founded the profession by promoting the distinction between the "mechanical engineer" and the "mechanic" based the distinction less upon the possession of technical knowledge than upon the exercise of significant supervisory authority. From the day of its founding, the ASME was little more than a social club run by and for the self-perpetuating oligarchy of a shop-culture elite who had become leaders of industry. Alexander Holley, one of the founders of the ASME, was both a professional engineer and an industrialist. The two careers, as he explained, went hand in hand. "The advantages of the association of business men and engineers in these societies are notorious. . . . These advantages are not only large membership, and hence large incomes to devote to publications and illustrations, but they lie chiefly in the direct business results of bringing professional knowledge, capital, and business talents together under the most favorable circumstances."[11] To the men who dominated the professional society of mechanical engineers, business success was the mark of professional success and business leaders were *ipso facto* prominent engineers. For the younger school-trained engineers, professional identity was hard to come by. Entrepreneurial opportunities had vanished and the independent consulting and industrial-leadership positions were firmly dominated by the shop-culture elite who controlled the professional society. As a result, a new type of professionalism emerged, one which emphasized academic credentials, scientific training, and formal promotion, within the corporate hierarchies, into management.

Since neither the electrical engineers nor the chemical engineers had any traditional cultures to contend with, the brand of professionalism which defined the engineer in terms of his position within the corporate hierarchy and his scientific training became the predominant one. When the electrical engineers split off from the ASME in 1884 to set up their own professional society, their discipline was still young and so too were most of its members. Mobility within the professional ranks was thus relatively open. This mobility, however, tended to be based upon promotion within the large corporations which dominated the electrical industry. For as Charles Steinmetz of GE pointed out, "a very large percentage of the prominent electrical engineers [were] associated with large manufacturing or large operating companies," and the majority of electrical-engineering graduates were absorbed into the largest corporations, GE and Westinghouse. At the same time, electrical engineering, as the product of both mechanical engineering and physics, was firmly rooted in science from the outset. There was no formidable "rule of thumb" tradition as in mechanical engineering, and so scientific training readily became a mark of professional identity.[12]

The same was true in chemical engineering; the product of mechanical engineering and chemistry, it was created to meet the demands of the expanding chemical and chemical-process industries. The industrial chemists—independent consultants and company officials—who pushed the American Chemical Society to establish the Division of Industrial and Engineering Chemistry in 1908 also set up the American Institute of Chemical Engineers (AIChE) in the same year to determine their own standards of professional competency.* Their experience within the industrial corporations had taught them that chemists alone, trained in science but with little knowledge of engineering principles or managerial techniques, could not meet the demands of the industry. What was needed was a new breed entirely, people who combined training in chemistry and physics with training in engineering and management, who were prepared to translate their understanding of chemical processes into the efficient design and profitable management of industrial plants on a large scale. "There must be a body of men supplied in increasing numbers," President Samuel P. Sadtler told the first AIChE convention, "who can take the technical charge of these industries, first as aides and ultimately as managers of the several works, qualified to continue the successful administration of the same and able to push them steadily to fuller development along safe and profitable lines."[13] The chemical-engineering profession was, in effect, made to order.

By the beginning of the twentieth century the United States had a healthy and rapidly expanding engineering profession. Excepting teachers, it was already the largest professional occupation in America,

---

*Although the ACS division initially included chemists working in the food-and-agricultural and analytical-chemistry fields, a group of prominent industrial chemists—led by such men as Arthur D. Little, W. D. Bigelow, L. H. Baekeland, and Willis R. Whitney —soon succeeded in having membership restricted to chemists working in industry. Richard K. Meade, a member of the ACS division and founder and editor of *The Chemical Engineer,* suggested to his colleagues that they start a separate organization as well. Meade, Little, William H. Walker (Little's former partner), and consultants Charles F. McKenna, William Booth, and J. C. Olsen were the organizers of the AIChE.

and during the next three decades its continued growth reflected the enormous development of modern industry. The approximately 45,000 engineers of 1900 had multiplied fivefold to 230,000 by 1930, and well over ninety percent of these were employed as technical workers and managers in industry. The greatest growth within the profession took place in the newer and most scientific fields: electrical and chemical engineering. By 1928 enrollment in electrical-engineering courses exceeded by fifty percent that of either mechanical or civil engineering, and enrollment in chemical engineering was already half that of mechanical engineering.[14]

Driven by the twofold dynamic of professionalism and industrial demand, the growth of the engineering profession outpaced that of the industrial workforce and the economically active population as a whole. This remarkable increase, however, did not significantly alter the social position or social complexion of engineers in America. The profession remained tiny and elite. By 1930 there were still only 45 engineers for every 10,000 workers in industry, and engineers accounted for but one half of one percent of the entire economically active population. All but one out of a thousand of these engineers were male, and three quarters of them were sons of middle-class professionals, proprietors, managers, or farm owners; most had enjoyed a college education, and the great majority came from native American, Anglo-Saxon, Protestant stock.[15]*

In theory, at least three types of engineering professionalism might have emerged by the turn of the century, each emphasizing a different

*Although the engineering profession in later years became a route of upward mobility for lower-class youths, it was at this time predominantly the preserve of middle- and upper-class men. It should also be emphasized that this enormously influential class of people was overwhelmingly male. Indeed, it would be no exaggeration to say that modern technology in America has always been an exclusively male domain. Thus, in addition to being a product of industrial capitalism, reflecting class control over the means of production, modern technology in America has been equally a product of a male-dominated culture reflecting male prerogatives and male preferences. As agents of modern technology, the engineers enjoyed the freedom of action reserved for the dominant class, and their work was an extension at once of their own freedom and of the forms of domination which made that freedom possible. Historians and others must keep this in mind when trying to account for and evaluate their prodigious achievements. On the changing social origins of engineers, see, for example, Carolyn Cummings Perrucci, "Engineering and the Class Structure," and Robert L. Eichhorn, "The Student Engineer," both in Robert Perucci and Joel Gerstl, eds., *The Engineers and the Social System* (New York: John Wiley and Sons, 1969); and Martin Trow, "Some Implications of the Social Origins of Engineers," in National Science Foundation, *Scientific Manpower,* 1958, 67–74.

aspect of the profession. One might have stressed the business-leadership function of engineering, giving rise to a type of businessmen's association. A second might have focused upon the role of engineers as corporate employees and encouraged trade-union organizations. A third might have emphasized the scientific nature of engineering and inspired a professional identity grounded upon the monopoly of esoteric knowledge, similar to that of lawyers, physicians, and scientists. In reality, which includes the engineer's self-conception as well as his actual social situation, strains from all three types survived into the twentieth century. Each was employed by the engineer, at one time or another, to define himself in relation to something or someone else: the scientist, the skilled worker, the layman, labor, the corporation, or the student. Not surprisingly, these conflicting identities within the mind of the professional engineer often generated confusion and delusion.

From the outset, all branches of engineering identified professionalism with business leadership. There were important differences, however, in the way business leadership was construed. The electrical and chemical engineers defined it in terms of management within the large corporations which dominated the science-based industries, but the tradition-bound civil, mining, and mechanical engineers included in their definition the roles of the independent consultant and the entrepreneur. With the expansion of machine industry and the demise of the shop culture, the tension created by these conflicting notions of business leadership became most acute among the mechanical engineers. In order to counter the corporate challenge to engineering prerogatives, for example, some of the older shop-culture mechanical engineers attempted to create a new identity for the engineer, one which tied him to the large bureaucratic corporation and yet enabled him to retain his power as an independent engineering expert within it. Although the scientific-management movement began within the large corporations (the father of scientific management, Frederick Taylor, worked for Midvale Steel and Bethlehem Steel), its origins actually lay outside the corporation, in the fading identity between the engineer and the machine-shop manager and entrepreneur. By attempting to centralize the management functions of the corporation within the planning rooms of the scientific-management engineers, the heirs to the shop culture sought to retain their traditional authority within the corporate structure.[16]

Thus, both the shop-culture mechanical engineers and the corporation-based electrical and chemical engineers contributed, if in different ways, to what ultimately became the dominant notion of business leadership for engineers: management within the large corporation. By 1920, moreover, with the abandonment of the original pretensions of the early scientific-management leaders, promotion within the corporate hierarchies of the science-based industries had become the *sine qua non* of professional advancement. In his massive study of engineering education, William Wickenden examined the career patterns of engineering graduates from 1884 to 1924 and observed "a healthy progression through technical work toward the responsibilities of management";[17] roughly two-thirds of the graduates had become managers within fifteen years after college. This marked trend, which Wickenden regarded with professional pride, reflected the unprecedented demand for technically trained managers in modern industry. At the same time, it clearly indicated that it had become "almost always necessary for an engineer to leave the engineering of materials and enter the engineering of men in order to become very successful financially and socially."[18]*

The emphasis upon management responsibility within the corporation as the mark of professional success undercut whatever support there might have been for trade-union activity by engineers. Such support had been weak from the start, of course, since any identification of engineering with skilled labor would have blurred the distinction between engineer and mechanic upon which the profession was grounded. The glorification of the management function of engineering reinforced opposition to unionism in general, and to engineering unionism in particular. As those charged with supervision of the industrial labor force, engineers found labor organizations difficult and disagreeable, and as professionals, they viewed unionism as a measure of mediocrity.

Mobility into corporate management, on the other hand, was for them evidence of their superiority and, in this respect at least, every corporate engineer was a staunch individualist. Since each engineer had to travel the road alone, isolated from the profession at large, and his success depended upon his performance as an individual, the collective solidarity upon which trade-unionism depended was never allowed to

---

*There have been recent efforts by some companies—the Bell Labs and General Electric, for example—to establish a dual advancement system which allows research scientists and engineers to remain as such and still advance their rank and salary. For an early discussion of this development by the personnel director of the Bell Labs, see John Mills, *The Engineer in Society* (New York: D. Van Nostrand Co., 1946), pp. 80–140. These developments do not require any significant alteration of the general conclusions drawn here, however; for the most part, engineers must still go into management in order to advance their positions.

develop. The same individualist orientation which tied success to corporate rather than strictly professional criteria also distinguished the engineering profession from the older legal and medical guilds, which often acted in unity like trade unions. If the physician were taken to typify the professional person, the engineer would fall far short of the mark in terms of both autonomy and professional clout. For most engineers, however, the potential conflict between professional integrity and subordinate corporate employment did not arise simply because their unique notion of professionalism was one which neatly embraced corporate position as the mark of status within the profession. If some bridled at the specter of business control over their profession, most engineers viewed the commercial or administrative success of the individual as the clearest sign of professional achievement, wholeheartedly subscribing to Charles Steinmetz's paradoxical notion that the corporation "was the most efficient means of making individual development possible in our present state of civilization."*[19]

Science, like management, gave engineers their identity. The mechanical engineers had emphasized the scientific nature of their discipline to distinguish themselves from mechanics and skilled craftsmen, and this emphasis increased over time as scientifically trained engineers made headway against the rule-of-thumb methods of the shop-culture tradition. The electrical and chemical engineers had, of course, placed

---

*Steinmetz's notion derived from his particular brand of socialism. An immigrant German Social Democrat (follower of Lassalle), Steinmetz spent all of his adult life in the employ of General Electric. He saw no contradiction, however, between his high-level employment in one of the country's largest capitalist corporations and his lifelong membership in the Socialist Party, U.S.A. Because of both his revisionist socialist political perspective and his personal experience at GE, Steinmetz came to view corporate development as the centralization of the means of production that would ultimately, and inevitably, usher in socialism. Considered the leading electrical engineer in the country, Steinmetz enjoyed a unique position within the corporation he served, which his mathematical genius had placed in the forefront of the electrical field. Aside from his work at GE, Steinmetz was a major force in the educational community. He was professor and for a time head of the electrical-engineering department at Union College, director of the educational activities for GE at Schenectady, and for many years president of the Schenectady Board of Education (a position he initially assumed under the socialist administration of Mayor George Lunn). In addition, in 1922—a year before his death—he ran unsuccessfully for the position of New York State Engineer on the Socialist-Farmer-Labor Party ticket. To date, there have been few noteworthy studies on Steinmetz, although a chapter by James Gilbert in his *Designing the Industrial State: The Intellectual Pursuit of Collectivism in America 1880–1940* is an important first step. Because of his unique career and perspective, reflecting the convergence of the socialist movement, corporate development, and the growth of modern technology, Steinmetz deserves more attention than he has received. His important book *America and the New Epoch* (New York, 1916) should be standard reading in American intellectual history.

heavy emphasis upon scientific methods and knowledge from the outset. By 1900, and at an accelerated rate thereafter, the popularization of science characteristic of the nineteenth century steadily disappeared in the wake of the very scientific and engineering professionalism which it had fostered.[20] While the systematic monopolization of scientific knowledge by the professionals increased the autonomy of scientists, however, it had the opposite effect upon engineers, tying them to the large corporation. The difference stemmed from the distinction between science and engineering. "Unlike science," Edwin Layton has observed, "technology cannot exist for its own sake."[21] The practice of engineering, as it gradually became "applied science" distinct from traditional craft, at the same time was ever more closely tied to the only organized social medium for the application of science, modern corporate industry. The wedding of science to the useful arts had become at once the professional habit of the engineer and the standard operating policy of the industrial corporation. If the engineer desired to teach or do scientific research, he was able to remain in the academy; but if he wanted to translate his scientific understanding into concrete achievements of social significance—that is, if he wanted to practice engineering—he was compelled to enter the employ of industry. Only the large industrial enterprise could afford to join effectively and efficiently the means of science to its socially useful ends. Moreover, the big corporations, because of their control over patents, had combined their capacity to command the industrial application of science with their exclusive legal right to do so. The industrial corporation, therefore, as it emerged as the locus of modern technology in America, became at the same time the habitat of the professional engineer.

The monopolization of "scientific engineering" by the professional engineers was thus the reverse side of the monopolization of the scientific engineers by science-based corporations.* The large corporations

---

*When industries increasingly turned to chemists and physicists for scientific expertise, after World War I and especially after World War II, the scientists also became monopolized, and with them their scientific knowledge. The percentage of chemists teaching in universities was approximately 10 percent in 1900; this dropped to 4.5 percent by 1920, with most of the difference being absorbed by private industry. With the expansion of universities from the 1920s on, and the establishment of industrial chemistry-research facilities within the universities in cooperation with industry, the universities gradually reabsorbed them. The shift toward industry was most dramatic among physicists. The ratio of teaching/nonteaching physicists dropped from 5.7 in 1900 to 0.4 in 1960 as they were increasingly employed by industry and, since 1940, by government. Because the professional identity of the scientist predated this industrial orientation, however, the scientist has had a somewhat more self-determined experience in industry. See Jay Gould, *The Technical Elite* (New York: Augustus M. Kelly, 1966), p. 172.

needed the technical knowledge which only the professionals could provide. On the other hand, the professional engineers required human organization and material resources in order to render their knowledge functional, and the large corporations had secured a monopoly on these. Not surprisingly, therefore, as the Wickenden study of engineering graduates showed, "the independent private practice of engineering [was] distinctly on the decrease and . . . engineers to a greater extent [were] going into the employ of corporate organizations, particularly of the large industries."[22]

As he strove to create a professional identity for himself, the engineer commonly tried to present himself to the public as "technology" itself, the great motive force of modern civilization. In doing so, however, he was forced to identify himself with corporate business, for that too was technology, as much a part of him as he was of it. The engineer's efforts to control the historical process which had created him, in order to render secure his position within it, thus complemented and reinforced the efforts of the science-based corporations to control the same process, which had created them as well.

As the engineers, in their various educational, professional, and industrial capacities, worked to monopolize and regulate their professional resources, they themselves became monopolized and regulated as industrial resources. Some among them recognized this trend and rebelled against it. The president of the ASCE, Onward Bates, warned in 1909 that the engineer was becoming simply "the tool of those whose aim it is to control men and to profit by their knowledge," and a writer in the *American Machinist* complained bitterly that GE and Westinghouse had so "commercialized engineering" that their control over the profession posed "almost insuperable barriers against new men and new ideas." The vast majority of engineers, however, were not overly concerned. As they increasingly tended to identify their professional status with their corporate status, their pursuit of professionalism became, at the same time, a crucial aspect of the corporate management of technical knowledge and technical people: the corporate control of technology.[23]

The engineers who undertook to regulate the historical process of modern technology in order to consolidate their professional position were drawn inescapably toward engineering education. DeVolson Wood, in his inaugural address as the first president of the Society for the Promotion of Engineering Education, noted how "in less than forty years about one hundred professional engineering schools have come into existence, graduating some twelve hundred annually." This growth, he observed, had been "spontaneous in character, without a central head or mutual conference." The new organization, he hoped, would change that. "If its efforts are properly directed, it may make of all of these schools a kind of university, though widely separated, . . . a kind of unity for accomplishing the best results in this line of education." H. F. J. Porter of Westinghouse agreed. "Each institution," he suggested, "is in reality a factory turning out engineers. . . . It is necessary . . . not only that a high grade but also that the same grade of men should be turned out from all the technical schools." The profession and the industries, he argued, needed "a uniform system of educating engineers."[24]

The engineering educators who met in Chicago in 1893 as Section E of the World Engineering Congress found that they shared a wide range of common interests which merited further discussion. When they established the SPEE the following year, the first association of college educators in the United States devoted exclusively to educational matters, the educators outnumbered practicing engineers ten to one; discussion focused upon the questions of academic respectability and the relative proportion of scientific and shop courses in the curricula. During the next two decades, however, while the number of teachers increased at a modest rate, there was a great growth in the number of nonteaching engineers and businessmen within the SPEE membership. The focus of discussion, moreover, shifted to such topics as "the efficiency of graduates," "adapting graduates to industry," changing the schools to meet the "specifications of industry," and preparing the engineering graduates for "business leadership."[25]

The combination of a status-minded community of engineering educators trying to enhance their prestige in the eyes of the classicists and of poorly endowed schools attempting in vain to keep pace with the rapidly changing industrial state of the art tended to produce graduates who were neither cultured gentlemen nor effective engineering practitioners. Perhaps most important, the schools turned out people imbued with a scientific zeal for untrammeled inquiry, the aristocratic arrogance of a college elite, or the entrepreneurial spirit of *laissez-faire* capitalism—people on the whole ill-adapted by temperament or training to employment in large authoritarian corporate organizations. The new reformers within the SPEE sought to change all that, to bring education into line with what they referred to as the new realities of modern industry.

Since, as Steinmetz later recalled, "the early corporate development of the electrical industry [had allowed it] to exert a considerable influence in shaping the curriculum of engineering colleges,"[26] the representatives of the electrical industries were the most forceful advocates of change within the SPEE. Dugald C. Jackson of MIT, a leading consultant for the electric-utilities companies, and Charles F. Scott of Yale, a Westinghouse engineer, initiated a plan for cooperation between the professional societies and the engineering educators which led to the first major study of engineering education in the United States, the Mann Report. Prepared for the SPEE and the Carnegie Foundation by physicist Charles R. Mann, the report reflected the new corporate industrial reorientation of engineering education. "For a number of years," Mann observed, "practicing engineers have felt that the instruction in colleges of engineering . . . was not organized to meet the demands of the profession."[27] The phrase "demands of the profession" was clearly intended, and understood, to mean the demands of the large industrial corporations and the engineers who served within them.

The "practicing engineers" who called for corporate, management-oriented training for engineers did so both in the interests of their profession and in the interests of the science-based corporations. Their approach to education was perhaps most succinctly articulated by Alfred H. White, professor of chemical engineering at the University of Michigan.

> There is some analogy between the college and the manufacturing plant which receives partially fabricated metal, shapes it and refines it somewhat, and turns it over to some other agency for further fabrication. The college receives raw material. . . . It must turn out a product which is saleable. . . . The type of curriculum is in the last analysis not set by the college but by the employer of the college graduate.[28]

Because the engineering schools produced at once professional engineers and the technical and managerial manpower required by corporate industry, leading engineers in industry looked to the schools both to meet their manpower needs and to recruit young men for their profession. The corporate and professional concerns of the "practicing engineers" were thus completely interwoven. Reform efforts within the schools aimed at gearing engineering education for the production of both efficient and loyal corporate employees and competent and dedicated "leaders of industry"; they were at once the work of a profession in search of power and recognition, and of corporate leaders in search of their subordinates and successors. It was through such reforms that engineering education in America became a major channel of corporate power.

The men who had begun to dominate the profession by the turn of the century were different in crucial ways from those who had pioneered before them. The transformation of the archetypal successful "professional engineer" closely paralleled that of the industrial leader from "tycoon" to "corporation manager" described by Paul Baran and Paul Sweezy in their study of corporate capitalism:

> In one respect [the corporation manager] represents a return to pre-tycoon days; his chief concern is once again the "surveillance and regulation of a given industrial process with which his livelihood is bound up." On the other hand, in another respect he is the antithesis of classical entrepreneur and tycoon alike: they are both individualists *par excellence,* while he is the leading species of the genus "organization man."
>
> There are many ways to describe the contrast between tycoon and modern manager. The former was the parent of the giant corporation, the latter is its child. The tycoon stood outside and above, dominating the corporation. The manager is an insider, dominated by it. . . . The loyalty [of the manager] is to the organization to which he belongs and through which he expresses himself. To the one the corporation was merely a means to enrichment; to the other the good of the company has become both an economic and an ethical end. The one stole from the corporation, the other steals for it.[29]

Perhaps no one better personified the new corporate approach to the profession and professional education than William E. Wickenden. Son of an Ohio civil engineer, Wickenden took a degree in electrical engineering and taught, under Dugald Jackson, at both Wisconsin and MIT. At MIT he teamed up with Magnus Alexander of GE to establish a cooperative program of study between GE at Lynn and MIT, and headed the curriculum committee. During the war he played an important role in the development of the Student Army Training Corps, and thereafter joined the Bell System, becoming successively personnel director of the Bell Telephone Labs, personnel-committee chairman of the Western Electric Company, and, in 1921, assistant vice-president of AT&T "in charge of recruiting and developing young men for technical positions." His work at AT&T was "to obtain the best and most promising men, and his field of research covered every technical school, scientific college and university in this country as well as abroad." In 1922, at the invitation of Charles F. Scott, who as president of SPEE the year before had set the stage for another major study of engineering education, Wickenden was appointed director of the newly created Board of Investigation and Coordination, and for six years headed the most comprehensive study of technical education in history. In 1929 Wickenden became president of the Case Institute of Technology in Cleveland, where he remained until retirement, and death, in 1947.[30]

Aside from his work in investigating engineering education in the 1920s, Wickenden's major contribution to the field was his influence in effecting the expansion of the curriculum along humanistic–social-science lines. Throughout his career he sought closer cooperation between industry and educational institutions, and strove to prepare the young engineer to be a "leader in business." A gifted speaker and a prominent professional in every capacity, Wickenden was one of the most important spokesmen for engineering professionalism in this century. His notion of professionalism was thoroughly corporate, and as an educator he sought to imbue young engineers with a similar spirit. Shortly before his death he began to write a "professional guide for junior engineers" which was subsequently published by the Engineers Council for Professional Development. In his "guide" Wickenden aptly articulated the requirements of the professional engineer, as laid down by his generation. He traced the path toward managerial and executive responsibility, and urged the student engineer to study human relations, read Elton Mayo's *The Social Problems of an Industrial Society* —"required reading for every young engineer"—and learn the "art of handling men." He even encouraged him to join unions "as an episode in [his] education," arguing that "membership will give him an opportunity to study the psychology, economics, politics, and tactics of unionism from within," and will prove a "laboratory experience" of great value when, as a manager, he tries to get "others to put their hearts into their jobs."[31] But nowhere does Wickenden better illustrate the ethos of his profession than in his instructions for those just "Beginning Professional Practice." The theme is the subtitle: "Subordinates Are Important: Be a Good One."

> In a sense, [the engineer] is like a soldier, he must have special qualifications for his arm of service, but his effectiveness . . . depends upon his team-play rather than upon his individualism. He must know how to follow before he is qualified to lead; he must learn to obey before he can give orders; and he must win his way to responsibility upward from the ranks.
>
> Of course, in a broad way you are working for society, the company, the department, your family, and yourself; but directly and primarily

you are working for and through your boss, and, as a rule, you will best serve your own and all other interests through this channel.

[Individualism on the part of the young engineer] very well may lead him into the pitfall of working for himself rather than for his organization. . . . If he is smart, he discovers that he is engaged in an almost unbelievably complex piece of teamwork rather than a race for individual recognition. . . . That is when he starts working for the organization. He has to find his place on the team; he has to learn the plays and the signals. . . .[32]

William Wickenden, like his colleagues in the professional-corporate-educational community of engineers, strove in the first thirty years of the century to get not only engineering students, but educators, industrialists, politicians, and working people "to find their place on the team," "to learn the plays and the signals." The team, however, was their corporate team; the plays and signals were their plays and signals. As leading spokesmen for the most dynamic sectors of corporate America, they aimed at creating a society geared to meet their scientific and corporate needs. The professional engineer, Wickenden instructed engineering students, "hopes to get on in life . . . by personal merit rather than by association with mass advancement" such as that offered by the unions. "The engineer," Wickenden—corporate executive, bank director, college president, and professional engineer—boasted, "is rarely class-conscious. When he is, it is usually a sign of defeatism and disillusionment."[33] Proud, dynamic, self-confident, a man supremely conscious of the purposes for which he and his associates labored, Wickenden belied his own boast.

Preservation Through Change

Corporate Engineers and Social Reform

If the process of social production in America was not automatic, neither was it blind. During the closing decades of the nineteenth century, the new institutions of science-based industry, scientific technical education, and professional engineering had gradually coalesced to form an integrated social matrix (composed of the corporations, the schools, the professional societies) within which that process could become visible to those who early participated in it. And it was within this circumscribed world that the first comprehensive, coherent view of modern technology emerged, a world-view of corporate reform. As science-minded corporate leaders strove to comprehend and meet the demands of industrial expansion and stability, as technical educators attempted to formulate the training requirements of the new industries within an academic setting, and as professional engineers tried at once to organize themselves and define their proper social function in the dual world of science and industry, corporate social production steadily became a conscious process and gained both direction and momentum. Attuned to the history that had produced them and their world, the corporate reformers of science-based industry now undertook to anticipate, and commit themselves to making, the history that would perpetuate both.

A collective portrait of these corporate reformers is not difficult to draw, since they were a remarkably homogeneous group. Like the majority of engineers, they were white males, an overwhelming proportion of whom were Anglo-Saxon Protestants of old-stock native American descent; the remainder came from families who had emigrated during the previous century from the British Isles or northern Europe, primarily Germany. The great majority were sons of prosperous middle-class businessmen, farmers, managers, teachers, lawyers, doctors, ministers, or engineers; had been born between 1860 and 1880; and had enjoyed a comfortable upbringing in the small towns of New England and the Midwest. Very few came from extremely wealthy families. Those who did were representative of what Christopher Lasch has called the "intellectual and moral rehabilitation of the ruling class"; having overcome the proclivities of their class toward leisure, sterile culture, escapism, and provincialism, they were able to use their hereditary wealth and class privileges to advantage in corporate industry.[1] Fewer still were born to poverty, this despite the recurring allusions of many in their later years to having worked their way up from nothing.

If any single characteristic best defined this group, it was the nature of their education. Most came from families in which a college education for a son was not only possible but probable (at a time when less than two percent of their generation went to college). All received a scientific training, in electrical, mechanical, or chemical engineering, physics, chemistry, or some combination of these, having attended such schools as Harvard, Yale, MIT, Columbia, and Princeton in the East; Michigan, Illinois, Purdue, and Wisconsin in the Midwest; and California, Stanford, or the University of Washington in the West. The largest proportion were trained in electrical engineering, and a significant majority received their education—as undergraduates, graduates, or postgraduate assistants—at MIT. Already by the turn of the century this educational hub of corporate engineering had begun to challenge Harvard as a breeding ground for the ruling elite in industrial America by bringing talent and capital together. (This is illustrated by the fact that in the 1920s the chief executives of General Motors, General Electric, Du Pont, and Goodyear—four of the largest corporations in the world — had been classmates at MIT a quarter-century earlier.)

In the course of their education and their subsequent work as employees of or consultants to the corporations, these men established the contacts and adopted the point of view which made them members of the elite professional group that promoted, oversaw, and directly profited from the scientific and industrial development of corporate America. Theirs was a relatively small circle of men whose common vision and purpose enabled them to shift easily between positions in industry, the universities, and the various scientific bureaus of government. When they met to discuss their work, they often did so in the committee rooms of the professional societies, the executive offices of the large corporations, or the deans' and presidents' offices of the universities. Most often they shared their insights over luncheon in such gentlemen's clubs as the University, Technology, Columbia, New Haven, Engineers, Chemists, Bankers, Century, Commonwealth, Union, or Cosmos. It was in drafting rooms like these that the designs for a new America were drawn.

Perhaps no one epitomized the spirit of these corporate reformers better than Magnus Alexander. A German-born electrical engineer who had worked as a designer for both Westinghouse and GE, Alexander early embarked upon social-reform projects which illustrated the breadth of vision shared by these men. As the first director of GE's education-and-personnel department at Lynn, Massachusetts, Alexander formulated the plan of cooperation between MIT and GE, founded and directed the apprenticeship training program, and supervised the technical courses for engineering-college graduates, all of which were pioneering ventures. Outside GE he became an active member of the Massachusetts commissions on old-age pensions and workmen's compensation, among the earliest in the country. He was a charter member of the National Association of Corporation Schools, the forerunner of the American Management Association, and vice-president of the National Society for the Promotion of Industrial Education. He participated in the affairs of the Society for the Promotion of Engineering Education and the American Institute of Electrical Engineers, worked with fellow electrical engineer Malcolm Rorty and economist Wesley Mitchell on some of the early work of the National Bureau of Economic Research, and prepared the first comprehensive study of the cost of labor turnover. In 1916 he launched his biggest project. Together with Loyall Osborne, chief engineer and later president of the Westinghouse Electric International Company, patent attorney Frederick P. Fish, former GE counsel and AT&T president, and Frank Vanderlip, president of the National City Bank of New York, he founded and directed the National Industrial Conference Board, the "research arm" of U.S. industry and the largest cooperative undertaking of American employers up to that time.

Alexander's career was indicative of the bold, confident, and pioneering spirit which moved him and his associates, and his varied concerns reflected the complex of economic, technological, social, and political problems with which they had to contend. In all of their undertakings they worked to counter the forces of instability inherent in the evolving capitalist economy, to ensure the continued prosperity of corporate industry, and to promote and regulate the scientific and technological progress upon which such prosperity depended. Driven by the dual requirements of corporate hegemony and technological development, they moved from the workshops of industry to join other farsighted business leaders, bankers, and corporate-minded social reformers in the universities and the government in the task of transforming America. In a sober and scientific manner they strove to analyze, rationalize, systematize, and coordinate the entire "social mechanism," to translate haphazard, uncertain, and disruptive social forces into manageable problems for efficient administration.

In October 1929, Magnus Alexander addressed the World Engineering Congress in Tokyo on "The Economic Evolution of the United States" and recounted the collective achievements of the corporate reformers. In a scholarly manner he traced the developments, "the refinement of technique, methods, and policies," which had characterized the work of the previous decades. He recalled how the industrial-merger movement had failed to stabilize production and domestic markets, and how bigness had proved initially unprofitable, and described the subsequent success of trade-association cooperation, government regulation, and scientific management. He stressed the importance of economic and social research, market analysis, industrial and governmental planning, labor-management cooperation, streamlined production management, and the "rationalization of the distributive process" in achieving social equilibrium under the corporate aegis. He proudly proclaimed the new guiding business principle "that productivity *creates* purchasing power," and described how higher wages, the enhancement of buying power through consumer credit, and the creation of demand through advertising and other merchandising techniques reinforce productivity and corporate prosperity.[2]

The phrase "time is money," Alexander observed, "once so widely and frequently heard, is now seldom used. As against the flair for magnitude, speed, ruthless competition and an element of gamble, which characterized American business life in the nineteenth century, there is today a tendency toward consolidation and association for greater efficiency, careful planning on the basis of diligent research, and much concern over the ramifications and effects of economic life upon social progress. . . . Whereas," he continued, "*laissez faire* and intensive individualism marked the economic life of the first half of the history of the United States, the emphasis is now shifting toward a voluntary assumption of social obligations . . . and cooperative effort in the common interest."[3]

At Tokyo, just before Black Thursday and the start of the worst depression in the nation's history, Magnus Alexander exuded the confidence that had prompted him to remark a few weeks earlier that "there is no reason why there should be any more panics." His optimism was rooted in the innovative and remarkably successful work of thirty years, during which time he had witnessed what appeared to him to be the creation of a new society, grounded upon corporate prosperity, in which the horrors of industrial warfare, economic instability, social inequity, poverty, and political strife had all but been eclipsed. "The United States," he proudly reported to his international audience, "shows definite signs that a period of a more settled existence has begun." The first three decades of the century, as they all well knew, had been anything but settled.[4]

The same nineteenth-century industrial capitalism that had spawned modern technology and professional engineering had also generated the corporations and, as well, the social turmoil that confronted them. The unprecedented economic expansion following the Civil War had given rise to intense competition between industrialists who vied for larger shares of the market.[5] This competition had the effect of depressing prices and thus compelled manufacturers both to increase production and to lower costs through the introduction of new machines and more efficient production methods. Enhanced productivity only aggravated the chronic problem of overproduction, which depressed prices further, and compelled manufacturers to expand their markets nationally and internationally to sell their wares. Competition between manufacturers was thus further intensified.

By the turn of the century the larger corporations which had survived this competition undertook to stabilize the chaotic economic situation, to keep capitalism alive by eliminating much of the economic freedom that had both made it so popular and driven it to the brink of disaster. In the place of economic freedom, they offered the goods, the abundant fruit of efficient and regulated industrial production. Toward this end, and in a country whose *laissez-faire* traditions precluded the establishment of legal cartels, these corporations expanded to control as much of the market and the production process as possible, to eliminate lesser competitors and coordinate the many aspects of each industry. By means of consolidations, mergers, trusts, holding companies, trade associations, and, ultimately, government regulatory agencies, they struggled to regulate production and stabilize prices. At the same time, they undertook to rationalize the sprawling empires which they had acquired in the process—to make them profitable as well as powerful by utilizing expensive plants at maximum capacity, fully realizing the potential of technological developments, and coordinating all the varied activities under their command. In the science-based industries this meant the rationalization of modern technology itself as well as the production processes it made possible: the standardization of scientific terminology and methods, the regularization of patent procedures, the efficient organization of scientific research, and the systematic production of technical manpower.

The corporate control over production, prices, and markets, coupled with an unparalleled productive capacity, made possible unprecedented profits and an astounding volume and variety of goods. The tremendous momentum of economic expansion of the previous half-century, however, had already begun to slow down, as plants reached maximum size for efficient manufacture, natural resources were depleted, the spur of railroad construction peaked and fell, and domestic markets became saturated. In such a context, the very success of corporate monopoly generated problems. Efficient large-scale production and the need to make full use of expensive plant and equipment led to overproduction when demand failed to keep pace, despite efforts to regulate output. As a result, new techniques of merchandising were required to create both demand and purchasing power, to produce the masses who would and could consume what was being mass-produced.[6] At the same time, artificially high prices, and thus profits, coupled with a declining rate of economic expansion, generated an unwieldy capital surplus that had somehow to be profitably invested; new areas of investment had to be found.[7] Aside from the burgeoning automotive industry, overseas ventures, and advertising—as well as the various social-welfare programs designed to ensure a modicum of social stability—none appeared so promising as research and development activities, scientific management schemes, and education.

The contradictions inherent in an economic system which must expand or perish thus defined corporate reform efforts. The tension generated by these contradictions, moreover, created a perpetually unstable economic setting for such efforts, one marked by the periodic "spurts and recessions"[8] following the bankers' panic of 1907, the boom of wartime expansion, the postwar depression of 1921, and the automobile-spurred prosperity of the 1920s. The imperatives, and perils, of the monopoly capital system, however, were only part of the picture. For the same process of industrialization that had given rise to the industrial capitalist and his vehicle of monopoly, the corporation, had also produced an industrial working class, the human machines whose labor produced the goods, the profits, and thus the capitalist system itself. In the closing decades of the nineteenth century this working population had also begun to organize—to struggle against the increasing capitalist monopolization of the means of production, and to secure a larger share of what was being produced.[9]

The emergence of the railroad brotherhoods and the American Federation of Labor out of the bitter confrontations of the 1870s and '80s gave "cohesion, confidence, and purposefulness"[10] to the labor movement. The brutal aftermath of the Haymarket Riot of 1886 reinforced the resolve of labor to organize, and the strikes of the 1890s—especially the Homestead and Pullman strikes—eclipsed those of the previous decades in terms of human cost. Encouraged by the growing popular antagonism toward big business, and by the progressive recognition of the legitimacy of labor demands, the strikes continued with greater intensity after the turn of the century: the United Mine Workers strikes of 1900, 1902, and 1912; the steel strike of 1901; the International Ladies Garment Workers Union strike in 1909; the IWW-supported strikes of the Western Federation of Miners in 1907 and against the Pressed Steel Car Company in 1909; and the free-speech fights at Spokane, Fresno, and San Diego between 1909 and 1912. In the second decade the struggle for union recognition, coupled with a movement for the radical restructuring of society, intensified even more, as union membership multiplied by sixfold the 1897 figure; unorganized workers, in addition, increasingly employed sabotage, work slowdowns, and similar means to achieve their objectives. The reports of the United States Commission on Industrial Relations, which investigated industrial strife between 1910 and 1915, reflected the enormous scope of industrial turbulence. As one historian has written,

> Regardless of cause, geographic location, type of industry or ethnic grouping, turbulence in industrial relations flared all over the United States. It rocked large cities and small towns, manufacturing areas and agrarian communities. . . . Industrialization had outdistanced American social attitudes and institutions. In many cases this led to a collapse of civil authority, to near anarchy and to military rule. . . . Americans on the eve of World War I lived in an age of industrial violence.[11]

The science-based industries were hardly immune to such labor conflicts; in 1913, for example, there was a major strike at the GE plant in Schenectady, in part a response to the firing of union organizers; in

1915 a major strike rocked the Westinghouse plants in East Pittsburgh and "the most prominent spokesmen of the strikers had Socialist Party, Socialist Labor Party, and I.W.W. affiliations."[12]

The strikes and industrial violence continued unabated during the World War, despite a mounting patriotism-spurred repression, on the one hand, and significant government efforts at conciliation, for wartime expediency, on the other. After the war, the industrial unrest flared to new heights, and events such as the Boston police strike, the Seattle general strike, and the United States Steel strike—all in 1919— drove the war-wearied nation to its first major government-sponsored purge of the Red Menace. Compounding the "labor problem"— as the corporate engineers and other industrial leaders viewed it— was the enormous tide of the "new immigration" between 1897 and 1914.

Immigration had declined considerably from the nineteenth-century peak in 1882 before an upsurge began three years before the turn of the century, one that would eventually bring over eight million people from foreign lands to American shores. This new influx of immigration differed from those which had preceded it in composition as well as magnitude. Over eighty percent by 1905 were from southern and eastern Europe, and were either Greek or Roman Catholic or Jewish; before 1883 ninety-five percent of immigrants had come from northern and western Europe, and had been primarily Protestant. The new immigration differed in other ways as well: it was predominantly composed of single males rather than families, and over a third of them— as compared with around three percent of previous immigrants—were illiterate; these newcomers generally had less interest in learning the language and ways of Americans, since they tended to consider the United States a temporary station from which they would return to their homelands after they had succeeded, or utterly failed, to "make their fortune in America."[13]

The "new immigration" played contradictory roles in the industrial situation of the period; coming in response to business prosperity— many of these immigrants had contracted with specific employers before they embarked—and emigrating in periods of business setback, they were used by employers to "keep wages low and develop management techniques."[14] They tended to settle in urban industrial areas rather than in the rural farmlands, and principally in New York, Massachusetts, and Pennsylvania, the centers of the mining, steel, and science-based industries. The engineer-managers of the last-named, overwhelmingly of either native American or northern European

stock,* thus had considerable experience with the new immigrants. At
GE alone, over two thousand Slavs worked in the Schenectady foundry
in 1903, under the direction of a New England Yankee and Worcester
graduate named Prince. The alien ways of the immigrants widened the
gap between the worker and the manager, and better enabled the latter
to treat the former as simply another object of "scientific" study. These
corporate engineers, in addition, played an important role in the Americanization movement—designed in part to orient workers to industrial
conditions—and many, like German-born Magnus Alexander, strongly
supported the ethnically prompted restrictive immigration legislation
of the 1920s.[15]

The new immigrants' activities did not always reinforce the position
of the employer against organized labor, however. They often "plunged
into battle" alongside American native labor against their employers,
and the growing Socialist Party commonly "united those of old American stock with newly arrived immigrants."[16] But whether actually
engaged in the pitched battles on the side of organized labor or employed as scabs and cheap labor to thwart union demands, the immigrants presented problems to their employers simply because of who
they were. All had brought with them into the industrial situation, as
Herbert Gutman has observed, "ways of work and other habits and
values" rooted in their own "pre-industrial cultures," habits and values
which were "not associated with industrial necessities and the industrial ethos." As a result, their arrival perpetuated the "recurrent tension
. . . between native and immigrant men and women fresh to the factory,
and the demands imposed upon them by the regularities and discipline
of factory labor" which had marked the American experience throughout the nineteenth century.[17] While corporate industry increasingly
dominated the economy of the society and the land was steadily abandoned for urban industrial employment, the new immigrants with their
"medley of foreign tongues and customs" and their "pre-industrial"
work habits continually revitalized the problems of habituating a society to industrial imperatives. Living in terms of their own cultural
heritage, which they sustained and which sustained them in a strange
and hostile environment, the new immigrants defied ready absorption
into the industrial process. Like other workers, they challenged and

*In his study of engineering education, Wickenden found that "comparatively few engi-
neering students seem to have derived from the wave of immigration from southern and
eastern Europe." *Report of the Investigation of Engineering Education* (Pittsburgh: Society for the Promotion of Engineering Education, 1930), I, 162.

thereby helped to shape that process, and thus the techniques and tasks of the engineer-managers who designed and ran it.*

The dynamics of the changing economic situation, the growing labor movement, and the swell of immigration gave rise to new perceptions of the good society, which found expression on the political level. By the turn of the century, various political forces sought to readjust the political economy to the evolving requirements and possibilities generated by industrial capitalism. The leaders of labor, small business, corporations, immigrant groups, traditional political parties, revolutionary parties, and the professions began to formulate the politics required to turn the gains in technics to their own advantage. First and foremost was the emergence of a popular movement for democratic and industrial reform. Dating back to the Populist surge of the 1890s and the writings of Henry George and Edward Bellamy, the reform movement gained new momentum throughout the country with the muckraking exposures of political and business corruption in high places. Reform leaders at the municipal, state, and national levels called for a progressive revitalization of the democratic spirit through the adoption of the initiative, recall, and referendum, and the direct election of U.S. senators. They demanded a redistribution of wealth and opportunity by means of welfare legislation, industrial "uplift" programs, and woman's suffrage, and sought a rebalancing of economic power through antitrust legislation, government regulation of business, and support of trade unions. At all levels of government, moreover, they worked to "clean up city hall," to make the "business of government" more orderly, efficient, and honest.

In addition to the progressive reform movement, which reached national proportions with the creation of the Progressive Party and

*Except for the familiar history of the American trade-union movement, little is known about how workers, through their resistance and opposition, helped give shape to the modern industrial process. The engineering and management of production rarely if ever involved simply the transfer of designs from drawing board to shop floor. New approaches were introduced and abandoned, or endlessly revised to better adapt them to the work situation, a context in which people with conflicting interests, rather than mere considerations of elegance or efficiency, determined the final outcome. Without a precise description of how this has happened, the history of technology must remain a one-sided, and hence distorted, account. Herbert Gutman and David Montgomery have begun to fill in this gap in our understanding. See, for example, Gutman's *Work, Culture, and Society in Industrializing America* (New York: Alfred A. Knopf, 1976), pp. 3–78, and two unpublished papers by Montgomery, "The 'New Unionism' and the Transformation of Workers' Consciousness in America, 1909–1922" (1972) and "Immigrant Workers and Scientific Management" (1973).

which aimed primarily at eliminating the horrors of industrial capitalism in order to make it work better, there existed both socialist and anarchist movements which aimed at eliminating the capitalist system altogether. Throughout the first two decades of the century an increasingly broad-based movement for socialism elected Socialist Party candidates to municipal offices in many industrial cities (among them Schenectady, New York, headquarters of GE) and, under the leadership of Eugene V. Debs, attained considerable strength on the national level. As James Weinstein has written, the socialist movement "was conscious of its traditions and was ideologically unified by a commitment to a socialist reorganization of society as the solution to the inequalities and corrupting social values it believed were inherent in American capitalism. Before 1920, the old Socialist Party had mass support at the polls, a widespread and vital press, a large following in the trade-union movement, and profound influence on the reformers and the reforms of the day."[18] In addition to the Socialist Party, the revolutionary International Workers of the World, the IWW, actively worked to achieve socialism in America. Under the leadership of men like Big Bill Haywood, the Wobblies attempted to organize the industrial workers outside the trade unions and to radicalize the skilled workers within them. With its antagonism toward the existing order, its willingness to employ sabotage, and its success among the unskilled and unorganized workers, the IWW did much to ignite the labor struggles of the period, and to arouse the anxiety and wrath of old-stock middle-class Americans, conservative trade-union leaders, and the managers of industry. It never elicited the revolutionary support necessary to challenge seriously the establishment parties, much less the capitalist system itself; it lacked a coherent and consistent vision of a new order; and it was crippled by increasing internal dissension. But the movement for socialism in America was nevertheless a force to be reckoned with. The specter of a socialist revolution, especially after the Bolshevik success in Russia, remained a staple of political rhetoric throughout the period, and this rhetoric reflected the genuine and pervasive fear which prompted continuous industrial and governmental surveillance and repression.

The business community actually split in response to the movements for socialism and democratic reform. The majority of small businessmen, many of whom sought collective action through various trade associations and the National Association of Manufacturers, fought vigorously against the labor unions, the closed shop, and government regulation of business, not to mention socialism. More farsighted big-business leaders, in contrast, aimed at absorbing moderate reform movements, anticipating or redirecting them, while at the same time isolating the proponents of more radical change. Espousing a general theme of cooperation, social harmony, and economic and political order, they stood in opposition to socialism, on the one hand, and the anarchy of unrestricted competition, on the other. These "corporate liberals,"* as they have been called, sought above all to reconcile traditional liberal democratic notions of individualism, self-reliance, free enterprise, and anti-statism with corporate-capitalist and scientific-technological demands for order, stability, and social efficiency. Emphasizing first one, then the other, they worked to regulate the corporate economy through the agencies of government, through private associations like chambers of commerce and trade organizations, and through such research agencies as the National Bureau of Economic Research, the Brookings Institution, and the National Industrial Conference Board. Through reform bodies like the National Civic Federation, they promoted social-welfare legislation in order to reduce the burdens and antagonism of working people, and strove to enlist the labor unions as voluntary partners in the corporate industrial system, thereby hoping to substitute orderly and predictable negotiation for industrial warfare.[19]

Engineers responded in different ways to the changing social situation from their vantage point within industry. Whatever the particular form of their reaction, they sought to test their strength in the political arena and thereby to define the potentials and limitations of their social function. The experience would provide further indication of the close relationship between engineering and institutionalized social power.

*For further elaboration of corporate liberalism, see Martin J. Sklar, "Woodrow Wilson and the Political Economy of Modern U.S. Liberalism," *Studies on the Left,* Vol. I, No. 3 (Fall 1960); James Weinstein, *The Corporate Ideal in the Liberal State* (Boston: Beacon Press, 1968); William Appleman Williams, *The Contours of American History* (Cleveland: World Publishing Company, 1961), pp. 343–469; James Gilbert, *Designing the Industrial State* (New York: Quadrangle, 1972); Ronald Radosh and Murray N. Rothbard, eds., *A New History of Leviathan* (New York: E. P. Dutton & Co., 1972); David Eakins, "The Development of Corporate Liberal Policy Research in the U.S., 1885–1965," unpublished Ph.D. dissertation, University of Wisconsin, 1966; and Jerry Israel, ed., *Building the Organizational Society* (New York: Free Press, 1972). For a recent critique of the concept of corporate liberalism, which argues that the significance of corporate liberals has been overdrawn by revisionist historians, see Kim McQuaid, "A Response to Industrialism: Liberal Businessmen and the Evolving Spectrum of Capitalist Reform, 1886–1960," unpublished Ph.D dissertation, Northwestern University, 1975. For a clear, concise analysis of the murky, always tenuous resolution between corporatism and liberalism and a brief history of corporate liberal attempts at statecraft, see Ellis Hawley's excellent unpublished paper "Techno-Corporatist Formulas in the Liberal State, 1920–1960: A Neglected Aspect of America's Search for a New Order."

In the first two decades of the century a number of engineers began to perceive a contradiction between socially beneficial technological progress and corporate control of the material and human means to that progress, between the possibilities of science and the demands of profit-making business. They began to question industry domination of their professional societies,* to seek public employment in government agencies on all levels, to offer their technical services to labor and radical movements, and to demand more power for themselves as engineers. In all of these efforts they tended to reject the engineering creed that the dollar had the last word, and to shift the balance of engineering priorities from profit to scientific integrity, social betterment, and political reform.

Emphasizing the scientific expertise of engineers, men such as Morris Cooke and Frederick Haynes Newell strove to democratize the professional societies of the mechanical and civil engineers and to redirect their energies from corporate to public service. They actively challenged industrial control over their profession through the establishment of new organizations such as the American Association of Engineers and the Federated American Engineering Societies. On the local level, civil engineers such as C. E. Drayer, F. W. Ballard, and other so-called Cleveland radicals—politically active members of the Cleveland Engineering Society—called for public control of utilities and "social engineering" in the public interest; Morris Cooke, as head of public works in the reform administration of Philadelphia Mayor Blankenberg, likewise attacked the privately owned utilities companies and tried to enlist the support of other public-spirited engineers.[20]

While "progressive" engineers aligned themselves with democratic and antibusiness reform movements, they also sought to enhance the engineer's status by encouraging reliance upon technical and administrative expertise in political decision-making. Some, like Cooke and Hollis Godfrey, had been close associates of Frederick Taylor and aimed toward acquiring more power for themselves outside of industry as they had, by means of scientific management, within industry. While a small number succeeded,† most were repudiated by their profession and some had to give up engineering practice altogether.

---

*Aside from being well represented in the executive committees of the professional societies, private companies commonly paid the membership dues for their employees.
†Godfrey, for example, was elected president of the Drexel Institute, largely as a result of his survey of technical education and manpower for the city of Philadelphia; Cooke eventually became head of the Rural Electrification Project during the New Deal; and Arthur Morgan, an outspoken flood-control engineer, was appointed first director of the TVA.

Most progressive engineers, however, conformed to the "corporate liberal" role of combining social reform with corporate requirements. The Cleveland radicals, for example, clearly fitted this mold; they were the most active members of that city's chamber of commerce and sought, above all, to streamline their city in order to attract large industry. The significant number of disenchanted young engineers who had rallied behind Newell and Drayer in the American Association of Engineers, moreover, did so in a period of business setbacks and unemployment; with the return of war-spurred prosperity, the Association collapsed. And the Federated American Engineering Societies, the most potent vehicle of progressive engineering reform, eventually elected as its president one of the foremost spokesmen of corporate America, Herbert Hoover. Tied as they were to the industrial organizations which controlled the means of professional practice, most progressive-minded engineers realized that antibusiness political activity grounded upon engineering expertise led inevitably to political elitism and impotence, technical impotence, or both; radical engineers, they understood, had to chose between being radical and being engineers.[21]

A handful chose to be both, and ended up being neither. Taylor's disciple Henry Gantt followed through the more radical implications of scientific management and called for the centralization of social power in the hands of technical experts; he set up the "New Machine" directed toward the radical restructuring of society, but this effort ended before it began, with his death in 1919. Like-minded colleagues Walter Rautenstrauch, Guido Marx, and the engineer-imposter Howard Scott formed the revolutionary Technical Alliance, with Thorstein Veblen at the New School for Social Research, and worked toward contradictory ends; while they sought to mobilize a "soviet of engineers" which would lead the working class in a social "overturn," they also envisioned a society run by science and engineers. Aside from Scott's brief flirtation with the IWW, the emphasis in practice was clearly placed on the latter. Thus, this most radical and isolated segment of the engineering community remained on the periphery, even when it surfaced briefly in the early days of the Great Depression as the media fad and political dead end, Technocracy.[22]

The corporate engineers of science-based industry were most like the sophisticated, class-conscious corporate liberals of the big-business and banking community in their response to the various social currents of the day. As the personification of the link between science and industry, however, they typically tended more toward techno-corporatism than liberalism (except perhaps when defending the prerogatives of the private industrial firm). They sought change in order to preserve, striving to meet by whatever means possible the dual imperatives of corporate growth and technological progress which defined their professional lives. Although they enjoyed the support of many of the most powerful figures in the industrial world, and through stock options, large salaries, and lucrative consultant practices directly shared a stake in such corporate growth, they were driven neither by simple self-interest nor by lust for power. Rather, they were moved by a shared dream, a compelling vision of an affluent, humane, tranquil, and powerful America. The strength of their commitment to it derived from their sense of history. They perceived themselves as revolutionaries striving to fulfill the promise of past human accomplishment; if they labored to gear the nation for a particular form of social organization, that of corporate capitalism, it was because they viewed such a system as both historically necessary and inevitable. As students of modern technology, moreover, they looked to science for their guide and defended their activities in the name of science. They were bold and resourceful largely because, to their mind, they moved with destiny in a world that made sense.

At the turn of the century these reformers embarked upon a far-reaching enterprise, to bring American society into line with technological advance and corporate growth. At a time of considerable social turmoil, marked by a chronically unstable economy, a widespread popular demand for industrial and political democracy, a militant labor movement, an unprecedented flood of immigration, and an indigenous movement for socialism, they set about to design the new social institutions and foster the social habits which this transformation required. Thus, in 1936, when President Franklin D. Roosevelt wrote a letter to the SPEE urging engineering educators to try to instill in young engineers more of a sense of social responsibility for technologically rooted social problems, Dean Andrey A. Potter of Purdue could respond cheerfully and with assurance that he and his colleagues were "fully appreciative of the responsibility of the engineer in bringing about a better balance between technological progress and social control." For thirty years they had worked to achieve what Magnus Alexander called the "benevolent circle" of prosperity. They brought to their task an intimate knowledge of modern industry and scientific technology that was unequaled by any of their adversaries, who sought to steer the country in altogether different directions, and they derived from it a pioneering élan, an unshakeable confidence of purpose. Aside from that, their only advantage lay in the fact that they entered the contest already on top.[23]

Part Two

Corporate Reform as Conscious Social Production

Laying the Foundation

Scientific and Industrial

Standardization

> We are the victims of looseness in our methods; of too much looseness
> in our ideas; of too much of that sort of spirit, born out of our rapid
> development. . . . Nothing can dignify this government more than to be
> the patron of and the establisher of absolutely correct scientific stan-
> dards and such legislation as will hold our people to faithfully regard
> and absolutely obey the requirements of law in adhesion to those true
> and correct standards.[1]
>
> —Lyman Gage

The corporate reformers from the engineering community moved
swiftly into the twentieth century, ahead of their contemporaries. By
the 1890s they had already begun to lay the foundation for it close to
home. In the rapid and uncontrolled growth of modern industry there
had evolved a great diversity of manufacturing techniques and industrial products, with each manufacturer devising his own and pitting
them against those of his competitors on the open market. The result
was a confusion and a duplication of effort unknown to later generations of Americans. The rule of the marketplace, survival of the fittest,
generally determined in what manner industry would proceed and what
shape its machinery, tools, and industrial products would take. As the
science-based industries emerged, they too had adopted this haphazard
mode of operating and, as a result, the market was flooded with competing types of electrical apparatus and chemical products peddled by the
new companies, in addition to the plethora of machined goods and
machinery itself. Since each company was concerned above all with
promoting its own products, and devised unique means of evaluating

their quality and performance, comparison between competing products was difficult. And since each manufacturer used its own specially machined parts, replacement by those of another was virtually impossible.

For the corporate reformers such a situation was untenable, from the standpoint both of economics and of engineering. Although they viewed the wide variety of consumer goods as evidence of the vitality of free enterprise, they regarded the same variety in industrial products as a serious obstacle to economic and technological development. The corporate consolidations which had been forged primarily to bring some order to the chaotic economic situation were to them the means of unprecedented scientific and technical development, giant engines of progress. With their massive investment in machinery, plants, and manpower, however, the coporations were economically viable only given maximum utilization of resources—that is, large-scale, continuous production. And such production was economically viable, in turn, only given large-scale consumption. But without some uniformity in the types and dimensions of industrial products which went into the production of consumer goods, and standard specifications for the performance of equipment and machinery, the interchangeability of parts and the regularization of manufacturing processes upon which large-scale production depended were impossible. And without widely recognized standards of quality and readily available means of servicing products after sale, large-scale consumption was impossible. Standardization in industry was thus the *sine qua non* of corporate prosperity and, since the corporations were the locus of technological innovation, of scientific progress as well. Just as competition in marketing had eventually threatened the survival of the very economic system which made it possible, so competition in production threatened to retard the technological development which it had heretofore promoted.

In the new science-based industrial corporations, the pressure for reform was most severe. Never before had there been such a demand for controls in production, for uniformity, precision, reproducibility, and predictability, as there was in the electrical and chemical industries. Just as they cohered in the theory and in the design, all the pieces of scientific production had to be made to fit in practice, to complement each other in a smooth, uninterrupted, predetermined process. The rationality of science thus defied the irrationality of the marketplace; like the corporations which thrived upon it, science itself demanded standardization. For the corporate reformers of science-based industry, trained in science and weaned upon large-scale enterprise, the challenge was made to order.

Modern industrial standardization presupposed scientific standardization, and this demanded precisely defined units of measure and accurate means of measuring them. The need for precision workmanship first became acute in the machine shop, the heart of modern industry. At the end of the eighteenth century, James Watt had encountered considerable difficulty trying to secure the right tools and competent workmen to bore engine cylinders with the uniformity demanded by his machine designs. Variations of one sixteenth of an inch in the bore of an engine cylinder were common in those days, and accuracy to one thirty-second of an inch was a sign of superior workmanship. During the nineteenth century, significant advances were made in mechanical worksmanship, owing largely to the development of measuring devices and refined techniques. The pioneering work of the Englishmen Charles Babbage, Henry Maudsley, and Joseph Whitworth and the American William Sellers in the refinement of machine-tool apparatus, gauges and standards, and the training of skilled machinists laid the basis for the modern machine-tool industry. By the start of the twentieth century, machinists could work metals to a tolerance of five one-thousandths of an inch and produce uniform and durable machine parts to precise industrial specifications. The mechanical engineers who emerged from the machine shops were thus keenly attuned to the industrial requirements for accurate standards. In 1884 the ASME created a Committee on Testing in an effort to coordinate standards for measurement and terminology, and in 1898 an American Section of the International Association for Testing Materials (incorporated four years later as the American Society for Testing Materials*) was established for similar purposes.[2]

The emergence of the science-based electrical and chemical industries heightened the demand for standards of measurement. "As any branch of industry grows more scientific," Dugald Jackson, chairman of MIT's electrical-engineering department, explained, "that is, as it comes to fully utilize organized and recorded knowledge in place of empiricism resting on the personal experience of individuals or groups of individuals, it becomes more and more conscious of the usefulness of precise methods and accurate measurement." The engineers pro-

*This name was subsequently changed to the American Society for Testing and Materials.

duced by these industries, which were grounded in physical science from the outset, thus tended toward precision in definitions and measurement. The electrical engineers, for example, established the standards committee of the AIEE in the 1890s to coordinate standardizing activities in the electrical industry. They also pushed for the inclusion of electrical and chemical standards in the Office of Weights and Measures of the U.S. Coast and Geodetic Survey, and fought vigorously for the adoption of the metric system. Industrial chemists set up a standards committee in the American Chemical Society around the same time for similar purposes. Above all, there was a need in these new industries for reliable instruments, calibrated by dependable standards, for the measurement not only of length, weight, and volume, but also pressure, heat, light, electricity, magnetism, and radioactivity, to determine the rating, output efficiency, and durability of machines, devices, and processes. Efforts on the part of engineers along these lines dovetailed with those of professional scientists, physicists and chemists, to standardize their terminology, establish accurate physical constants, and systematize methods of analysis.[3]

Physical scientists played an important role in the quest for standards, and their work laid the groundwork for subsequent engineering and industrial standardization. The physicist A. A. Michelson, for example, America's first Nobel Prize recipient, calculated the speed of light by developing accurate methods of determining it. By the 1890s he had come up with an apparatus, which he called the interferential refractometer, for the measurement of physical lengths in terms of an incorruptible standard, the wavelengths of homogeneous light. William A. Noyes, a brilliant analytical chemist at Rose Polytech, gained an international reputation for his pioneering work of developing and systematizing the standard analytical methods and specifications for chemicals. He later became first chief chemist of the U.S. Bureau of Standards. The man who perhaps contributed more than any other toward standardization in science, however, was Henry S. Pritchett.

The son of an astronomer, Pritchett himself became one of the most prominent astronomers in the country and, at Washington University in St. Louis, the teacher of the country's most extensive course in the discipline. Working with the U.S. Coast and Geodetic Survey, he developed a transit instrument to make time observations and calculate longitudes which formed the basis for the topographic maps of the U.S. Geological Survey. His transit instrument was used to establish standard time in the central time zone when it was inaugurated in 1883, and for several years thereafter "his observatory was the source of the exact time which facilitated uniform time for station clocks and trainmen's watches along . . . 50,000 miles of railway in the Mississippi Valley."[4] In 1897 Pritchett became superintendent of the Coast Survey and, three years later, the president of MIT as well. In 1905 he submitted a plan for faculty retirement allowances to Andrew Carnegie which led to the creation of the Carnegie Foundation for the Advancement of Teaching, with Pritchett as president. As a scientist with a great interest in the scientific developments in engineering, industry, and education, Pritchett fully recognized the necessity for centralized standards other than those for time. When Fred Halsey delivered a paper at MIT expressing the machine-tool industry's opposition to the metric system, for example, Pritchett denounced him and argued for governmental standardization.[5] Pritchett endorsed a centralized standardizing mechanism for both industrial and scientific reasons, and thought such development was vital to the nation's defense against industrial competition from Germany. Germany, with its Physikalische-Technische Reichsanstalt, had a well-established national physical laboratory to which many American industries and scientists were compelled to resort for crucial standardizing services. This, Pritchett and many industrialists and scientists felt, placed the United States at a distinct competitive disadvantage. America needed its own national physical laboratory.[6]

With the discovery of large discrepancies in weights and measures in various customs houses, the U.S. Congress had established, in 1836, an Office of Weights and Measures to provide some regulation and uniformity. While legally a bureau of the Treasury Department, the office in practice operated as a part of the U.S. Coast and Geodetic Survey. It had limited funds, lacked the authority to issue verification certificates, and did not have the facilities to determine either chemical or electrical standards. As late as 1886 the head of the office, Charles Sanders Peirce—the first philosopher of pragmatism—complained that "the Office of Weights and Measures at present is a very slight affair, I am sorry to say." The National Academy of Sciences, at its annual meeting in 1900, concurred: "the facilities at the disposal of the government and of the scientific men of the country for the standardization of apparatus used in scientific research and in the arts are now either absent or entirely inadequate, so that it becomes necessary in most instances to send such apparatus abroad for comparison."[7]

Pritchett, appointed director of the Coast and Geodetic Survey in 1897, was determined to bring the government agency into line with changing industrial and scientific needs. "It was clear," as he later explained, "that the advancing industrial needs of the country were already beginning to cause a demand for standards other than those of weight and measure." He persuaded Samuel Stratton, another physicist (and later MIT president) who had simplified Michelson's apparatus for measuring light, to leave the University of Chicago temporarily and help him reorganize the Office of Weights and Measures, to work on "a plan for its enlargement into a more efficient bureau of standards, which might perform in some measure for the country the work carried on by the *Reichsanstalt* in Germany." As he and others would later use the German model for the restructuring of American technical and vocational education, so Pritchett used the German example as the model for "a standardizing bureau adapted to American science and to American manufacture."[8]

Stratton prepared a comprehensive report on the necessity of a well-equipped Bureau of Standards, and outlined plans to establish it. He was persuaded to push the plan through Congress, and to direct the bureau when it was established, by Pritchett's and Stratton's longtime Chicago friend Frank Vanderlip. Vanderlip, who became vice-president of the National City Bank of New York in 1901, was at this time Assistant Secretary of the Treasury; since the Office of Weights and Measures came under the Treasury Department, Vanderlip's support was critical. This was also true in the case of Vanderlip's boss, Lyman Gage, Secretary of the Treasury. A prominent Chicago banker, Gage had been one of the primary promoters of Chicago's World Columbian Exposition of 1893, and was the president of the Chicago Civic Federation, the forerunner of the corporate-liberal National Civic Federation. Gage and the other three men encouraged Congressman James H. Southard, who had earlier introduced a bill for metric legislation, to sponsor the bill for the creation of the bureau. The bill, strongly endorsed by the National Academy of Sciences, the American Association for the Advancement of Science, the American Physical Society, the American Chemical Society, and the American Institute of Electrical Engineers, was passed overwhelmingly and the bureau was established in 1902, with Stratton at its head. The next year it became part of the new Department of Commerce and Labor.[9]

Pritchett indicated at the time that the strongest government support for the bureau came from Gage, a man "who appreciated not only the direct commercial results of the measure, but also its indirect moral effect."[10] Testifying before the Senate committee considering the bill, Gage eloquently spelled out the dual role of standards in the search for a new social order:

There is another side to this which occurs to me. It may appear to many to have a more sentimental than practical value, but it gives the proposition, to my mind, great force, and that is what might be called the moral aspect of this question; that recognition by the government of an absolute standard, to which fidelity in all relations of life affected by that standard is required. We are the victims of looseness in our methods; of too much looseness in our ideas; of too much of that sort of spirit, born out of our rapid development, perhaps, of a disregard or a lack of comprehension of the binding sanction of accuracy in every relation of life. . . . Nothing can dignify this government more than to be the patron of and the establisher of absolutely correct scientific standards and such legislation as will hold our people to faithfully regard and absolutely obey the requirements of law in adhesion to those true and correct standards.[11]

The spirit of standardization thus promised to weld science to power, and to lend to the might of legal and moral authority the legitimacy of scientific truth; the creation of the Bureau of Standards was a symbolic and actual step in that direction.

Under Stratton's direction, and with W. A. Noyes as chief chemist, the bureau made "enormous contributions to science and to industry." "Compared with the old Office of Weights and Measures the new Bureau of Standards was an aggressive and expansionist outfit," its growth reflecting "the rapid penetration of science into technology in the United States in this period." By 1916 it had firmly established itself as a "direct link between government and industry." Its work involved standards of measurement, the determination of standard constants, standards of performance, quality, and practice, which included the formulation of industrial safety codes. For a fee, it verified instruments submitted to it, and occasionally conducted narrowly defined industrial research on a limited basis. In addition, in affiliation with the American Society for Testing Materials, the bureau tested the capacities of materials, machinery, and equipment. The bureau's elaborate facilities—by 1920 it had the largest precision-testing machine in the world—enabled it "to determine the commercial practicability of a process, and yet with much greater ease than would be the case in a full-sized plant." The centralized Bureau of Standards thus facilitated industrial technical development with but moderate expense and little duplication of effort by industry. Evaluating the work of the bureau twenty years after its creation, Henry Pritchett surmised that "perhaps no government bureau ever attained in twenty years so great a development or one which has so allied it to the problems of the industrial and scientific development of the country at large."[12]

Scientific standardization paved the way for industrial standardization, and here too the reformers from the science-based industries played a commanding role. Of course, it could well be argued that industrial standardization dates back to Eli Whitney, who standardized interchangeable parts for the manufacture of muskets. Standardization, however, was much more difficult to effect within an industry than within one company, since it required cooperation among rivals. Probably the first major step was taken by William Sellers, the dean of the Philadelphia machine-tool industry. In 1864 Sellers, then president of the widely respected Franklin Institute, successfully pushed for the machine-tool industry's adoption of his system of standard screw threads, a system which became known as the American Standard. Within twenty years the Standard had been extended to bolts, taps, and cap screws. The railroads also pioneered in intercompany standardization, with the adoption of the standard gauge for track and—through the work of the Master Car Builders, an intercompany association—the adoption of standard coupling designs and dimensions of rolling stock.[13]

The professional mechanical engineers who emerged from the machine shops and the railroad yards made the major contributions to intercompany standardization. Working through the ASME and the ASTM, they were best able to transcend the narrow self-interest of any specific company and view the industry as a whole. They promoted standardization primarily in the interest of industrial efficiency and safety, devising methods of testing materials and establishing standard specifications for steam boilers. Although they agreed upon the need for industrial standards, however, the mechanical engineers often split, roughly along the lines of the opposing shop and school cultures, over the best methods of achieving them. The shop-culture engineers, like Sellers, emphasized standardization for shop efficiency, but most, as entrepreneurs or high-ranking managers in manufacturing corporations, opposed the imposition of standards from outside their particular domain. They advocated standardized screw threads, gear teeth, bolts, nuts, and various machine processes, but believed in the primacy of private enterprise, and that the manufacturer who dominated a particular market would determine the standards for that market. The school-culture engineers, on the other hand, advocated some centralized authority, preferably governmental, which would set absolute standards to which all industries must conform. They assumed, of course, that school-trained engineers in government would set those standards.

This conflict over the methods of standardization crystallized now and then in the controversy over adoption of the metric system.*[14]

The differences among mechanical engineers which often hampered cooperative efforts at industrial standardization did not exist among electrical and chemical engineers, owing to their close attachment to physical science.[†] In the electrical industry, moreover, the domination

*The metric system, dominant in the European world of science, lent itself to convenient standardization and facile calculation. Machine-shop owners, however, who had invested in apparatus which conformed to the English system, were against any change, partly for financial reasons. Their opposition was also based upon the realization that metric adoption could destroy in a moment one of the few advantages which the shop man had over the technical graduate: the ability to make relatively complicated calculations accurately in seconds, the fruit of years of experience. As early as 1881, therefore, under the leadership of prominent shop-culture engineers Coleman Sellers and Henry R. Worthington, the majority of ASME voted that "the Society deprecates any legislation tending to make obligatory the introduction of the metric system of measurement into our industrial establishments." In the 1890s, however, there was a big push for the metric system among professional scientists and the most scientific of the new industries and engineering fields; "most of the newer industries of the United States were eager to take advantage of the metric system's simplicity."

Among those pushing for the new system were Andrew Carnegie, Thomas Edison, George Westinghouse, Alexander Graham Bell, and Henry Ford. In 1896 Congressman Southard, who headed the House Committee on Coinage, Weights and Measures, introduced a bill to make metric measure mandatory in all government departments and federal contracts. Scientists, educators, and the electrical and chemical industries joined forces behind the bill and passage appeared certain. The shop-culture elite of the ASME, and the metal-working and machine-tool industries which they represented, headed the opposition. A committee was established in ASME that same year, composed of shop-culture leaders, specifically to prepare material for use "in opposition to legislation to make the metric system and its use compulsory" in the United States; led by Fred Halsey, editor of the *American Machinist,* this opposition bitterly attacked supporters of the bill and labeled them "socialists." Southard temporarily withdrew his bill in order to marshal more support, but the opposition ultimately succeeded in eliminating the threat of metric legislation.

In the 1920s the issue was revived with still greater support from scientists, educators, many manufacturers, physicians, pharmacists, as well as chemical, radio, and electrical engineers; by that time, however, many of the industries that had supported the conversion at the turn of the century had invested heavily in tools and machines in inch measure and thus stood to lose considerable sums. Their opposition, bolstered by a National Industrial Conference Board estimate that the conversion would cost over $200 per employee, once again dashed metric reform efforts. See Monte Calvert, *The Mechanical Engineer in America, 1830–1910* (Baltimore: Johns Hopkins University Press, 1967), pp. 169–86; "Proceedings of the Hartford Meeting," *Transactions of the American Society of Mechanical Engineers,* II (1881), 9; John Perry, *The Story of Standards* (New York: Funk and Wagnalls Co., 1955), pp. 78, 87–100.

[†]As Dugald Jackson pointed out, "the remarkably large influence that electrical engineering has exerted on the American engineering industries in general seems to have arisen partially as an influence of the sources and early relations of electrical engineering in and with physical science and physicists. This gave to electrical engineering a tendency to precision in definitions and measurement from the earliest days." "The Relation of Standards and of Means for Accurate Measurement of Effective Development of Industrial Production," typescript, Jackson Papers, MIT Archives.

of manufacturing and electrical engineering by a few large corporations made intercompany standardization a relatively straightforward affair. Charles E. Skinner directed the internal standardization program at Westinghouse and as an active member of the AIEE Standards Committee developed the standards and specifications for electrical equipment which were eventually adopted nationally and internationally. As chairman of the American Engineering Standards Committee in the late 1920s, Skinner recalled that "there were very few manufacturers engaged in the production of electrical apparatus so that relatively few voltages, frequencies, and types of systems found their way into service."[15] This unique situation made it possible also to standardize lamp bases, generators, and filaments, and, perhaps most important, to develop the standard means of rating electrical equipment.

In the early days of the electrical industry, as Comfort Adams, then head of the AIEE standards committee, later explained, "no agreement existed among the various manufacturers as to what constituted a ten-horsepower motor or a thousand-kilowatt generator. In competitive selling a salesman could claim that a ten-horsepower motor could carry twice that load." The standard rating of electrical equipment thus early became a major goal in the electrical industry. Through the effective cooperation of the research staffs under B. C. Lamme and Skinner at Westinghouse and Steinmetz at GE, however, and with the prestige of the AIEE standards committee behind the effort, such standards were soon developed and adopted throughout the industry.* In addition, other agencies were created to promote industrial standardization outside the companies and the AIEE. The Electric Power Club established standards for wires and cables, switches, circuit-breakers, meters, and control apparatus, and the National Electric Light Association's Electrical Apparatus Committee extended AIEE standards to areas of particular importance to operating companies, such as voltages for transformers, and uniform service rules for motors.[16]

In the chemical industry, which was much more diverse and fragmented than the electrical, the development of intercompany standardization was more difficult. At the prompting of chemical engineers like Arthur D. Little and William H. Walker, both the Division of Industrial Chemists of the ACS and the fledgling AIChE set up committees to establish standard terminology and to promote the use of uniform

---

*According to Stephen Dizard of MIT's political-science department, who is just completing a study of the Bell System, the standards for the telephone industry were similarly developed by AT&T and were subsequently adopted by the Federal Communication Commission for the industry as a whole.

specifications and analytical methods devised by Noyes, William F. Hillebrand, and their colleagues at the Bureau of Standards. "In the early days of the committees on definitions and specifications, manufacturers practically ignored the committees' requests for data."[17] Standardization in the chemical industry thus had to await the corporate consolidations which followed the war. Only then, and with the widespread adoption of Arthur Little's key concept of "unit operations" as the basis of chemical-engineering practice, did it finally make significant headway.

The automotive industry also made significant contributions to intercompany standardization, and had perhaps the widest impact upon American industry as a whole. In 1906 the Association of Licensed Automobile Manufacturers, those who were licensed to operate under the Selden patent, adopted standards for automobile screws and nuts. The major drive for standards in the industry, however, came four years later, from the automotive engineers who in 1905 had organized the Society of Automobile Engineers (SAE). In 1910, under the leadership of Howard E. Coffin, an automotive engineer and vice-president of the Hudson Motor Car Company, the SAE launched a movement to achieve intercompany standardization throughout the industry. The immediate impetus came from the small manufacturers who dominated the SAE in the early days. This was a period of severe economic contraction in the industry, and small manufacturers found themselves wholly dependent upon the fortunes of the particular parts manufacturers who supplied them with their own, uniquely designed materials. The work of the SAE aimed at strengthening the position of the small manufacturer by standardizing parts and rationalizing purchasing specifications. The engineers from the larger companies like Ford and General Motors had little to do with the SAE at this time. However, they were busy carrying out similar programs internally while their companies acquired their own parts manufacturers. By the 1920s, after such internal reorganization and standardization had been completed, the General Motors engineers became the most active members of the SAE Standards Committee. Because the automotive industry was the major consumer of industrial products, the work of the SAE automatically extended the drive for standards into many other industries. Predictably, emphasis was placed on uniform specifications for chemical products (lacquers, finishes), petroleum products, rubber products, steel, nonferrous metals, and machined products of all kinds, as well as many types of electrical apparatus.[18]

By the first decade of the twentieth century, every engineering society in America had established its own standards committee. Before 1918, however, only the work of the ASTM cut across all branches of engineering, and there was thus a pressing need for some central agency to coordinate all of these activities. In that year, therefore, the American Engineering Standards Committee was created. It was, as the AESC secretary later observed, "the agency through which industrial standardization in this country is passing from the second to the third stage, namely, from the standardization by association, societies, and government agencies, to standardization on a national scale."[19] The prime mover behind the creation of the AESC was Comfort A. Adams. A Case Institute electrical engineer whose forebears came to Plymouth in 1621, Adams had worked as a designer for the Brush Electric Company before becoming an engineering educator at Harvard and MIT, and an industrial consultant. In 1919 he became the first dean of the Harvard Engineering School, a position he left to chair the Division of Engineering and Industrial Research for the newly established National Research Council.

In 1910, Adams was made chairman of the AIEE standards committee and immediately recognized that the institute's standards overlapped and conflicted with the requirements published by other professional engineering societies. At his suggestion, therefore, the AIEE took the initiative in calling several conferences of the societies with the purpose of coordinating standardization activities. The outcome of this cooperation was the AESC, the founding members of which were the mechanical, electrical, and mining engineering societies and the ASTM. Joining the engineering societies, at their invitation, were the U.S. Departments of Commerce, War, and the Navy, which lent to the AESC a quasi-governmental status. The following year membership was opened to trade associations, private companies, professional societies like the ACS, the AIChE, and the SAE, and the Departments of Interior and Agriculture. In 1928, just a decade after its founding, the AESC became the American Standards Association. The change was made, Comfort Adams explained, because it had become "obvious that the idea of a national clearinghouse for engineering standards alone was too narrow and impractical in view of the growing need for national standards in nearly all spheres of economic activity."[20]

By the time Americans began to gear their industrial resources for World War I, there was thus already in existence a healthy and growing "standardization movement." The war stimulated further growth.

Probably the most important figure behind the wartime standardization effort was Howard Coffin of the SAE. When the Naval Consulting Board was organized in the summer of 1915 through the joint effort of the engineering societies, Coffin, representing the SAE, was made chairman of the Committee on Production, Organization, Manufacture, and Standardization. In this new capacity he expanded his horizons to include the entire industrial plant of the nation, carrying out a nationwide inventory of plants and initiating what became known as the industrial preparedness campaign. When the campaign grew beyond the capacity of the Naval Board Preparedness Committee, Coffin, together with Hollis Godfrey of the Drexel Institute, succeeded in securing legislation for the establishment of a Council of National Defense. As a member of the advisory committee of the council, and head of the subcommittee on munitions and manufacturing, Coffin carried further the work of coordinating industry and standardizing its operations. When the council was superseded by the War Industries Board, Coffin was given the responsibility for establishing uniformity in the manufacture of materials, machinery, and parts, and advanced these ends considerably.[21]

The war provided great stimulus to the standardization movement and gave rise to the related drive for product "simplification," aimed at reducing product variety. Government-sponsored programs eliminated much duplication of effort during wartime, and the programs were carried beyond the war through the efforts of management-oriented engineers. The famous Waste in Industry Report, prepared jointly by the Federated American Engineering Societies under Herbert Hoover and the American Engineering Council, emphasized the importance of both standardization and simplification for reducing industrial inefficiency.[22]

Once he became Secretary of Commerce, Hoover used his position to further the cause of standardization in industry, which now included simplification as well. He directed the Bureau of Standards to "set up new divisions to promote the adoption of commercial standards and simplified practices." (He did so, incidentally, over the objection of Samuel Stratton, who argued that the bureau had not been created for such strictly commercial purposes.) The new Division of Simplified Practices ultimately became the "medium through which producers, distributors, and consumers could agree upon simplification of production by reducing the number of sizes and models of products." Working in cooperation with the U.S. Chamber of Commerce and the AESC, Hoover established an operating procedure for simplification of products; it involved thorough study of the particular problem area, followed by a meeting of representatives of manufacturers, distributors, and consumers to secure agreement to the elimination of certain types and sizes of products. As a matter of course, however, the poorly organized consumers and smaller competitors generally gave way to the interests of the stronger manufacturers. Leon P. Alford, one of the authors of the Waste in Industry study and a prominent management engineer, saw no problem with this. In his review of technical changes in manufacturing industries for the Hoover Commission study of Recent Economic Changes, he happily estimated that the Division of Simplified Practices had achieved a ninety-eight-percent reduction of varieties in some areas. The result was the elimination of a great deal of what Magnus Alexander called "wasteful competition."[23]

If scientific standardization laid the basis for industrial standardization, the latter provided the foundation for the rationalization of production. By the 1920s, moreover, the standardization movement had reached into the realms of accounting, distribution, and consumption. The most important extension of the concept of standardization was into the area of "personnel," management's scientific term for standardized labor. The scientific management of labor followed directly, in the minds of the engineers, from the standardization of materials and machinery. While standardization was the "elimination of waste in materials," Magnus Alexander observed, scientific management was "the elimination of waste in people."[24]

It was no coincidence that Frederick Taylor, the father of scientific management, spent as much time systematizing the methods of cutting metals as he did formulating his principles of shop management.* Each was the complement of the other. He systematized and standardized the processes of production in order to concentrate control over them in the hands of management; and he formulated his strategy of shop management in order to make maximum use of the newly rationalized operations. The crux of the matter was the standardization of human work activity and, ultimately, of human beings themselves. As Dugald Jackson pointed out, standardized manufacture demanded "that the

*In his work at Midvale Steel, Taylor received the encouragement and support of the president of the company, who was none other than William Sellers, creator of the American Standard for screw threads. The two executive heads of the company, moreover, were chemists trained at Yale's Sheffield Scientific School. See Bruce Sinclair, "At the Turn of the Screw," *Technology and Culture,* XI (1969), 26; Edward C. Kirkland, *Industry Comes of Age* (New York: Holt, Rinehart and Winston, 1961), p. 177.

operations of groups of employees and machines . . . be associatively joined, and that individual whims . . . be restrained. . . . The disciplinary relations within the manufacturing organization must be definite and strict." Dexter S. Kimball, manager of GE's Pittsfield plant, Dean of Engineering at Cornell University, and a leader in industrial management, had this insight in mind when he noted with some urgency that "the extension of the principles of standardization to the human element in production is a most important and growing field of activity."[25]

The Corporation as Inventor

Patent-Law Reform
and Patent Monopoly

The patent system was established, I believe, to protect the lone inventor. In this it has not succeeded. . . . The patent system protects the institutions which favor invention.[1]

—E. F. W. Alexanderson

When he intimated his opinion of those particular inventions and discoveries which had most facilitated other inventions and discoveries, Abraham Lincoln included, along with the art of writing and printing and the discovery of America, "the introduction of patent laws." For it was these laws, he explained to his Springfield audience in 1860, that had "added the fuel of interest to the fire of genius,"[2] by conferring the protection of monopoly over an invention to the "true inventor" exclusively, thereby directly rewarding the inventive spirit. When Lincoln's oft-quoted words were inscribed in stone above the doors of the new Patent Office in 1932, however, they no longer conveyed either the intent or the *de facto* practice of the system administered therein. "In his day, Abraham Lincoln could well say that 'the patent system added the fuel of interest to the power of genius,' " one observer concluded after the hearings of the Senate Patent Committee in 1949. "Today it would be more correct to say that the patent system adds another instrument of control to the well-stocked arsenal of monopoly interests . . . it is the corporations, not their scientists, that are the beneficiaries of patent privileges." The mass of evidence of the corporate use of patents to circumvent antitrust laws which was collected in the testimony before the Temporary National Economic Commission in the early 1930s prompted another writer to concur. "It would require more than twenty years of Rip Van Winkle oblivion to events of this world," he wrote, "to miss the fact that the overwhelming proportion of significant inventions now come out of scientific laboratories, and that these . . . are institutions which have largely if not wholly removed—by deliberate intent—the pecuniary reward for the inventor."[3]

These latter-day critics were but echoing the warnings of those who had much earlier witnessed the transformation of the patent system. Within a half-century after Abraham Lincoln offered his glowing evaluation of it, the American patent system had undergone a dramatic change; rather than promoting invention through protection of the inventor, the patent system had come to protect and reward the monopolizer of inventors, the science-based industrial corporations. "It is well known that patents in the United States are bought up in large numbers for the purpose of suppressing competition," one commentator observed in the *Iron Trade Review* of 1915. He noted that the monopoly of an industry by means of patent control constituted a "monopoly of monopolies" and "a patent on the very industry" as a whole. Such control of patents, he warned, with the resultant capacity for direction and suppression of invention itself, "strangles the sciences and the useful arts, and contributes liberally to illegitimate commercial schemes."[4] As it gave rise to "monopoly of monopolies," the patent system gradually fostered the corporate control of the process of invention itself and thus facilitated the commercially expedient retardation, as well as promotion, of invention.

The framers of the Constitution, who formulated the basis of the American patent system, had deliberately sought to avert this possibility, and in doing so they had departed significantly from the practice of their time. Letters patent had been issued in England as early as the sixteenth century, in accordance with the principles of unwritten common law, and the first patent statute was passed by Parliament in 1623. The patent system of England, however—which was adopted by the colonies and the states under the Articles of Confederation—was geared toward the promotion of new industries by granting monopolies to importers of inventions and processes as well as to inventors themselves. Reward went to the "man who introduces or improves the manufacture and not alone the man who originated the improvement." The writers of the American Constitution, however, "introduced a radically new idea into the view of the function and scope of a patent system."[5]

A proposition placed before the Constitutional Convention would have empowered Congress "to establish public institutions, rewards and immunities for the promotion of agriculture, commerce and manufactures." This was rejected by the convention in favor of one which authorized Congress "to promote the progress of science and the useful arts, by securing for limited times to authors and inventors the exclusive right to their respective writings and discoveries." "For the first time in the world," the nation's most prominent patent lawyer explained in 1909, "the framers of our constitution laid the entire stress . . . on the recognition and reward of inventive thought."[6] The system thus focused upon the inventor, who alone could receive a patent for a particular invention and subsequently either work under it, sell it in part or whole, or grant licenses for its exclusive or nonexclusive use. Legally, there was "no limit to the inventor's absolute control of the thing" covered by the patent, and it could be a process, method, machine, manufacture, a composition of matter, or any improvement of them; ideas or principles, mathematical formulae, laws of nature, or philosophical abstractions "not embodied in concrete form," however, could not be patented. "In the beginning," one student of the patent system has written, "it was easy to fit the definition of invention to the simple economic conditions that prevailed. Any invention was ordinarily the creation of one individual. . . . Our patent system was designed to stimulate the individual to invent by giving him the right to exclude others from making, using, and selling his invention."[7] In the eyes of the designers of the system, moreover, the rewards were not the inventor's alone; the patent right was "really a just reward for service rendered to the community," and the community benefited as well.[8]

The patent system was created for the mutual benefit of the inventor and of society, to which he disclosed his invention in return for patent protection. It was assumed that the patentee would certainly develop the invention for commercial use under the protection granted him; indeed, "the whole historical background of granting monopolies for promoting the progress of the useful arts gives no sanction to the suppression of inventions." As a means of fostering commercial progress, the patent system did not "sanction a monopoly of kindred or competitive patents or a restraint of trade other than in the particular thing which the patent covers"; the Constitution, moreover, provided for patent monopolies "for limited times only"* to guard against any long-term restraint of trade. Between 1790 and the latter part of the nineteenth century, however, the role of patents in American commercial development underwent significant changes. These affected the

---

*The life of a patent eventually became set at seventeen years.

methods by which patents were issued and to whom they were issued, as well as their use once granted.[9]

The first United States Patent Law, of 1790, was administered by Thomas Jefferson and his colleagues under very strict standards, and relatively few patents were issued. Three years later a more relaxed system was adopted whereby "anyone who swore to the originality of his invention and paid the stipulated fees could secure a patent," its validity being decided by the courts. In 1836 this second law was repealed and a Patent Office was created. The 1836 Patent Act "marked the beginning of our present patent system," based upon the "examination system" involving scrutiny of each patent application. The fantastic growth in the number of patent applications thereafter had begun by the end of the nineteenth century to place a great strain upon the rather meager resources and small staff of examiners in the Patent Office.[10]

Praising the admirable foresight of the Founding Fathers, Frederick Fish explained how they had "adopted a new theory that men are encouraged to invent by the certainty of reward; if the fire of genius is fed with the fuel of interest, the industries will take care of themselves." As a former president of AT&T and a GE counsel, Fish could declare with assurance that the industries had done so, and in ways never imagined by the framers of the Constitution: "as business units became larger, patent-owning corporations supplanted inventors in the exploitation of patents." The inventor, the original focus of the patent system, tended increasingly to "abandon" his patent in exchange for corporate security; he either sold or licensed his patent rights to industrial corporations or assigned them to the company of which he became an employee, bartering his genius for a salary. In addition, by means of patent control gained through purchase, consolidation, patent pools, and cross-licensing agreements, as well as by regulated patent production through systematic industrial research, the corporations steadily expanded their "monopoly of monopolies." Although the first patent pool, among manufacturers of sewing-machine parts, was established as early as 1856, it was not until the end of the century that corporations clearly became the dominant factor in patent exploitation. In 1885 twelve percent of patents were issued to corporations; by 1950 "at least three-fourths of patents [were] assigned to corporations." The change in the focus of the patent system, from the protection of the inventor to the protection of the corporation which either employed the inventor or purchased his patents, was succinctly phrased by E. F. W. Alexanderson, a Swedish immigrant who became one of GE's early leading research engineers. "The patent system was established, I believe," he said, "to protect the lone inventor. In this it has not succeeded . . . the patent system protects the institutions which favor invention."[11]

The growth of the corporations, and the intensification of their control through trusts, holding companies, mergers and consolidations, and the community of interest created by intercorporate shareholding and interlocking directorates generated a counterdevelopment within American society: the antitrust laws. The patent system which conferred legal monopolies to inventors came into increasing conflict with antimonopoly legislation as corporations replaced lone inventors as the primary holders of patents, and used patents to create monopolies. The conflict surfaced in court interpretations of the patent monopolies in the light of the Sherman and, after 1914, the Clayton antitrust acts. Section three of the latter explicitly declared the illegality of monopoly based upon sales regulation of patented machinery and products, and was largely a response to the monopoly held by the United Shoe Machinery Company. The interpretations of corporate patent practice varied from court to court, but the judicial history of patent monopolies falls roughly into three periods. Between the signing of the Constitution and the first decade of the twentieth century there was either disregard for, or approval of, monopolies based upon patents. For two decades thereafter there was a gradual tightening of restrictions, although as late as 1926, in an important precedent-setting case involving General Electric, the Supreme Court "emphasized the right of a patent owner to license manufacturers with restrictions as to price." In the third period, beginning "about 1940, indifference and leniency in general gave way to a more aggressive prosecution and court decisions and decrees which reflect[ed], as never before, the purpose of the Sherman Act." The period under examination here, that of 1900–1929, was one of comparatively little judicial restriction of corporate patent monopoly and the market control it made possible. The tremendous strides made along these lines in this period, in addition, were of such proportions as to render subsequent judicial and legislative efforts to check corporate monopoly through patent control too little too late.[12]

The novel American patent system, designed to protect the inventor by granting him a monopoly over his creations, had by the turn of the century fostered the development of "institutions" that demanded a controlled promotion of the "progress of science and the useful arts," one that conformed to the exigencies of corporate stability and prosperity. The science-based industries, based upon patent monopolies from the outset, thus sought to redefine the patent system as yet another means to corporate ends. In particular, they aimed to bend the system in ways which would enable them to circumvent the antitrust laws. Their efforts included intercorporate agreements, industrial research and regulated patent production, and reform of the patent-system apparatus. Edwin J. Prindle, a mechanical engineer and patent lawyer, was active in all three areas.

In numerous articles Prindle outlined the means of securing patent monopolies to bypass the antitrust laws; methods of securing patents from inventors, and employee-inventors; and the legislative means of streamlining the patent system along corporate lines. An early member of the American Patent Law Association, which was founded in 1897, Prindle pursued a career which involved him in countless court cases in the defense of patent-holding corporations, and provided him with the opportunity of formulating, along with Frederick Fish and the other members of the National Research Council Committee on Patents, the bill which authorized the revamping of the Patent Office in the early 1920s. In a widely read series in *Engineering Magazine* in 1906, entitled *Patents as a Factor in a Manufacturing Business*, Prindle clearly spelled out the possible uses of the patent system for purposes of corporate monopoly. In offering his suggestions, he indicated that they arose out of his own experience at the patent bar and as a practicing engineer, as well as from the successful experiences of the pioneers in such undertakings: Bell Telephone, GE, Westinghouse, and the United Shoe Machinery Company.

> Patents are the best and most effective means of controlling competition. They occasionally give absolute command of the market, enabling their owner to name the price without regard to cost of production. . . . Patents are the only legal form of absolute monopoly. In a recent court decision the court said, "within his domain, the patentee is czar . . . cries of restraint of trade and impairment of the freedom of sales are unavailing, because for the promotion of the useful arts the constitution and statutes authorize this very monopoly."
>
> The power which a patentee has to dictate the conditions under which his monopoly may be exercised has been used to form trade agreements throughout practically entire industries, and if the purpose of the combination is primarily to secure benefit from the patent monopoly, the combination is legitimate. Under such combinations there can be effective agreements as to prices to be maintained . . . ; the output for each member of the combination can be specified and enforced . . . and many other benefits which were sought to be secured by trade combinations made by simple agreements can be added. Such trade combinations under patents are the only valid and enforceable trade combinations that can be made in the United States.[13]

Prindle proceeded to outline methods of prolonging monopolies and expanding them through ownership of auxiliary patents. "If a patent can't be secured on a product," he suggested that "it should be secured on processes for making the product." And "if none of these ways is feasible, it should be considered whether or not the product cannot be tied up in some way with a patent on some other product, process, or machine." As a patent lawyer, Prindle understood that "a patent is valid only when granted in the name of the inventor," and he emphasized the importance, for corporations, of securing the patent rights of their employees.[14] He alluded to a long series of cases in which corporations were unsuccessful in their attempts to gain control of patented inventions of their employees because they had failed to contract with them specifically for such privileges. He thus strongly argued that

> It is desirable to have a contract with every employee who is at all likely to make inventions which relate to the business of the employer . . . the courts will sustain such contracts, even though they contain no further provision for return for the inventions than the payment of the ordinary salary. . . .[15]

Prindle was aware that he was deliberately subverting the intent of the patent system. In citing cases where employees had refused to give up "their rights" guaranteed by the Constitution, he emphasized the importance of using "psychology" to obtain the patent rights of employees, acknowledging thereby that "rights" were in fact being lost, and that at least some employees were fully aware of it. "The difficulty of inducing the employees to sign such a contract," he noted, "will be reduced if the officers of the company will set the example by signing such a contract." Quite clearly, Prindle understood that what he was proposing—the compulsory signing of employment contracts which automatically assigned employee patent rights to the employer— amounted to confiscation, and something that neither he nor his readers would have liked to have happen to them. Prindle thus acknowledged that the signing of the contract by the officer to set a "reasonable" example was in fact reasonable for the officer alone, since for him it was "a mere matter of form, as [he] is frequently a man who is either not inventive or one who is glad to take his returns in the form of dividends from the stock."[16] For corporate employees in the science-based industries, however, this matter of form would become standard and compulsory procedure.

The methods outlined by Prindle were in part the fruit of long experience in the electrical industry. The largest corporations of the industry—GE and AT&T—had years before mapped out the patent territory and begun to refine their tactics for mastering it. Although patent-control measures became widespread in the automobile, rubber, steel, chemical, and other industries, the stories of AT&T and GE in particular and the electrical industry in general provide the earliest examples of how they operated.

AT&T, as has already been noted, was incorporated as the consolidation of the various Bell System interests in 1900. By that time the Bell System had already substantially occupied the field, since it had been "successful in every contest involving the original patents."[17] Having anticipated the expiration of those patents, the Bell companies had, as President Theodore N. Vail phrased it, "surrounded the business with all the auxiliary protection that was possible."[18] As a result, AT&T found itself in an excellent position to stifle and harass competitors through patent-infringement suits. The success of these procedures was explained by an AT&T patent lawyer:

> It appears to me that the policy of bringing suit for infringement on apparatus patents is an excellent one because it keeps the concerns which attempt opposition in a nervous and excited condition since they never know where the next attack may be made, and since it keeps them all the time changing their machines and causes them ultimately, in order that they may not be sued, to adopt inefficient forms of apparatus.[19]

Among the patents secured by AT&T were those which underlay AT&T's monopoly of long-distance telephony—Michael Pupin's patent on loading coils, and the Cooper-Hewitt patents on the mercury-arc repeater. AT&T purchased Pupin's rights in 1900 and secured exclusive domestic rights for the Cooper-Hewitt patents in 1907. AT&T was also able to obtain the rights for Lee De Forest's three-element vacuum tube in 1913, half a dozen years after it was first patented. By gaining control of the De Forest invention, "which is the heart of radio broadcasting, wireless reception and amplification of long-distance conduction of electrical waves, whether used in radio, telegraph, or telephone communications," AT&T secured a key position in the nascent radio industry.[20]

Between the time of its organization as the Bell Patent Association in 1875 and the creation of the Federal Communications Commission in 1934, the Bell System "remained free from federal regulation."[21] By 1935, through licensing agreements, mergers, purchases, and research, it had increased its patent holdings from two—the original Bell patents —to 9,255, which included some of the most important inventions in telephony and radio. In addition, through the various radio-patent pool agreements of the 1920s, AT&T had effectively consolidated its position relative to the other giants in the industry. An FCC investigation of the telephone industry and extensive study of the patent system led Floyd L. Vaughan to conclude that

> By amassing thousands of patents on inventions in the whole field of communication . . . American Telephone dominates the telephone and also controls "the exploitation of potentially competitive and emerging forms of communication." It thus excludes others from its field and avoids being excluded by them. Would-be rivals may enter and remain only as licensees under restricted conditions. It pre-empts for itself new frontiers of technology for exploitation in the future and, in the meantime, protects what is already developed. It keeps itself in a commanding position for the exchange of patent rights. In short, it employs patents to maintain its dominance . . . in communication.[22]

The experience of General Electric was similar. GE was formed in 1892 as the consolidation of the assets, and especially patents, of the Edison and Thomson-Houston interests. Beginning in 1896 with the establishment of the GE-Westinghouse Board of Patent Control, "the first important example of collusion in acquiring patent rights,"[23] GE "followed the conscious policy of funnelling into its control all patents held by its licensees and touching any phase of the [incandescent lighting] industry."[24] A Tariff Commission report on incandescent electric lamps indicated that "since that time, through the purchase and consolidation of numerous companies, through the purchase of patents and through its own research organization . . . GE has acquired most of the important patents covering electric lamps, their parts and machinery and processes for making them."[25] Subsequent court decisions further indicated the success of GE's patent policies. A favorable Supreme Court decision of 1926 noted that GE's control of the patents covering the manufacture of tungsten filaments, and the use of gas in light bulbs to increase light intensity—Just and Hanaman (1912), Coolidge (1913), and Langmuir (1916)—"secure to GE the monopoly of the manufacture of making, using, and vending" of the means of "the making of the modern electric light."[26] In 1949 a considerably less sympathetic New Jersey district court found that GE's "offensive of patents" had led to a situation in which "it individually monopolizes patents employed in the incandescent electric lamp industry." In the opinion of the court,

> General Electric's apparently impregnable position was a formidable barrier to anyone who contemplated entering the lamp manufacturing field and this, coupled with the knowledge that it controlled the manufacture of lamp bases, lamp manufacturing machinery, along with a tight block on the supply of glass,* created a situation sufficient to deter entry. The link to unlawful monopoly is apparent from the fact that upon expiration of the lawful patent monopoly in 1933, there was no new entry into the field.[27]

The individual policies of GE and AT&T were carefully designed to gain and prolong monopolies over patents vital to their industry. Toward this end, they employed such methods as incomplete disclosure of information in patent applications, the use of trademarks, the outright suppression or delayed introduction of patented apparatus,† the compulsory assignment of employee patents to the company, and the deliberate production of auxiliary patents. Perhaps of greater significance than all of these combined, however, were the agreements through which the efforts of individual companies were coordinated in the interest of all, and insulation from national and international competition in particular fields was secured. The radio-patent pool agreements of the 1920s, among such odd bedfellows as AT&T, GE, RCA, United Fruit, American Marconi, and Westinghouse, provide an illuminating example.[28]

By the beginning of World War I, a number of companies had arrived at a stalemate with regard to radio development, due to mutual patent interferences. During the war, when the government guaranteed to protect the companies from infringement suits, research in radio proceeded at a rapid pace. The close of the war, however, brought with it a renewed deadlock. "Ownership of the various patents pertaining to vacuum tubes and circuits by different concerns prevented the manufacture of an improved tube for radio use."[29] In addition to domestic competition, there was a very real possibility that control over radio

*GE and the Corning Glass Company had an agreement whereby they jointly manufactured all bulbs for lamps made in the United States. Floyd L. Vaughan, *The United States Patent System* (Norman: University of Oklahoma Press, 1956), p. 121.

†AT&T held the patent on the combined handset telephone twelve years before introducing it; GE followed a similar course with fluorescent lamps. Vaughan, *The United States Patent System,* pp. 75, 76. N. R. Danielian, *AT&T* (New York: Vanguard Press, 1939), p. 102.

might be secured by the British Marconi Company, which was trying at the time to obtain rights to the necessary GE-controlled Alexanderson alternator.

In light of this threat to American supremacy of the airwaves, Woodrow Wilson and a number of armed-forces representatives prevailed upon GE to withhold the necessary patent rights and set up instead an American-owned company to control radio. In late 1919, GE thus established the Radio Corporation of America; it purchased the stock of the British-controlled American Marconi Company and transferred its assets, along with the Alexanderson and other GE-owned patents, to RCA. The industry-wide impasse nevertheless remained, and "the only solution to the conflicts was to declare a truce: get together and draw up an agreement defining the rights of the various squatters on the frontiers of science." The truce was declared between AT&T and GE in the license agreement of July 1, 1920, and within the following year, through collateral agreements, the other companies in the patent conflict joined the radio-patent pool.[30]

The consolidation of radio patents, which numbered some thousand, "divided up the telephone, electrical, and radio fields and established supremacy in each through exclusive licenses." At the same time, the agreements kept all who were not party to them out of the radio field. AT&T, for example, retained "exclusive rights under its patents in two-way telephone services, both wire and radio," while GE and Westinghouse received "all patents held by AT&T and RCA relating to the lamp industry, in return for their patent rights applicable to telephone and radio." "Each one, in effect, kept and obtained the patent rights in its particular field."[*][31] Perhaps the best description of the intent behind the agreements was offered by J. E. Otterson, general commercial manager of the Western Electric Company, in 1927:

> The regulation of the relationship between two such large interests as the AT&T Company and the GE Company and the prevention of invasion of their respective fields is accomplished by mutual adjustment within . . . the "no-man's land" lying between . . . where the offensive of the parties as related to these competitive activities is recognized as

---

[*]In all, there were eight parties to the agreement, and they divided themselves into two groups; the "Telephone Group" consisted of AT&T and its manufacturing subsidiary, Western Electric Company; the "Electrical Group" or the "Radio Group" consisted of GE, Westinghouse, RCA, United Fruit Company (which owned certain radio patents), the Wireless Specialty Apparatus Company, and the Tropical Radio Telegraph Company.

a natural defense against invasion of the major fields. Licenses, rights, opportunities and privileges in connection with these competitive activities are traded off against each other and inter-changed in such manner as to create a proper balance and satisfactory relationship between the parties in the major fields.

"The contract," Otterson explained,

is an example of the character of arrangement that may develop out of an effort on the part of two large interests to avoid an invasion of their respective fields and a destructive conflict of interests. It was through trading off rights in connection with these competitive activities that an adjustment between the two interests was reached and the two major fields left intact.[32]

What was actually left "intact," of course, were not the fields themselves, but rather the corporate control over the fields, the "spheres of influence" of the large companies. Through agreements like this within the country, and the establishment of international cartels to regulate the global "field," the large corporations of the electrical industry sought to dominate not only markets for their products but the manufacture of those products as well. Their "Napoleonic concept of industrial warfare, with inventions and patents as the soldiers of fortune,"[33] served them well, as the court's evaluation of GE activities in 1949 clearly indicated; GE, it concluded,

paced its industrial achievements with efforts to insulate itself from competition. It developed a tremendous patent framework and sought to stretch the monopoly acquired by patents far beyond the intendment of those grants. It constructed a great network of agreements and licenses, national and international in scope, which had the effect of locking the door of the United States to any challenge to its supremacy . . . arising from business enterprise indigenous to this country or put forth by foreign manufacturers.[34]

The corporate control of patents by means of intercorporate agreements was inextricably coupled with the research arm of the "patent offensive." As a basis for systematic patent production and monopolization, industrial research played an increasingly important role, from the turn of the century on, in the development of bargaining power for such agreements. Although the primary objective of research was to find solutions to immediate technical problems, another objective was "to anticipate inventive trends and take out patents to keep open the road of technical progress and business expansion."[35]

Scientific research at AT&T began in 1907 when J. J. Carty became chief engineer and brought into the domain of AT&T the research which his predecessor Hammond Hayes had left to the students of MIT and Harvard College. The growing importance of research within the company was highlighted against the backdrop of a severe economic downturn. J. P. Morgan formally took over the reins of the company the same year and, in the wake of the bankers' panic, called for greater efficiency and increased standardization of equipment in order to cut costs. By 1910, however, there were 192 engineers doing development work under Carty, with annual expenditures of half a million dollars. By 1916 this number had increased to 959, with expenditures of $1.5 million, and by 1930 AT&T was spending $25 million for research. Between 1916 and 1935 the engineering department of the Western Electric Company and the Bell Laboratories combined, spent $250 million on engineering and research; this sum, as one historian of AT&T has estimated, "far exceeded the total operating budget, for instance, of Harvard University for the same period."[36]

A major impetus behind this expansion was the early recognition, by Carty and his superiors, of the potentials of wireless, coupled with the need to develop a "repeater" for long-distance telephony. That the promise of radio, in terms of AT&T prosperity and competitive strength, was a prime motivation behind the rapid expansion of research was suggested by Carty himself, in a memorandum of 1909.

At the present time scientists in Germany, France, Italy, and a number of able experimenters in America are at work upon the problem of wireless telephony. While this branch of the art seems at present to be rather remote in its prospects of success, a most powerful impetus would be given to it if a suitable telephone repeater [vacuum-tube amplifier] were available. Whoever c.n supply and control the necessary telephone repeater will exert a dominating influence in the art of wireless telephony when it is developed. The lack of such a repeater . . . and the number of people at work upon [it] . . . created a situation which may result in some of these outsiders developing a telephone repeater before we have obtained one ourselves, unless we adopt vigorous measures from now on. A successful telephone repeater . . . might put us in a position of control with respect to the art of wireless telephony should it turn out to be a factor of importance.[37]

Work on this problem preoccupied the research branch of the company as soon as it was established. As Frank Jewett, head of the Bell Labs, recalled in 1932, "it was early clear to the AT&T Company . . . that a full, thorough, and complete understanding of radio must be had at all times if the art of telephony . . . was to be advanced and the money invested in that service safeguarded."[38] Such research, in which the prime motivation was commercial dominance through patent offense and defense, proved quite successful; although AT&T had to purchase the De Forest vacuum tube, the research done in its laboratories provided a still greater bargaining position in the radio-patent pool agreements after the war. Perhaps the clearest statement of intent behind such industrial research was provided by J. E. Otterson in his memorandum of 1927:

> A primary purpose of the AT&T Company is the defense and maintenance of its position in the telephone field. . . . Undertakings and policies must be made to conform to the accomplishment of this purpose. The AT&T Company is surrounded by potentially competitive interests which may in some manner or degree intrude upon the telephone field. The problem is to prevent this intrusion.
>
> It seems obvious that the best defense is to continue activities in "no man's land" and to maintain such a strong engineering, patent, and commercial situation in connection with these competitive activities as to always have something to trade against the accomplishment of other parties. . . . It seems essential to . . . maintain an active offensive in the "no man's land" lying between it and potentially competitive interests. . . . The nearer the trading can be carried to the major field of our competitors the more advantageous the trading position we are in. . . . Ability to stop the owner of a fundamental and controlling patent from realizing the full fruits of his patent by the ownership of necessary secondary patents may easily put one in position to trade where money alone may be of little value.[39]

While research and patent warfare on the part of large "institutions which promote invention" provided them with a competitive edge in their bargaining among themselves, it completely overwhelmed the independent inventor whom the patent system was originally designed to protect. Lone inventors could either try to fight for their rights within "no man's land" or join the dominant forces which occupied the fields around it. Out of frustration and survival instinct, they increasingly flocked to corporate employment in exchange for security and abandoned their patent privileges in the process.[40]

When inventors decided to fight for an independent position, they were faced with frustration on three fronts: trying to obtain a patent; trying to sell or develop it; and trying to see it put to use after it was sold. Because of the conscious policies and extensive resources of the larger corporations, lone inventors were obstructed at every turn. As Floyd Vaughan observed,

If the inventor sells his patent rights at all, it is usually for a lump sum rather than for royalties. If he develops his own invention, which is seldom, he must seek the capital of others. Most of his inventions are never sold or developed at all. In any case he usually receives little or nothing. As the obstacles of the inventor have grown, patents, to an increasing extent, have stimulated him through delusion rather than reward.

Obtaining the patent was perhaps hardest of all:

It is a common practice, especially of large companies well-financed and equipped with technicians and patent lawyers, to take out every possible patent in their fields and thus block any would-be intruder. If an outsider seeks a patent in this domain, he must find out in some instances about hundreds of patents on kindred ideas and avoid them. Creative minds may be compelled to spend more time in obtaining or avoiding patents than in solving a problem.[41]

By means of interference and infringement suits, the corporations were well able, and equally inclined, to harass patent applicants and cause them to abandon their claims. According to the president of the Thomas Edison Company, Edison himself had spent more money in obtaining patents, litigating them, and preventing infringements than he had received from them. Lee De Forest, while successful in defending his claims and selling some rights to AT&T for a sizable sum, was pushed into bankruptcy as a result of other patent litigation. In 1917 B. A. Behrend awarded Nikola Tesla the coveted Edison Medal of the AIEE with the words: "Were we to seize and to eliminate from our industrial world the results of Mr. Tesla's work, the wheels of industry would cease to turn, our electrical cars and trains would stop, our towns would be dark, our mills would be dead and idle. Yea, so far-reaching is this work that it has become the warp and woof of industry." Tesla himself, however, as Alexanderson later recalled, "was a frustrated inventor and had to spend his old age in impoverished retirement."[42]

Edison, De Forest, and Tesla, whatever the cost, had been able to translate their inventions into commercial developments, and to sell them. The majority of inventors were less fortunate. While perhaps able to obtain patents, they were unable either to interest investors in them or to sell them to established companies. At a House committee hearing on the pooling of patents, one witness testified that "the greater the contribution, the more certain is it to be denied recognition by the entrenched corporations and their servile laboratory staffs. And the lack of such recognition . . . [in part] explains the shameful spectacle of every single one of the world's great inventions having been forced to be idle until outside competition had forced their adoption despite the cunning and conspiracy of the great corporations in that field—and often only after the inventor was no longer here to receive his due reward."[43] The individual corporations were usually resistant to significant change simply because of their established dominance in the field; moreover, pooling agreements often required the sharing of important new developments with competitors, and this canceled out any motivation of an individual company to invest in a new invention. For similar reasons, if a company did invest in something new, more than likely it was in order to obtain some future bargaining advantage, and thus the immediate use of the patent was suppressed.

In 1912 Louis Brandeis concluded that "these great organizations are constitutionally unprogressive." But in terms of the human cost of independent invention, they were more than that. "Edwin Armstrong," wrote E. F. W. Alexanderson, "was one of the few inventors who made a fortune out of his patents. But the end was sad. His patent litigations got so on his nerves that he committed suicide." Alexanderson, whose alternator was the focus of attention and GE's strongest asset in the radio-patent pool agreements, provided some insight into life outside the pool. Both he and Owen D. Young, chairman of the board of GE, had received a number of threatening letters. "An inventor," Alexanderson recalled, "claimed that he had a patent which was being infringed by my patent. The threats were directed primarily at Mr. Young. He was the head of the Company. The letter writer said that it costs only a thousand dollars to get somebody killed and he had the thousand dollars. General Electric had to fight a suit in patent court to prove that his patent . . . was different from my patent. I never met the inventor . . . but those who did meet him said that he was a simple, mild-mannered man."[44]

The frustrations of independent invention led the majority of inventors into the research laboratories of the large corporations; in the process, invention itself was transformed. "Team research in the laboratory of the large corporation has largely displaced the inventive activity of the individual. The assembly line of invention, like that of manufacturing, is dominant today. The improvements of various workmen and technicians are put together, under the guidance of lawyers and business managers, so that patents can be acquired which will provide for dominating a field of production more completely."[45] Inventors became employees in corporations to spare themselves the hardships of going it alone. Their patents were thereby handled by corporation-paid patent lawyers and their inventions were made commercially viable at company expense. Corporate employment thus eliminated the problem of lawsuits, and in addition provided well-equipped laboratories, libraries, and technical assistance for research. The nature of their actual work, however, had changed. "Work was often done under high pressure. The employee-inventor was expected to direct his efforts along lines in accord with the company's commercial policies and not to spend time fooling around with any interesting idea that appealed to him. He was expected to produce results of definite commercial value and not to take too long about it." The "collectivization" of invention done in the research laboratory presupposed the specialization of each task: "company inventors were usually organized into departments or sections; they were assigned definite projects to work on and problems to solve," and the various efforts so assigned were assembled only by management.[46]

The new role of the "employee-inventor" further reinforced the changes in the *de facto* patent system that had created him. By incorporating within them the material and human means of technological development, the large corporations effectively eliminated the threat of "outsiders." As one historian of the patent system, Floyd Vaughan, observed, "By controlling the only market for improvement patents and by controlling the factory operations of laboratories where new and pertinent ideas were most likely to occur, a company could command the stream of inventive thought."[47]

According to the United States patent system, "no one except the true inventor can obtain a valid patent." By employing the technical experts capable of producing inventions, the corporations were also obtaining the legally necessary vehicles for the accumulation of corporate patents. At first a number of corporations provided limited compensation to an employee for a patentable invention; GE, for example, rewarded the lucky employee with a dollar.[48] In time, however, employees became required to assign all patent rights to their employer, as part of the employment contract, in return for their salaries. In addition, "to make sure that an employee does not conceive a bright idea and then leave the company to develop it on his own . . . most such employment contracts contain a 'trailing clause.' By means of such a clause, the company can claim inventions not only during employment but also for a period—a year, for instance—after employment has ended."[49] The Bell Laboratories at first compensated employees for patents beyond their salaries, but, as Frank Jewett explained, such incentive allowances encouraged the worker to work for himself rather than his employer, and in competition with his coworkers.

> The incentive was to get as many patents that could pass the Patent Office as possible. An invention was made. It could be covered by one strong patent or it could be covered by a dozen minor patents. It was to the company's advantage to have one strong patent, but it was to the employees' advantage to have a dozen minor patents. . . . It created a situation where men would not work with each other . . . yet the problem which was before us was a problem which required team action; . . . so some way had to be found to get over that.[50]

The Bell System's solution was the one Prindle had suggested in 1906: the elimination of patent reward for employees. The "fuel of interest" was completely divorced from the "fire of genius" and, as one writer put it, "the heroic age of American invention" had come to an end. It was true that corporate employees, while no longer able to exploit the fruit of their own inventiveness, nor even to exercise that inventiveness fully, were nevertheless able to eat regularly, " a consideration not to be sneered at." In addition, they had, as Alexanderson noted, "a safe and resourceful place to work." The safety within the corporation, however, had been achieved at the expense of, and in deliberate violation of, whatever safety there was outside it. More and more independent inventors were forced to "abandon their patents" as they would a sinking ship, and seek refuge on the shore from which the ship was being bombarded. After hearing the testimony before the Senate Patent Committee in the early 1940s, Bernhard J. Stern concluded that

> Genius is not nourished, for when the research worker joins the staff of an industrial laboratory, he relinquishes his right to patent the fruits of his researches to the corporation which employs him. If, on the other hand, he remains a member of that almost extinct tribe of solo inventors, he is usually powerless to compete with the industrial giants that control credit, technological facilities, and the markets, and he is generally unable to develop his patent in the face of the expenses of infringement suits.

"No one except the true inventor can obtain a valid patent," Frederick Fish, patent attorney and corporation president, assured his audience at the annual meeting of the AIEE in 1909. "In so far as there is any foundation for the contention that under modern conditions, the inventor himself does not get all that he should for his work, the basis for the contention is not the patent system or the law, but the social and industrial conditions which prevail."[51]

The Patent Office faced a continuously increasing flood of applications during the last two decades of the nineteenth century, and this intensified in the twentieth. The patents issued to individuals increased substantially between 1900 and 1916, but thereafter the role of the lone inventor declined as the corporate apparatus for patent control—strengthened during the war—became firmly established. As a result, according to the testimony of the Commissioner of Patents before the Temporary National Economic Committee, the proportion of patents issued to large corporations (with assets over $50 million) grew from 3 to 17.2 percent of the total.[52] In addition to the *de facto* transformation of the patent system brought about by industrial research and corporate patent-control practices, the actual apparatus of the Patent Office and the legal apparatus surrounding it underwent a significant change in this period. These changes, like the others, owed much to the efforts of engineers and the corporate spokesmen of the science-based industries.

In 1870 Congress appropriated funds for the codification of existing patent laws and for the streamlining of the Patent Office, and authorized the Supreme Court of the District of Columbia to hear appeals on its decisions. By the turn of the century, however, the Patent Office had again become seriously inadequate to the demands placed upon it by expanding industry. A spate of articles in technical and trade journals attacked the Patent Office as "A Big Handicap to Industry" and decried the "Abuses of Our Patent System" and "Our Antiquated Patent System." In response to these industrial demands, the American Bar Association in 1896 initiated legislative proposals aimed at the greater efficiency of the patent system through the elimination of costly delays and drawn-out litigation. The following year a group of patent lawyers, including Frederick Fish and Edwin Prindle, established the American Patent Law Association to promote these ends in the name of professionalism and industrial progress. They sought higher standards of competence (through educational requirements) of patent lawyers and Patent Office examiners, correspondingly higher salaries for Patent Office staff, and greatly expanded facilities to handle the flood of cases. In 1899 a Classification Division was established within the agency to classify patents and thereby facilitate determination of the novelty of applications. But by 1905, with the addition of trademarks to the list of Patent Office responsibilities, the inadequacy of the existing apparatus became more acute than ever.[53]

In 1909 Fish delivered his address to the American Institute of Electrical Engineers on "The Patent System in Its Relation to Industrial Development" and strongly argued the case for patent-system reform. He called for greater efficiency in the handling of applications and the increased salaries, competence, and size of the Patent Office staff that it required. In addition, he emphasized the importance of reforms in the judicial aspects of the patent system. He said that more expedient and uniform handling of patent cases by the courts would reduce the "uncertainty" which patent litigation usually entailed.

> The uncertainty in the application of the law to the facts . . . makes it a difficult matter to foresee how a patent will be viewed and construed by the courts and leads to more or less conflict in judicial decisions. . . . It is not at all impossible that the establishment of a single Appellate Court to take the place of the nine independent Courts of Appeal now existing would go a long way towards helping the situation. Such a court would surely bring about harmony of judicial decisions to an extent that does not now exist.[54]

The efforts of patent lawyers like Fish and Prindle were aimed toward streamlining the patent system in accordance with the new "social and industrial conditions which prevail." The increased demand for patent services was undeniable: the one-millionth patent was issued in 1911, the two-millionth only twenty-four years later.[55] But in addition to this increased burden in numbers of applications, corporate requirements of efficiency and predictability in patent processing and litigation demanded reform of the system. These men and their colleagues thus sought to standardize both patent procedures and the legal evaluations of Patent Office practice. At the same time, they pushed for stiffer educational requirements (both legal and technical) of Patent Office personnel, to assure greater competence; they were all too well aware that a great proportion of patents authorized by the Office was subsequently declared invalid by the courts. They similarly pushed for tougher requirements for entry into the legal bar, and sought to establish a harmony of judicial decisions through centralization of the appeals mechanism. In both areas they hoped to replace existing personnel with better-educated people who would be more competent to handle and promote the increasing complexity of the patent system, and who, because of their education, would be more apt to think the way the reformers themselves did.

L. H. Baekeland, the Belgian-born chemist, promoted patent reform for similar reasons, but approached the problem from a different perspective, one common to inventors who had succeeded in the industrial world. The same year that Fish spoke to the AIEE, Baekeland addressed his colleagues in the industrial division of the American Chemical Society on the uses and abuses of the U.S. Patent System from the inventor's standpoint. Baekeland, however, was no ordinary inventor. He was to the fledgling chemical industry what Edison had been to the electrical, ably combining inventive genius with a flair for industrial promotion. Baekeland had spent his early career in Belgium as a professor of chemistry and an industrial chemist. In 1889 he came to America on a traveling scholarship and, having made contact with the largest photo-supply house in the country, A. & H. T. Anthony and Company, he decided to remain to pursue his research on photographic paper and film. Two years later he quit Anthony and set up the Nepera Chemical Company to manufacture and promote his invention, the photographic paper known as Velox. After considerable success, he sold the company to Kodak in 1899. He then turned his attention to electrochemistry and patented an apparatus for regenerating electrolytes while consulting for the Hooker Electrochemical Company. Baekeland's major discovery, however, came in 1909, when he announced his invention of "bakelite," a heat- and chemical-resistant synthetic phenol resin which was to have wide application in the electrical, radio, automotive, aviation, and other industries. Through this work Baekeland contributed tremendously to the development of the burgeoning phenol-based lacquer and finish industries and almost overnight gave birth to the modern plastics industry. In 1910 he organized the General Bakelite Company to exploit his invention, and by 1922 it had undergone such rapid expansion that a holding company, the Bakelite Corporation, was organized to consolidate the different interests of the enterprise. Baekeland remained president until 1939, at which time the company became part of the Union Carbide and Carbon Corporation.

Baekeland's career symbolized the tendency inherent in the patent system at a time of industrial consolidation, the subtle inversion whereby the system which protects the inventor becomes the system which thwarts him. An inventor of genius who had enjoyed extreme good fortune, Baekeland admired the patent system as a social institution superbly designed to promote individual creativity. At the same time, as a capitalist and industrialist concerned with the fortunes of his rapidly expanding company, Baekeland saw the system as the foundation for science-based industrial expansion and monopoly, the bedrock of corporate prosperity. In his own mind, Baekeland conceived of the patent system as the friend of both inventor and corporation. He did not fully recognize that equal treatment for both under the law naturally favored the latter. In 1909 Baekeland thus called for legal reforms that would enhance rather than stifle the blessings of the system:

> I could hardly suggest an improvement in the U.S. patent law without curtailing the privileges and interests of the poor inventor. On the other hand, it is very unfortunate that although the laws for filing and registering a patent ... are almost all that can be desired, ... before the courts ... the poor inventor is entirely at the mercy of a legalized system of piracy. ... This game is so successfully played that I know of rich companies here in the U.S. whose main method of procedure is to frighten, bulldoze, and ruin financially the unfortunate inventor who happens to have a patent which he is not willing to concede to them on their own terms; that is to say, for next to nothing. ... Thus has it come about that an otherwise liberal patent law intended for the protection of the poor inventor has become a drastic method for building up powerful privileges in the interest of big capitalistic combinations.[56]

Baekeland's answer to the plight of the inventor was the same as that offered by Fish in the interest of corporate patent policies: more efficient and expedient procedures for patent litigation. "The sooner we have a special and adequate patent court," he told his fellow chemists, "to which all patent litigation can be referred and which can operate without the absurd delays and abominable expenses now involved in suits, the sooner will cease the arrogant frustrations of the generous efforts of those who framed the patent laws." Three years later, in his presidential address to the American Institute of Chemical Engineers, Baekeland called for a special court of appeals in Washington to act upon questions of patent validity.[57]

By the second decade of the century the injustices of the patent system were being scrutinized by political reformers who concurred with Baekeland about the nature of the abuses but seriously questioned whether the tide could be turned simply by streamlining court procedures. Rather, they sought drastic revision of the patent system itself with the aim of reversing the monopolistic trend which it fostered. The Oldfield Bills of 1912 were indicative of such sentiment. They proposed compulsory working of patents under penalty of government licensing to counter the suppression of patents by corporations, and called for the elimination of the "product patent,"* which underlay patent monopoly in the chemical industry.[58]

---

*A patent on a new combination of matter which prohibited such synthesis even by an entirely different process.

The hearings on the bills elicited testimony from inventors, scientists, manufacturers, and patent lawyers. Frederick Fish, as counsel for GE, argued strongly against compulsory working of patents. "If the patent law is changed so that the result of such work cannot be controlled," Fish declared, " . . . the progress of the arts will be stayed." As far as the suppression of patents was concerned, Fish maintained that "there is not a particle of evidence before the committee . . . which indicates that this is a matter of the slightest consequence." Thomas Edison concurred, testifying that he knew of no instances of suppression, and Baekeland, in an address to the AIChE, strongly opposed the compulsory-working clause and the elimination of product patents, calling instead for the public inspection of patent claims. The suspicion that the large corporations were using the patent system against the public interest nevertheless persisted. One writer in the *Iron Trade Review* insisted that "by the process of suppressed invention, this country, instead of enjoying the most efficient means of carrying on many activities, is today using appliances devised ten to twenty years ago." But the corporate interests prevailed. After a joint meeting, the AIChE, Chemists Club, AIEE, AIM&ME, ASME, and the Patent Law Association successfully petitioned Congress to defer the patent legislation pending further study, and the Oldfield Bills died in committee.[59]

Shortly thereafter Thomas Ewing, grandson of the first Secretary of the Interior, was appointed Commissioner of Patents. A prominent patent lawyer, Ewing had handled the patent applications of Frank Sprague on electric railway traction and Michael Pupin on long-distance telephony; in addition, he had served as tutor in science at the Columbia School of Mines. Upon his appointment, he declared that "a man had the right to say what the resale price of his patent should be," thereby endorsing the tradition of market monopoly based upon patent control. A strong advocate of patent reform, Ewing worked with the Economy and Efficiency Commission on the Patent Office to seek ways of streamlining and expediting patent operations. In 1917 he became head of the Patent Board of the War and Navy Departments, an agency which encouraged patent pools in the munitions and aircraft industries as an aid to wartime technological development.*[60]

The Patent Board and other wartime experiences, besides paving the way for the radio and aircraft and automobile patent pools, exposed the glaring inadequacies of the patent system and thus further stimulated

---

*Another wartime boost to patent consolidation, this time in the chemical industry, was the creation of the Chemical Foundation. See chap. 1, p. 16.

reform efforts. The Patent Office Society was formed in 1917 "to promote and foster a true appreciation of the American Patent System"[61] and enlist support for reform measures. More importantly, a new centralized agency emerged during the war as a major vehicle for the promotion of engineering and industrial interests: the National Research Council. At Ewing's request, and as one of its first postwar projects, the NRC created a Patent Committee specifically charged with the preparation of legislation for patent reform. Acting chairman for the committee was Baekeland, and its members included Ewing, William F. Durand, a Stanford mechanical-engineering professor who held many important patents in the nascent aeronautical field, Robert A. Millikan, Nobel Prize physicist and consultant for AT&T, Samuel Stratton, head of the Bureau of Standards, Fish, and Prindle.

The NRC committee proposals expanded on an unsuccessful reform bill of 1915. They included measures for increasing the efficiency of Patent Office operations, enlarging the staff, raising salaries, and ensuring a higher degree of competence through educational requirements. In addition, the committee proposed that the courts be authorized to provide compensation for loss due to patent infringement, that the Patent Office be made independent of the Department of Interior, and that a single patent-appeals court be created to harmonize the court decisions pertaining to patents. "We shall never have a uniform and definite patent law, consistently applied," the committee argued, "until we have a single court of Patent Appeals independent of local sentiment, realizing responsibility to fix the principles of the law and enforcing an harmonious application of these principles on the lower courts." The proposals were first introduced as part of the Nolan Bill of 1921, which was enacted to alleviate the confused international patent situation created by the war. Most of them—those relating to the streamlining of the office, higher qualifications of staff, and higher salaries—were passed as the Lampert Patent Office Bill of 1922.[62]

The Lampert Bill, which had received strong endorsement from the engineering community—the Engineering Council declared its official support, and Prindle, on behalf of the NRC committee, had propagandized for it among engineers—coincided with another, quite different piece of legislation. The Stanley Bills sought the compulsory licensing, by the United States government, of U.S. patents held by foreign companies for the sole purpose of suppressing manufacture in this country of inventions made abroad. The bills required the compulsory "working of patents" within two years after issuance, under penalty of government licensing. The lawyers representing American corporate interests fought against this measure, despite their concern over international competition and the seeming attractiveness of the bill for domestic enterprise. Edwin Prindle, as chairman of the Patent Committee of the American Chemical Society, explained the opposition: "There are many more American-owned foreign patents than foreign-owned American patents," and the Stanley Bills could provoke countermeasures by other countries which would threaten America's international patent position. Of equal importance Prindle argued that it could serve as a "wedge" in the patent system which might lead to compulsory working of patents by Americans as well as aliens. Passage of the bill, he argued, would thus constitute a grave threat to the patent-control practices he himself had promoted so forcefully in 1906.[63]

The Lampert Bill was passed and the Stanley Bills never emerged from the Senate committee. The reform of the Patent Office, however, had only begun. In 1925 the expanded office was transferred from the Interior Department to the Department of Commerce, where it, like the Bureau of Standards, became an important agency of industrial prosperity and efficiency under the direction of Herbert Hoover.[64] An editor of the *Scientific American* assured his readers that the change was for the better. Herbert Hoover, he wrote, "knows that the preeminent position of American industry is largely based upon the protection afforded by patents."[65] Hoover further streamlined the Patent Office: the time allowed for response to office actions was reduced from one year to six months. Shortly thereafter, in 1929, the Court of Customs Appeals, renamed the Court of Customs and Patent Appeals, assumed "new duties of hearing patent appeal cases direct from the Patent Office."*[66]

The successful reform efforts between 1900 and 1929, while they had to be buttressed by subsequent measures in the wake of ever-expanding patent demands, brought the American patent system more closely into line with the needs of corporate industry. They set the basis for a "formalism" in the handling of patents which progressively eliminated the individual inventor, who, unlike the large corporations with their well-staffed legal departments, was not equipped to cope with its in-

---

*The creation of the court, it should be pointed out, did not signal the entire realization of Fish's early proposal for a single court of patent appeals. Thereafter patent applicants had the option of appealing their cases either to the new court, which judged the merits of the application only and heard no witnesses, or to the district courts and civil action, as had been the case. The fight for a single court of patent appeals thus continued. See, for example, Otto S. Schairer, *The Patent Problem from the Viewpoint of Industry* (New York: National Industrial Conference Board, January 19, 1939).

tricacies and complexities. Rather than having been the creation of some autonomous "bureaucratization," however, the new formalism was, as Robert Lynd noted, "overwhelmingly the result of the shrewd, meticulous tactics of corporation lawyers"[67] and well served the interests of both the corporations and the lawyers. The changes in the mechanism of the patent system, coupled with the emergence of corporate patent monopoly and industrial research, presented a formidable obstacle to the individual inventor whom Lincoln had in mind when he uttered the phrases that were inscribed above the doors of the Patent Office in 1932.

The significance of the transformation, however, was not recognized by Conway Coe, Commissioner of Patents, as he testified before the Temporary National Economic Commission less than a decade later. After defending the great benefits of patent monopolies and rejecting the notion that they could conflict with the antitrust laws, Coe concluded his testimony:

> In our estimate of the patent system ... we cannot disregard its spiritual influence in our national life and destinies. ... Our patent system has developed in our people a creative faculty [which] has proved signally useful in solving some of the great problems that have arisen in our task of preserving and perpetuating our democratic form of government.[68]

The committee which heard his testimony, along with that of many others, substantially agreed with Coe's appraisal of the system. It was for this reason that they grieved over its demise, something which Coe himself was unable to see. "The control over applied science which business holds," they concluded, "is the key to the explanation of its dominant position in the process of government." Another interpreter of the testimony was sociologist Robert Lynd, a contributor to the Herbert Hoover–sponsored study of *Recent Social Trends* and the co-author of *Middletown.* "The problem we face today," he wrote, "is that, in an era that increasingly lives by science and technology, business control over science and its application to human needs, gives to private business effective control over all the institutions of democracy, including the state itself."[69]

7

Science for Industry

The Organization of Industrial
and University Research

Science like charity should begin at home, and has done so very imperfectly. Science has been arranging, classifying, methodizing, simplifying everything except itself. . . . It has organized itself very imperfectly.

Scientific men are only recently realizing that the effective power of a great number of scientific men may be increased by organization just as the effective power of a great number of laborers may be increased by military discipline. . . . The prizes of industrial and commercial leadership will fall to the nation which organizes its scientific forces most effectively.[1]

—Elihu Root

Patents petrified the process of science, and the frozen fragments of genius became weapons in the armories of science-based industry. Control over science by way of patent monopoly had a serious shortcoming, however, in that advantage could be seized only after the fact. During the nineteenth century, engineers focused their attention upon translating the haphazard discoveries of university-based scientists into patentable processes and products; rarely did they concern themselves with the actual production of scientific discovery itself. But in the first years of the new century their attention shifted. With the introduction of organized research laboratories in industry, and the unprecedented effort to integrate universities within the industrial structure, the corporate engineers undertook to anticipate scientific discovery, to guarantee and regulate the supply of what had become the lifeblood of modern industry.

As early as the 1830s various industrial firms had on occasion employed university scientists to do research, but these were rare and isolated instances. Samuel Dana conducted investigations of the chemical processes involved in textile manufacture for the Merrimack Manufacturing Company; Benjamin Dudley, another chemist, set up a small lab to study the composition of steel rails for the Pennsylvania Railroad; a German chemist named Fricke did metallurgical research for Andrew Carnegie; and Professor Silliman of Yale analyzed samples of Pennsylvania crude for the infant petroleum industry. Until the turn of the century, however, industrial research remained essentially the unorganized effort of individuals.[2]

The electrical and chemical industries were the first to establish research as a systematic part of the business. The primary purpose of these early laboratories, and of those which later sprang up in the nascent rubber, petroleum, automotive, and pharmaceutical industries, was applied research and development (process-testing, routine chemical analysis, cost-cutting, and quality control). In time, however, some of the giant companies which dominated these industries extended their activities into the area of fundamental scientific research. Nearly all of the basic research done by industry, as well as the bulk of applied research, was restricted to large firms with ample financial resources, since they alone were able to provide researchers with a relatively stable working situation and adequate facilities. The smaller companies turned to the new cooperative laboratories of trade associations, government bureaus, private consulting companies, and the universities for most of their applied and all of their basic research.

Before 1900 there was very little organized research in American industry, but by 1930 industrial research had become a major economic activity. In a 1928 survey of nearly six hundred manufacturing companies, 52 percent of the firms reported that research was a company activity, 7 percent stated that they had established testing laboratories, 29 percent were supporting cooperative research activities of trade associations, engineering societies, universities, or endowed fellowships, and 11 percent of those doing little or no research indicated that they intended to introduce research work.[3] In less than three decades American industry had clearly become infatuated with scientific research. The major reason for this rush to science is not hard to fathom: there was money in it.

While many sectors of industry turned to science only when they became convinced that they would reap ample returns on their investment, the electrical and chemical industries had depended upon science from the outset. Thus it was they who set the example which the others would follow. Their dependence upon science compelled the leaders of these industries to become pioneers in industrial research. More important, perhaps, it forced them to look outside of industry for solutions to the problems they faced within it. Thus they came to conceive of their own particular activities as integral parts of an all-embracing industrial system, a system composed of distinct but coordinated units: industry, the universities, and government. For these men, the historical, philosophical, and geographical distinctions between the workshop and the laboratory, the industries and the universities, applied and fundamental research, were collapsing in practice, in the daily functioning of the industrial system.

"In the last analysis," J. J. Carty, AT&T's chief engineer declared, "the distinction between pure scientific research and industrial research is [merely] one of motive."[4] The motive behind the former is the disinterested search for truth; the motive behind the latter is the intent to maximize utility and profit. The two realms were not nearly so contradictory or mutually exclusive as it might seem, however, since the disinterested search for truth could be incorporated *within* the search for profit. The scientist could continue to delve into the mysteries of the universe without concerning himself with practical or pecuniary matters, so long as his discoveries could readily be translated by others into the means of capitalistic industrial development. Agreement over motives was not required. All that was required was satisfactory coordination between means and ends, and, for that, the proper organization.

During the first three decades of the twentieth century, therefore, the corporate engineers undertook to organize and harness science to industry. Their work evolved in three overlapping phases. The first involved the establishment of organized research laboratories within the industrial corporations, as integral parts of the enterprise. The second concerned the active support of, and cooperation with, research agencies outside of the corporations: trade-association laboratories, research foundations, government bureaus, and, most important, the science and engineering departments of the universities. The third saw the national coordination of these myriad research activities, primarily through the National Research Council, in support of corporate industry. The first two developments began roughly around the turn of the century; the third surfaced during World War I.

The first research laboratory in American industry, the GE Laboratory, was formally established in 1900, but had its origins in the work of Thomas Edison and Elihu Thomson. AT&T began organized research under Jewett's leadership in 1907 after a quarter-century of isolated activity by J. J. Carty. Westinghouse set up a research department in 1903 under C. E. Skinner, but it did not really become a going concern, with separate facilities, until 1916. GE and AT&T were the real pioneers in electrical research, and they have dominated industrial research in America ever since.

"It is the fashion to call this the age of industry," noted Henry Ford in the 1920s. "Rather, we should call it the age of Edison. For he is the founder of modern industry in this country." Norbert Weiner concurred, adding that "Edison's greatest invention was that of the industrial research laboratory. . . . The GE Company, the Westinghouse interests and the Bell Telephone Labs followed in his footsteps, employing scientists by hundreds where Edison employed them by tens."

Edison's laboratory had focused less upon fundamental discovery than upon practical and commercial development, and the work of his laboratory became the work of the early Edison companies and, ultimately, the work of General Electric. Elihu Thomson, co-founder of the Thomson-Houston company which had merged with Edison's interests to form GE, remained with the new company as its scientific sage. Thomson described the original lab as "a space set aside from a portion of the manufacturing and testing department, where with a few tools and perhaps one or two workmen, devices and new appliances were constructed in the form of working models, which were there to be refined and immediately put into manufacture." Basic research at GE before the turn of the century was thus hardly a routine matter, but as the effects of the 1893 depression eased somewhat, the company became charged with a renewed optimism and a spirit of innovation. E. Wilbur Rice, one of Thomson's early associates and vice-president and technical director at GE, together with Thomson, Steinmetz, and Albert G. Davis, head of the patent department, began to push for expanded facilities devoted to basic research. In 1901 the annual report to the stockholders stated that "it has been deemed wise during the past year to establish a laboratory to be devoted exclusively to original research. It is hoped by this means that many profitable fields may be discovered."[5]

Rice succeeded in persuading Willis R. Whitney, a professor of physical chemistry at MIT, to come to Schenectady and set up the laboratory. Rice encouraged Whitney to bring his own research to GE and work on it there, but Whitney commuted between Boston and Schenectady for three years before leaving MIT altogether to devote his full energies to the new GE Lab. "Mr. Rice's idea, from the very first,"

Whitney later recalled, "was to develop a laboratory for research in pure science. He wished it set sufficiently apart in the company organization to be free from the responsibilities of current problems of the company."[6] Whitney proceeded to lure W. D. Coolidge—an electrical engineer and physicist who had worked with him at MIT—away from the Institute, and to make him assistant director of the GE Lab; Whitney encouraged Coolidge to bring his own research with him, as he himself had been encouraged by Rice.

Under Whitney's direction the GE Lab expanded at a rapid rate. In 1901 there were 8 persons on the staff; by 1906 there were 102; by 1920 there were 301; and by 1929 there were 555. Whitney's interest in basic research, however, was not altogether satisfied. He assured Irving Langmuir, who brought GE its first Nobel Prize in 1932, that he was "not just interested in practical results. The laboratory will have to pay for itself, but, in addition, it should make a contribution to the advancement of science and knowledge. . . . I want some men in our lab who are contributing to fundamental scientific discovery." He complained, however, that the detachment from company operations that Rice had sought "has been impossible to maintain; the rule in GE has been to give calls for assistance from the engineers and production men precedence over all else . . . if they involve . . . possible loss to the company or unsatisfactorily meeting a customer's needs."[7] Although GE probably laid greater emphasis upon individual initiative, and less on directed projects, than most other industrial laboratories, science was nevertheless a handmaiden to corporate interests.

Applied research on telephony had always been done at AT&T, if in a rather random fashion. Research activities were divided both geographically and according to the particular focus of investigation. Work on transmission devices was done in Boston, in the laboratories which grew out of Bell's original workshop, while the development of switching and signaling devices was carried on in the "experimental labs" of the Western Electric Company in Chicago and New York. There was relatively little coordination of these various activities until 1907, the year Carty, who had been with the Bell organization in Boston since 1879, became chief engineer of AT&T. Under his direction the various operations were combined to form the engineering department of the Western Electric Company.[8]

The lab conducted some basic research, but the prime objective was always utility, in the most immediate sense of the term. Carty explained this to the members of the Chamber of Commerce of the United States:

These laboratories are devoted to a severely practical purpose. They are organized on a strictly business basis, and the work conducted in them is directed to no other purpose than improving and extending and conducting in a more economical manner the service which we render to the public. . . . The criterion which we apply to the work conducted in these labs is that of practical utility. Unless the work promises practical results it is not undertaken, and unless as a whole the work yields practical results it cannot and should not be continued. The practical question is, "Does this kind of scientific research pay?"[9]

The hard-nosed attitude of Carty was as much a display of corporate responsibility as anything else, and was intended to lend credibility to what appeared to be a rather uncertain enterprise.

It was not very long, however, before Carty's cautious approach led to bolder beginnings. Important theoretical work was already being done in Boston by George Campbell, the MIT-, Harvard-, and European-trained physicist who would later develop the electric wave filter, and it was Campbell who first brought Frank B. Jewett into the Bell fold. Jewett, the son of a civil engineer who constructed the California division of the Atchison, Topeka & Santa Fe Railroad, was a graduate of Throop Polytech (the forerunner of the California Institute of Technology) and took a Ph.D. in physics under A. A. Michelson at the University of Chicago. After receiving his doctorate, however, Jewett decided to forgo a promising university career and instead joined MIT's electrical-engineering department to prepare for a career in industry. In the eyes of Michelson, Jewett later recalled, "I was prostituting my training and my ideals . . . when I entered industrial life."[10] Leaving MIT, Jewett became a consultant and then Transmission and Protection Engineer at AT&T.

In the winter of 1910–11, Jewett brought together a group of scientists to form a separate research department for the study of long-distance telephony. The next year, when he became assistant chief engineer of the Western Electric Company (the manufacturing subsidiary of AT&T), Jewett coordinated the activities of his group with those of another research group, which had been set up by physicist Edwin H. Colpitts of the engineering department, and thus established the nucleus of the Bell Labs. "The industry had outgrown its ability to progress wholly on the basis of random invention," Jewett, who became president of Bell Labs, later explained. "It had also outgrown the second stage in which inventive ability and genius was teamed up with engineering skills, skills of the trained engineer. It had reached a stage in which it was clear that some other kind of attack on many problems had to be made."[11] The new attack was launched in the form of fundamental scientific research.

The Bell Labs, which were incorporated under that name in 1925, expanded at a tremendous rate and attracted many university scientists. In 1916, just four years after the creation of a separate research department, the annual expenditures amounted to $2.2 million; by 1930 they had multiplied tenfold. The great attraction of the Bell Labs for university-based scientists was suggested by W. Rupert McLaurin, an historian of the electrical industry. "In the emphasis placed on fundamental research, the Bell Telephone Labs resemble some of our leading university research centers. Until World War II, however, the Labs were far bigger and had a much larger budget for scientific research than any single university in the country." By 1925 the Bell Labs employed over 3600 people; in 1937 C. J. Davisson was the first Bell-sponsored winner of a Nobel Prize in physics.[12]

By far the most ambitious research effort in the new American chemical industry was that undertaken by Du Pont. In 1888 the company had established a small laboratory to test smokeless powder, and this work was extended during the next decade and a half to include high explosives such as dynamite. In 1902 the company was purchased by Pierre, Coleman, and Alfred Du Pont, all MIT alumni, and the new management decided upon an extensive research program as a means of expanding company interests. That year, as part of a general consolidation of company operations, the Eastern Laboratory was organized at Repauno, New Jersey, to conduct research on dyes and organic chemicals. Charles L. Reese, a chemist trained at Johns Hopkins and Heidelberg, was called upon to direct the research. In 1911 the lab became the center of the new Du Pont chemical department, with Reese at its head. It became the largest in the industry and by 1925 employed over 1200 chemists. Under Reese's leadership the company branched out into such new areas as synthetic organic chemicals, heavy chemicals, textiles, paints and pigments, cellophane, lacquers, and varnishes. In 1927 Du Pont started a program of large-scale fundamental research, primarily in the chemistry of polymerization. Under the direction of C. M. A. Stine, another chemist with Johns Hopkins and European training, who succeeded Reese as chemical director, this basic research drive paved the way for the development of nylon and other synthetic fibers as well as synthetic plastics and rubber.[13]

Other chemical corporations followed Du Pont's successful example. Dow, which had been established by a university-trained chemist, as early as 1901 created a laboratory to study the electrolytic process among other things. At the outset of World War I the company assembled a group of organic chemists to conduct extensive investigation on phenol derivatives and dyestuff manufacture. Eastman Kodak, the manufacturer of photographic materials and equipment, had relied heavily upon the work of chemical consultants from the very beginning. In 1912 George Eastman succeeded in securing the services of C. E. Kenneth Mees, a brilliant British chemist, physicist, and inventor, who set up the Eastman Laboratories of Kodak in Rochester. The Kodak laboratory quickly became one of the finest and largest in the industry and during World War I was the chief domestic supplier of organic chemicals used in research.

American Cyanamid organized a research department for routine analysis and testing in 1909 and a formal research lab three years later. During the war the company assembled a special research staff to study cyanamid derivatives and in the 1920s expanded its research operations significantly with the acquisition of the Lederle Laboratories and other firms. In 1930 the Stamford research center was established, with the prominent Columbia University chemist M. C. Whitaker as director of research. Union Carbide consolidated the research activities of its various subsidiaries in 1921; fundamental research was begun under the direction of the synthetic-organic chemist George O. Curme, who, as research fellow for the Prest-O-Lite Company, had done pioneering work on acetylene.[14]

By the 1920s other major industries had begun to follow the example of the electrical and chemical companies, often at the prompting of scientifically trained men who had some experience in those companies and had learned the value of industrial research. William M. Burton had organized an analytical lab for Standard Oil (Indiana) in 1890 to study the cracking process, but it was not until the 1920s that the petroleum industry undertook research on a significant scale. Among the earliest efforts were those of Standard (New Jersey), Shell, Gulf, and the Universal Oil Products Corporation. Charles Goodrich, a university-trained chemist, set up the first rubber-industry laboratory in the 1890s, and Paul Litchfield, an MIT chemical engineer, instituted the Goodyear research and development lab in 1908. Arthur D. Little, the dynamic promoter of industrial chemical research, founded the research department for General Motors in 1911. In 1920 GM acquired Charles F. Kettering's Dayton Engineering Labs (DELCO), and five years later the two labs were merged to form the GM Research Laboratory. Yale chemist John Johnston meanwhile directed the fledgling research work at U.S. Steel, while fellow chemists John A. Mathews,

E. C. Sullivan, and William Bassett headed the laboratories of Crucible Steel, Corning Glass Works, and the American Brass Company, respectively. During the 1920s the pharmaceutical industry embarked upon wide-ranging industrial research, with the Abbott Laboratories, Eli Lilly, Parke, Davis and Company, and E. R. Squibb leading the way.[15]

The giant firms which dominated the electrical and chemical industries pioneered in placing research work in industry on an organized basis. In doing so, they sought to institutionalize the foresight of those men who had laid the scientific foundations for the new industries, to transform what heretofore had been the result of random discovery and ingenious invention into the routine product of a carefully managed process. Systematic research, described by Frank Jewett as "cooperative effort under control," lessened their dependence upon the vagaries of genius; it made possible instead the creation of what GE's Philip Alger called "synthetic genii"—many specialists assembled as a team and "held together by bonds of sympathy and understanding, as well as by the company management."[16]

The research laboratories, above all, gave to the corporations command over the flow of scientific investigation. In the nineteenth century, scientific ideas had given rise to industrial manufacture; now the industrial corporations undertook to manufacture scientific ideas. "The Bell system," Jewett boasted, "is a completely integrated affair in which, from the inception of an idea through its development, its manufacture, its installation, and its operation, to the end of its life when it goes on the junk heap, the whole thing is under common command."[17] It was this process of research and development, followed by commercial application, that prompted Joseph Schumpeter to assert that the large industrial corporations "create what they exploit."[18] He neglected to add, however, that they exploit to create.

As the industrial research laboratories grew in size, the role of the scientists within them came more and more to resemble that of the workmen on the production line and science became essentially a management problem. The industrial laboratory was quite different from its university counterpart, which supplied it with scientific personnel. Whereas the university researcher was relatively free to chart his own paths and define his own problems (however meager his resources), the industrial researcher was more commonly a soldier under management command, participating with others in a collective attack on scientific truth. E. B. Craft of the Bell Labs described the operations which he oversaw. "I might say that each of these main [laboratory] departments is organized on the functional basis with a military type of organization so far as responsibility and supervision are concerned." The primary objective of Craft and his associates was to make the best use of "the human material which is available."[19]

> Perhaps the outstanding characteristic of this organization, the one that sets it apart a little from others, is its conduct of research and development by a group method of attack . . . the result is the necessity of a high degree of specialization. So in all of these technical departments we have specialists, chemists, metallurgists, physicists, engineers, statisticians, mathematicians, men who are trained and skilled in their particular branches of science and engineering. Their activities are so coordinated by means of this organization, that their best brains can be brought to bear upon any specific problem. . . . When a problem is put up to the Labs for solution, it is divided into its elements and each element is assigned to that group of specialists who know the most about that particular field but they all cooperate and make their contribution to the solution of the problem as a whole.[20]

The "group method," first introduced on a large scale at the Bell Labs, eventually became the standard operating procedure of industrial research. At the outset, however, it created as many problems as it solved. "We have had our troubles in working out this scheme," Craft later explained.

> In the first place it might appear that it would tend to destroy the initiative of the individual; that it would make it difficult to properly assign the credit and give the reward to the individual worker. These are all problems of administration that have had to be worked out. First of all we must establish in the individuals a state of mind which leads them to really believe that their best results are attained through cooperation with others.[21]

The organization of industrial research thus gave rise to a new field of expertise—research management. If science was to be effectively controlled, scientists had to be effectively controlled; the means to such control was the fostering of a spirit of cooperation among researchers second only to a spirit of loyalty to the corporation. Jewett took the first steps in this direction around 1912. Before this time Bell engineers were awarded $100 dollars for each new patent, a policy which Jewett perceived as counterproductive since it fostered individual rather than cooperative effort. Jewett eliminated such incentives altogether, labeling them anachronistic, vestiges of the fading "era of the inventor."[22]

In their place the laboratory adopted elaborate procedures for recognizing individual accomplishment in lieu of rewarding it. "We can analyze the results the same way that we analyze the problem in the beginning," Craft explained, "and [thus] assign to the individual the particular contribution that he has made. This is all very carefully recorded in our laboratory notebooks, a complete record of all the work that is done, and these are turned over to the patent organization and they determine who the inventors are." Those research workers not fortunate enough to be recognized by patent lawyers as inventors were recognized in other ways, such as through the publication of their work in technical journals. In this regard the management encouraged joint authorship of articles by perhaps half a dozen workers, so as to allow as many as possible to "share in the glory of the achievement."[23]

The managers of the large industrial research laboratories thus sought to resolve the needs of individual workers with the corporate imperatives of their "military" operations. They offered incentives to boost the ego of the individual without cost to the company, and emphasized the spirit of loyalty and cooperation in order to elicit his best efforts. In so laying the groundwork for research management, however, they overlooked an important point, one which has plagued management ever since. True cooperation, the spirit of collective activity, presupposes individual autonomy and intention. The workers in these laboratories, however, neither initiated their cooperative undertakings out of a sense of their own needs, nor directed them toward realization of some commonly defined purpose. Rather, they entered into cooperation not by will but as a compulsory aspect of their employment, by assignment. Thus, their activity did not reflect a spirit of cooperative investigation so much as one of collective subservience.

If the large industrial research laboratories undercut the individual aspirations of the scientific cadre within their firms, their work also had a chilling effect on independent research efforts outside of them. As the Temporary National Economic Committee discovered in the 1930s, "industrial research ... gives a competitive advantage to those firms able to pursue it." Those firms able to pursue it, however (especially if by "it" is meant fundamental research), were the larger ones. Thirteen companies—less than one percent of those reporting research activities in 1938—employed more than one third of all research workers. One half of industrial-laboratory personnel were employed by forty-five large laboratories, all but nine of which were owned or controlled by companies among the nation's two hundred leading nonfinancial corporations. "There is probably no other basic function of general economic activity," the TNEC concluded, "so dominated by a few enormous concerns." The success of industrial research laboratories thus gave a competitive advantage to the already dominant firms, and further increased the concentration of economic power. At the same time, it blinded many within them to any possible alternatives. At the hearings conducted by the TNEC on the question of the concentration of economic power, William Coolidge, director of the GE Labs, was asked by the chairman if he was "satisfied that the natural individual is sufficiently protected under the present system against competition from huge collective enterprises." "I think so," Coolidge replied. "We are always on the lookout for new inventions (I mean our company is), whether they come from our lab or whether they may be from the outside."[24]

The industrial laboratories of the large electrical-manufacturing and chemical corporations provided the nation with its first experience in large-scale organized scientific research. The lesson provided by their undeniable commercial success, coupled with the almost missionary promotional efforts of consultant chemists and research-oriented engineers, prompted many companies in all areas of manufacture to try industrial research. Research laboratories, however, with elaborate facilities and highly trained staff, were expensive to maintain, and only the largest companies could afford them. And even in those giant enterprises which were able to support some fundamental as well as applied research (notably Du Pont, GE, and the Bell System), in-house industrial laboratories were not able to meet all of their research needs. As Frank Jewett explained in a paper delivered to the American Association for the Advancement of Science, "those most obviously dependent on science have organized research laboratories whose sole function it is to search out every nook of the scientific forest for timber that can be used." But such industrial efforts could not by themselves "advance the frontiers of knowledge at a rate commensurate with the demands for industrial advancement."[25]

Among the earliest efforts taken to supplement the work of the in-house industrial laboratories involved the establishment of trade-association laboratories and semiprivate or private nonprofit research institutes. A prominent example of the trade-association laboratory was Nela Park. The National Electric Lamp Association was founded by John B. Crouse, a Cleveland lamp manufacturer, to serve the collective needs of the electric-lamp industry. In 1907 the association established the Nela Park research facilities in Cleveland as a cooperative engineering and research center, and it soon became the authoritative center of research activities concerning all problems of interest to the industry. The association subsequently established a school of electric-illumination engineering and Nela Park became known as the "university of light." In 1912 control over Crouse's National Electric Light Company was secured by GE and Nela Park became the National Lamp Works of GE. William Enfield, an electrical-engineering student of A. A. Potter at Kansas State and an MIT graduate, became director of Nela Park two years later. In 1913 the National Canners Association Laboratory was organized under the leadership of a professor of chemistry from Oregon State College, William D. Bigelow, and other cooperative laboratories were subsequently established by the Portland Cement Association, the Tanning Institute of Scientific Research (at the University of Cincinnati), the American Textile Research Institute, and the American Institute of Baking. (In the wake of the Volstead Act, the AIB was fortunate enough to obtain control over the laboratories established by leading American brewers.) Yet another research outfit was established by the Horological Institute, and its energies were directed toward the promotion of perhaps the most important industrial science of all, that of timekeeping and watchmaking.[26]

In addition to the trade-association labs, there were private and semiprivate research institutes, which were usually affiliated with universities. The earliest and most important of these centers of research was the Mellon Institute of the University of Pittsburgh. The man most responsible for its creation was the Canadian-born chemist Robert Kennedy Duncan. With the exception of Arthur D. Little, no single person did more to promote industrial chemical research than Duncan. After completing his graduate work at Clark University, Columbia, and various European universities, Duncan became a professor of chemistry and a prolific journalist of popular science, writing for such publications as the New York *Evening Post* and *McClure's.* Immediately following the tremendous success of his book *The Chemistry of Commerce,* he was commissioned by *Harper's Magazine* to make a study of the relationship between chemistry and industry in Europe. In the course of his study Duncan was impressed again, as he had been as a student, by the German system of cooperation between universities and manufacturers, and became convinced that such cooperation could improve the efficiency of the American chemical industry.[27]

Returning to the University of Kansas in 1907, Duncan established the Industrial Fellowship system, through which manufacturers were able directly to support research personnel in the university laboratories and to define their research work. The purpose of the system, as Duncan later explained, was to enable the universities to keep abreast of the changes in the industries and to enable the manufacturer to obtain needed expertise. The manufacturer "does not know how to treat a man of science in his factory in giving him either power or trust. . . . This consideration is responsible for much of the failure of American factory research."[28] According to the fellowship plan, a member of the chemistry staff was appointed for a two-year period to work exclusively on a problem defined by the sponsoring company, which would underwrite the cost. Any discoveries made during the fellowship period became the property of the company, and all patents were assigned to it. The fellow was required to submit periodic progress reports to the company and was permitted to publish his work in a manner which did not, "in the opinion of the company, injure its interests." From Duncan's perspective, the advantages of the fellowship plan were twofold. The university gained increased opportunities to fulfill its function of promoting research, three hours' teaching a week by a company-supported faculty member, and the "catalytic influence"[29] of industrial contacts. The manufacturer meanwhile obtained the use of adequate library facilities, enormously increased laboratory resources, greater opportunities for consultation among the fellow's colleagues, freedom from the problematic supervision over research work, and the services of a man educated to the company's particular needs. At or before the expiration of the fellowship, moreover, the company was entitled to secure the full-time services of the fellow for three years.

Others shared Duncan's perspective. Among those who recognized the practical success of the industrial fellowship program were Andrew and Richard B. Mellon of the Pittsburgh banking, mining, and petroleum interests. The Mellons invited Duncan to set up a similar system at the University of Pittsburgh and in 1913 endowed the Mellon Institute of Industrial Research, with its own financial resources, building, and management, to continue the program. Less than a year later Duncan, who became first director of the institute, died and was succeeded by one of the earliest fellows, the petroleum chemist Raymond F. Bacon. The success of the Mellon Institute led to the establishment of similar organizations elsewhere during the next two decades. Among the largest were the Batelle Memorial Institute in Columbus, Ohio (1929), and the Purdue Research Foundation (1930) in West Lafayette, Indiana. As early as 1918 Du Pont began to sponsor industrial fellowships, patterned after the Duncan idea, at several universities, and the

Chemists Club established the Victor G. Bloede Fellowship program. In a similar vein, the University of Chicago, at the prompting of chemist Julius Steiglitz, created industrial fellowships with the aim of shaping chemical courses and defining research activities to bring about closer cooperation with business.[30]

Private contracting provided yet another means of outside support for industrial research. C. F. Burgess, a professor of both electrical and chemical engineering at the University of Wisconsin, set up the C. F. Burgess Laboratories, Inc., in 1910; Carleton Ellis organized the New Jersey Testing Labs in 1912; and the Hirsch brothers instituted the Hirsch Laboratories in New York City in 1920. Among the most prominent independent consultants in the country, all of whom operated out of various colleges, were John E. Teeple, William M. Grosvenor, M. C. Whitaker, Carl S. Miner, William Hoskins, and Samuel P. Sadtler. Easily the foremost proponent of independent consulting and research contracting, however, was Arthur D. Little.[31]

Born in Boston in 1863, Little attended private schools in Portland, Maine, and New York City before entering MIT to take up the new special course in chemical engineering. At MIT Little made his mark as the founder of *Technology Review,* but left before graduation to become the chemist and superintendent of the Richmond Paper Company in Rumford, Rhode Island, the first sulfite mill for wood pulp in the United States. In this capacity he became one of the country's leading experts on the sulfite process of papermaking, and went on to design and operate pulp mills in Wisconsin and North Carolina. In 1886 Little opened an office in Boston as an independent chemical consultant; after a few false starts he teamed up with a young instructor at MIT, William H. Walker, to form the firm of Little and Walker. In 1905, when Walker returned to MIT (where he eventually became first director of MIT's Division of Industrial Cooperation), Little incorporated his business as Arthur D. Little, Inc., which was to dominate contract research in the United States for over a generation.[32]

Little's specialty was cellulose chemistry (papermaking and fiber treatment), but he soon expanded his activities into areas such as the electrolytic manufacture of chlorates, petroleum refining, leather tanning, and the production of artificial silk. More important perhaps than his actual research activities was Little's tireless promotion of industrial research. As a founder of the Industrial Section of the ACS (and later ACS president), and the prime mover behind the establishment of the AIChE, Little probably did more than any other individual to convince the chemistry profession of the importance of industrial research and to place the new field of chemical engineering on a sound basis. At the same time, he worked constantly to convince industrialists of the dollar value of scientific research. "Research, whether fundamental or applied," Little preached, "is the lifeblood of chemical engineering," and thus of the industries dependent upon it. Little was, above all, a master salesman of research—the "voice of research," as Maurice Holland, director of the National Research Council, called him. As the consultant for such industrial giants as the United Shoe Machinery Company, the General Chemical Corporation, General Motors, and International Paper, Little practiced what he preached, all the while alerting industrialists to what he called "the handwriting on the wall," the inevitable scientific transformation of modern industry.[33]

The expanding facilities of the federal government offered another solution to the industrial research problem. The conservation programs of Frederick Haynes Newell, a civil engineer and director of the Reclamation Service, the expanded Bureau of the Census under Francis A. Walker, a president of MIT, and his successors, and the new Bureau of Mines, established in 1910, combined to provide extensive surveys of the natural resources upon which industry depended. At the same time, research and the training of researchers became a fundamental function of the Geological Survey, the Bureau of Standards, and an expanded Smithsonian Institution. In 1913 a new organic act enlarged the scope of the Bureau of Mines to include "such fundamental inquiries and investigations as will lead to increasing safety, efficiency, and economy in the mining industry."[34] The next year a Petroleum Division was instituted "to provide production research for an industry that had in its rapid growth provided little research of its own,"[35] and the following year Congress created the National Advisory Committee for Aeronautics, which functioned "to direct and conduct research and experiment in aeronautics."[36] These various agencies were certainly "the servants of industry as well as of the government,"[37] and they cooperated with each other in their work. In 1925 the Bureau of Mines was transferred along with the Patent Office to the Department of Commerce, already the home of the Bureau of Standards, the Coast and Geodetic Survey, and the Bureau of the Census, and under the direction of Herbert Hoover the work of all of these agencies was further coordinated and directed toward the aid of industry.

The bulk of research appropriations in the federal government remained minimal, however. While they facilitated the centralization and coordination of industrial efforts throughout the country and acted as information clearinghouses, the governmental bureaus themselves did relatively little research. The largest appropriations for research were those for agriculture, not industry. As late as 1937 only 2 percent of government expenditures went for research (as compared to 25 percent in some universities and 4 percent in some industrial concerns) and well over half of that went toward agricultural and social-science statistical research. The development of interest in research within the military departments, the Army and the Navy, had been slow before World War I, but it accelerated at a feverish pace during preparedness and the war itself. While this stimulation had profound and lasting effects on American science and technology, and led to increasing military sponsorship of industrial research, appropriations to the military "plummeted" after the war, and the major research agency to emerge from it, the National Research Council, had to rely predominantly upon private and industrial funding for its operations.[38]

Private foundations were yet another means of meeting the research needs of industry. The Carnegie Institution was endowed with $10 million by Andrew Carnegie in 1902 "to cooperate with other institutions throughout the United States in encouraging investigation, research and discovery; in showing the application of knowledge to the improvement of mankind; in providing such buildings, laboratories, books and apparatus as may be needed; and in affording instruction of an advanced character to properly qualified students."[39] The first president of the institution was Daniel Coit Gilman of Johns Hopkins; he was succeeded after only two years by engineer-scientist Robert S. Woodward, assistant to Henry Pritchett at the Coast and Geodetic Survey and a former professor of mechanics and dean of science at Columbia. The chairman of the Institution's board of trustees was corporation lawyer Elihu Root, a staunch supporter of organized industrial research. A few years later Carnegie set up another foundation, at the suggestion of Henry Pritchett, whereby university faculty could receive retirement allowances. In addition to establishing a pension system, the Carnegie Foundation for the Advancement of Teaching, with Pritchett as director, broke new ground in the investigation of educational facilities and sponsored, among others, Abraham Flexner's revolutionary study of medical education and Charles Mann's study of engineering education.

In 1901 John D. Rockefeller set up the Institute for Medical Research, and followed that two years later with the General Education Board for the promotion of education within the United States. The Rockefeller Foundation proper was established in 1913 and consolidated with the Laura Spelman Rockefeller Foundation in 1928. Although under the direction of Simon Flexner the Rockefeller money was used to sustain basic research in physics and chemistry as a function of medical researches, the Rockefeller organization did not directly support basic scientific research of the kind needed in industry until the 1930s. It did, however, perform considerable service in that area indirectly through the funding of the National Research Fellowship of the National Research Council. "That event in 1919," wrote a former Rockefeller Foundation president, "marks the beginning of Foundation aid to the natural sciences."[40]

The Engineering Foundation was instituted in 1914 by Ambrose Swasey, a leading Cleveland machine-tool manufacturer and devotee of science.* This foundation was administered by the Engineering Foundation Board, composed of representatives from ASME, AIEE, and AIME; the board was a department of the United Engineering Societies and functioned "as an instrumentality of the Founder Societies . . . for the stimulation, direction, and support of research."[41] After World War I the Engineering Foundation, under the direction of W. F. M. Goss, former Purdue dean of engineering, and men such as Jewett of AT&T, President E. Wilbur Rice of GE, Arthur D. Little, and Gano Dunn of the J. G. White Engineering Company, assumed the financial responsibility for the National Research Council.

The Chemical Foundation was another privately endowed institution which promoted and supported scientific research for industry. Established during the war by executive order as the repository and licensing agency for seized German patents, it was created specifically to stimulate the development of and serve as a holding company for the American chemical industry. When Alien Property Custodian A. Mitchell Palmer became Attorney General, his assistant, Francis P. Garvan, assumed the directorship of the foundation. Under Garvan's dynamic leadership, and the strong support of chemical companies and the ACS, the foundation quickly became a major focus of activity within the industry. Aside from its important role as owner and dispenser of 4500 patents, the foundation assumed the major responsibility for public relations for the growing industry. In this capacity it had distributed some thirty million pieces of educational literature by 1935, and supported the journalistic work of chemist E. E. Slosson, author of the highly successful book *Creative Chemistry: A Description of Recent*

*Swasey's company, Warner and Swasey, produced the telescopes for George Ellery Hale and other leading astronomers—Pritchett included—at a loss, because of his concern for scientific progress.

Achievements in the Chemical Industries, and editor of Science Service.* Probably the most important aspect of the foundation's work,
however, was its encouragement and financial support for chemical
education at all levels and research in universities and colleges. In all
of its activities the overriding mission of the foundation was to effect
close cooperation between the educational institutions, the major research centers in the country, and the industry.[42]

The research activities of trade associations, semiprivate institutes,
independent contractors, government bureaus, and private foundations
provided an essential service and subsidy to the expanding science-
based industries. They were not, however, equipped to meet all of the
scientific requirements of modern industry. Only the nation's colleges
and universities, with their unequaled research facilities and trained
personnel, were prepared for this. Given the proper organization and
spirit of cooperation, the universities could provide an unlimited
amount of applied research for industry. More important, as the traditional site of fundamental scientific research, they could support the
basic investigations upon which industrial research was grounded, investigations which only a handful of the largest corporations could
afford to underwrite. "The very existence of the great research programs of industry," William Wickenden maintained, "is predicated
upon the existence of a vast army of free, disinterested and even impractical researchers at work in the laboratories of colleges and universities." Long the source of the random discoveries upon which the
science-based industries grew, such university research could be enlarged and upgraded and tied in with the related work of applied
scientists if only the universities were, as MIT's Dugald Jackson urged,
"integrated as research centers within the industrial structure."[43]

Finally, the universities could do what no other research agencies,
within or without industry, could do: they could reproduce themselves.
Whereas industrial research, as Jewett remarked, was "man-consum-
ing," university research was "man-producing." As the centers of scientific education as well as research, the universities alone were
equipped to provide "the continuous supply of well-trained workers . . .
to meet the increasing needs of industry." Thus, the universities, potential suppliers of applied research, fundamental research, and research
manpower, were the key to science-based industrial development.

*For a discussion of Slosson's Science Service, see Ronald Tobey's *The American
Ideology of National Science* (University of Pittsburgh, 1971), pp. 62–95.

"American industry finds it increasingly profitable to become interested in, and to aid by means of money and counsel, research in universities," Dugald Jackson reported. "The more influential of the men of the technical industries have come to recognize the desirability of cooperating in the joint processes of education and industry."[44]

Because of the enormous practical achievements of the industrial research laboratories and the meager resources of the universities, J. J. Carty explained in his AIEE presidential address, "it has been suggested that perhaps the theater of scientific research might be shifted from the university to the great industrial laboratories. . . . But we can dismiss this suggestion as being unworthy . . . the natural home of pure science and pure scientific research is to be found in the university, from which it cannot pass."[45] Carty thus stressed the importance of industrial support for universities. The money for university research

> should come from the industries themselves, which owe such a heavy debt to science. While it cannot be shown that the contribution of any one manufacturer or corporation to a particular purely scientific research will bring any return to the contributor . . . it is certain that contributions by the manufacturers in general and by the industrial corporations to pure scientific research, as a whole, will in the long run bring manifold returns through the medium of industrial research.[46]

In addition to the research itself, Carty concluded, "the time has come when our technical schools must supply in largely increasing numbers men thoroughly grounded in the scientific method of investigation for the work of industrial research."[47]

Frank Jewett enlarged on these themes, which became increasingly common in technical and university circles. He urged that the balance be shifted back to the universities; that the universities, properly funded and staffed, could indeed become the answer to industrial needs "as a whole" and "in the long run."

> Not only must [universities] advance the frontiers of knowledge at a rate commensurate with our demands for industrial advancement, but they must, at the same time, develop the scientifically trained personnel required to carry on the work of the industries as well as to carry their own work. It is a well-recognized fact that within recent years industry has made extremely heavy demands upon the faculties of the universities by reason of their ability to offer greater monetary rewards, and frequently better facilities for research.
> We now find ourselves confronted with the need of increasing the bargaining powers of universities and the attractiveness of academic

positions. In this matter the industries have a clear-cut obligation to the universities, an obligation which they cannot avoid without themselves being the chief sufferers. It is an obligation which rests upon all industries alike, for in the final analysis ... what benefits one industry, benefits the others. That thoughtful men in all walks of life are coming to see the vital need of a proper coordination of the nation's scientific interests is a happy augury for the future.[48]

A growing number of industrial spokesmen, scientists, and university educators joined Carty, Jewett, Whitney, and other corporate engineers in their promotion of university research and industrial support of education. Cooperation between universities and industry became the urgent message of the science-based industries, the engineering profession, and technical-school educators from roughly 1906 on; it gained attention and support through the efforts of engineers and scientists during the war and, after it, through the National Research Council. Not all who endorsed this movement, however, were as aware of its probable implications as were the corporate engineers. They well understood that what Jewett called "the stimulation of scientific research in a more diverse fashion through the universities and higher educational institutions" was one way of retooling American higher educational "processes" to meet industrial demands: not only would it involve a major reorientation of the universities themselves—something various interests within them sought for professional reasons—but, viewed from outside them, from the vantage point of corporate industry, it promised to provide a nationwide resource for further corporate expansion and social control. The return on the investment in academic science and education, moreover, while benefiting all industries, would not necessarily benefit them all equally. The ample rewards for such self-interested philanthropy would flow back not in random fashion, but rather according to "the spontaneous economic distribution of the benefits of science." What this meant, Jewett explained, was that the benefits of science "reach practically all industries, and in proportion to the size and the importance of each."*[49]

The growing need within industries for scientific research, and the drive toward cooperation with educational institutions to secure it,

*It was for this reason that firms which did not dominate an industry the way AT&T did—and thus could not guarantee that the fruits of their investment would be returned to them rather than to competitors—were reluctant to support pure research in universities. For further discussion on this point, see Lance E. Davis and Daniel J. Kevles, "The National Research Fund: A Case Study in the Industrial Support of Academic Science," *Minerva*, XII (April 1974), 214–20.

paralleled the development of research within the universities. The influence of German universities—the number of American students in Germany peaked in the mid-1890s—had resulted in the creation of Johns Hopkins University and Clark University, and the establishment of research-oriented graduate studies at Harvard, Columbia, Chicago, and Wisconsin. Daniel Coit Gilman, an advocate of utilitarian educational reform since his support of the Morrill Act in the 1850s, became the first president of Hopkins. Clark, the country's only university devoted solely to graduate studies, had as its "major aim . . . the promotion of pure science." Its first president, G. Stanley Hall, came to Clark from Hopkins to set up a "purer" Hopkins. Beset by financial difficulties from the outset, however, Clark did not realize its original aims as an all-graduate college.[50]

The men who headed Harvard, Chicago, and Columbia were also interested in developing facilities for research "as a means of gaining or retaining an 'up-to-date' reputation for their institutions." By 1906 a statistical analysis prepared by James McKeen Cattell, editor of *Science,* indicated that Americans produced "from one-seventh to one-tenth of the world's scientific research"; in the natural sciences the country had received its first Nobel Prize and boasted such prominent scientists as Stratton, Michelson, Remsen, Millikan, Noyes, Pritchett, G. N. Lewis, Hale, and Pupin.[51]

In the first decade of the new century the efforts of industrial leaders to reshape the educational institutions into a valuable industrial resource of both research and manpower coincided with university efforts to extend the services they provided. The University Extension movement, which originated in the 1870s and 1890s at such schools as Harvard and Chicago, at first involved only the teaching of traditional courses in art, history, literature, and the like to a wider community. For educators like William Rainey Harper of the University of Chicago it was a way of spreading the intellectual culture so long a monopoly of the elite. Between 1890 and the first years of the twentieth century the movement suffered "fifteen lean years" of inactivity. When it was revived, however, at institutions like the University of Illinois, it had taken on an entirely new meaning. Unlike the experience of the University of Chicago before the 1890s, it had become, as at Wisconsin, "intentionally utilitarian."[52]

"The first official recognition of the responsibilities of the university towards the industries of the state and country," noted C. Russ Richards, dean of the Engineering College at the University of Illinois, "was given by the Board of Trustees of the University of Illinois on December 8, 1903."[53] On that date was established the first university "engineering experiment station" in the country. The Hatch Act of 1887 had provided the necessary federal appropriations for the establishment of "agricultural experiment stations" at land-grant institutions throughout the country; while it thus greatly stimulated research for agriculture, and further promoted the intentions of the Morrill Act, it did not so encourage the development of the other aspect of the Morrill Act, "the promotion of the mechanic arts." As the needs of industry became greater relative to those of agriculture, the demand for government support for industry and engineering grew accordingly. In the federal government, the creation of the Bureau of Standards and the ultimate transfer of such industry-oriented bureaus from the Department of Interior to that of Commerce—where they did not have to compete with agriculture for funds—reflected this development. Within the land-grant engineering schools the demand for government support took shape in the creation of engineering experiment stations to provide for industry the same services that the agricultural experiment stations offered farmers. Richards, a prominent mechanical engineer, was among the leaders of this movement. The engineering experiment station, he argued,

> affords opportunities for the establishment of relations which will be of mutual advantage to the industries and to the institution ... the result of a free interchange of knowledge of importance to a particular industry will benefit every concern in the industry by standardizing and improving the product and through the establishment in the minds of the public of greater confidence in the products of industry. While engineers or special industries may be primarily benefitted through the extension of scientific knowledge in a particular field, the public is undoubtedly benefitted because commodities which it needs may be produced more economically and of better quality.[54]

A professor at Illinois, writing in the *Proceedings* of the AIEE, explained that the new educational experiment was "conducted as an institution of scientific research," that "its purposes are the stimulation and elevation of engineering education and the investigation of problems of special importance to professional engineers and to the manufacturing, railway, mining and industrial interests of the state and country." The engineering experiment stations promised to provide the necessary research, the supply of scientific information to industry (through regular publication in bulletins), and the training of engineers

in scientific research. As a result, and through the efforts of industrialists and educators alike, they grew in number. Support came from
industrial or state sources. Those at Iowa University and Pennsylvania
State were receiving special appropriations from the state legislatures.
Others at Kansas State, Kansas University, Illinois, and Cornell relied
upon private funds.[55]

The engineering-experiment-station movement gained momentum
with the revival of the extension movement, particularly at the University of Wisconsin. "In the entire history of university extension," wrote
one student of the subject, "no event had more critical importance than
the re-establishment of the Extension Division of the University of
Wisconsin by President Charles R. Van Hise and Dean Louis E. Reber
in 1906–07." Van Hise, an engineering graduate of Wisconsin, persuaded Reber to leave his post as dean of engineering and director of
the experiment station at Penn State to set in motion what became
known as the Wisconsin Idea. The two men sought to extend the
educational services of the university to the community of Wisconsin,
not only in the traditional "cultural offerings" but in the more practical
areas as well. Not surprisingly, "the engineering instructors were the
first full-time instructors employed in the Extension Division." Because
they sought to offer advanced training rather than basic vocational
training in their extension program, Van Hise and Reber also helped
to create the State Industrial Education Board, the state commission
for the promotion of a state vocational-school system. Through their
efforts, the university was relieved of providing vocational training and
was free to focus instead upon teaching and research for industry as
well as agriculture.[56]

The Land Grant College Engineering Association was created, with
Anson Marston of Iowa as president and A. A. Potter of Kansas State
as secretary, in part to promote the engineering-experiment-station
movement; in 1915 Van Hise helped set up the National University
Extension Association to promote extension education as well. Despite
the swell of support for industry-oriented educational reform, however,
there remained considerable resistance to such publicly funded support
of industry, both within and without the universities. Thus, in 1906,
with the Adams Act, and again in 1914, with the Smith-Lever Act, the
agricultural experiment stations received additional federal appropriations without even a suggestion of similar support for the engineering
and industrial interests. After the passage of the Smith-Lever Act,
some university and industrial engineers tried to correct this situation.[57]

Andrey A. Potter, an MIT electrical engineer, had come to Kansas State University from GE's Schenectady plant, at the behest of his boss, to help establish a course in electrical engineering. (In return, GE promised to provide him with consulting work to supplement a small academic salary.) Potter soon developed a keen interest in power farming and a great admiration for the achievements of the agricultural experiment stations. As an engineering educator with close ties to industry—he was later dean of engineering and president of the local Chamber of Commerce—he saw great potential in the establishment of similar institutions for the benefit of industry. The demand from industry had already resulted in the institution of engineering experiment stations in about twenty states, and he felt strongly that such effort should receive federal support "analogous to the aid provided by the Hatch Act of 1887."[58] Research extension programs were being established in England, Germany, Austria, France, Russia, Sweden, and Switzerland, and Potter, together with many leaders in engineering and industry, demanded that the United States at least keep pace with these developments.

In 1916 Potter drew up a bill for the creation and support of engineering experiment stations at all land-grant colleges, and secured the active support of his former chemistry professor at MIT, Willis Whitney. Whitney, besides being director of the General Electric Labs, was chairman of the Committee on Chemistry and Physics of the country's first national scientific advisory board (the United States Naval Consulting Board) and thus had considerable influence within the scientific community. Potter and Whitney were able to elicit the support of Senator Francis G. Newlands of Nevada, who agreed to introduce in Congress Potter's bill, which authorized the creation of and support for

> experiment stations to conduct original researches, to verify experiments, and to compile data in engineering and in the other branches of the mechanical arts as applied to the interests of the people of the United States and particularly of such as are engaged in the industries.[59]

In addition to providing actual university facilities and personnel for industrial research, the bill promised to encourage the scientific training of engineers through experience with both industrial problems and scientific research and the dissemination of scientific information, through the "publication of bulletins giving results of investigations at least once in six months."

Whitney remembered how he and A. A. Noyes had "chipped in their own money to start research in physical chemistry" at MIT and knew well how "most of our American teachers and research men had to go abroad for their first experience in research." He thus felt that "this experiment station scheme should contribute a good foundation for scientific cooperation in our country" and allow the industries, GE among them, to "take advantage of the enormous amount of available apparatus in the colleges."* He circulated a letter to elicit support for the measure, and the response was overwhelming. The nature of his mailing list had something to do with the response. P. G. Nutting, president of the Association for the Advancement of Applied Optics, avidly endorsed the proposal. "The contemplated distribution of labs," he wrote, "will provide more intimate relations between research and the interests most likely to profit by it." Arthur D. Little was no less candid in his appraisal of the plan; he thought Potter's inclusion of such services as water supply, sewage treatment, waste management, flood protection, road building, and transportation in the proposal, while possibly gaining support for it, might confuse the issue. "Perhaps if the stations were called industrial research stations," he suggested, "their purpose would be more clearly indicated." Among the strong supporters of the bill were prominent engineers and scientists from Yale, Harvard, Johns Hopkins, the University of Chicago, the University of Wisconsin, and many land-grant colleges.

The bill also received support from the American Academy for the Advancement of Science,† various government bureaus, the Chemists Club of New York, the foundations and various trade associations, and the research directors of industrial laboratories. All this support, however, could not get the Newlands Bill through Congress; many legislators were concerned over the possible cutback of agricultural appropriations which might have resulted. Many leaders within the land-grant universities could not condone what they viewed as the

---

*Whitney was an active member of the ACS Committee on Cooperation Between the Universities and the Industries. Joining him on the committee were W. A. Noyes of the Bureau of Standards, R. F. Bacon of the Mellon Institute, John Johnston of Yale and United States Steel, Julius Steiglitz of the University of Chicago, W. H. Nichols, founder of the Allied Chemical Corporation, and such prominent research chemists as G. N. Lewis, T. W. Richards, and Alexander Smith.

†In 1913 the AAAS established a Committee of One Hundred to promote "cooperation between the industries and the universities"; this committee, which had already begun to prepare an inventory of researchers, projects, and research support in America, strongly endorsed the Bill. James McKeen Cattell, editor of *Science,* chaired the committee.

obvious public subsidization of private industry. A number of professional consulting engineers were also opposed to what amounted to government-supported competition.[60]

In time, the growing strength of the university engineers, who wanted to use the schools as the base for their professional practice, coupled with their assurances of cooperation with independent consultants, overcame this initial resistance. Even if it did not succeed in its immediate purpose, the Newlands Bill drive did convince the land-grant colleges of the importance of scientific research, and industry spokesmen continued to campaign for university and government assistance. By 1937 there were thirty-eight land-grant-college engineering experiment stations in full operation in the United States, "spending over a million dollars annually from state and local sources on research."[61]

Probably the fullest expression of the idea of industry-education cooperation in research was realized at the Massachusetts Institute of Technology. During the first three decades of the century all four MIT presidents were enthusiastic advocates of such cooperation, and their efforts set the trend for decades to come. The period began with the presidency of Henry Pritchett and closed with that of Samuel Stratton. In addition, the governing body of the MIT Corporation was composed of men who shared their enthusiasm: Frederick Fish,* Elihu Thomson, Charles Stone, Edwin Webster, Pierre Du Pont, Coleman Du Pont, George Eastman, Gerard Swope, Arthur D. Little, Willis Whitney, and Frank Vanderlip. The most dramatic achievements in industrial cooperation were those of the electrical and chemical engineering departments, headed by Dugald Jackson and William H. Walker, respectively.

Instruction in electrical engineering began at MIT in 1882, in the physics department; two years later, it was given its own number—Course VI—and in 1902 a separate department. By 1891 it enjoyed the highest enrollment at MIT, something it would sustain through the 1920's. In 1887 Thomas Edison donated materials, machines, and dynamos for departmental instruction, and additional equipment was secured from Westinghouse. These links with industry were reinforced by the impressive flow of department graduates into prominent industrial positions. Among the earliest

---

*When Pritchett resigned as MIT president in 1907, the position was offered to Fish, who had just stepped down as president of AT&T. Fish rejected the offer in order to pursue his legal career, but remained on the executive committee of the MIT Corporation.

were Frank Pickernell, chief engineer at AT&T, 1895–99; Edwin Webster and Charles Stone, who submitted a joint thesis and later institutionalized their collaboration as Stone and Webster, Inc.; Calvin Rice, a GE engineer who served as the very influential secretary of ASME from 1906 to 1934; Alfred P. Sloan, president of General Motors; Gerard Swope, president of GE; and William Coolidge, director of the GE Laboratory.[62.]

MIT had made a move toward research early in the century, with the establishment by A. A. Noyes and Whitney of the Physical Chemistry Research Laboratory in 1903 and the Graduate School of Engineering Research that same year. On the latter occasion President Pritchett clearly indicated the larger motivations behind such a new course:

> The Germans need fear in the industrial world neither the Englishman nor the Frenchman, only the American. The time has now come when the American engineer must be capable, not only of the most modern practice, but also of conducting investigation and research. . . . This is the first effort of any technical school in the country to offer research work distinctive from that of the colleges, and directed toward engineering subjects.[63]

The cooperation in research in the electrical-engineering department was thus part of a larger research effort of MIT; Dugald Jackson was a prime mover in this effort. He himself epitomized the convergence of interests of technical educators, professional engineers, and leaders of the science-based industries. Jackson had been chief engineer for the Sprague Electric Railway and Motor Company and the Chicago office of the Edison GE Company before setting up the country's second electrical-engineering department at the University of Wisconsin (the first was at Missouri). As a founding member and president of the SPEE (1906–7), and president of the AIEE (1910–11) he played a critical role in the development of modern engineering professionalism and engineering education. A power consultant who was involved in some of the largest power and railway-electrification projects of his time, Jackson had become a leading defender of efficient management and the private ownership of public utilities. It was he, for example, who defended the Philadelphia utilities against the progressive attacks of Morris Cooke. Vannevar Bush, the pioneer developer of computers, first presidential science advisor, and prime mover behind the establishment of the National Science Foundation, had been a young colleague of Jackson's at MIT. He later recalled that whenever Jackson chose to instruct Bush's introductory course in electrical engineering, the students complained that they were "learning a great deal about public utility companies and their management" but little about the subject proper of the course. When Jackson was brought to MIT by Fish and Pritchett in 1907, he was encouraged to open a Boston office in order to maintain his consulting practice and, at the same time, to help bring about closer ties between industry and education.[64]

Another medium for industrial participation in departmental development was established in the same period, the Visiting and Advisory Committee of the electrical engineering department. The committee, which was very active throughout the period under study, was at first composed of Elihu Thomson of GE (who also served as part-time teacher), Hammond V. Hayes of AT&T, Charles F. Scott, chief engineer at Westinghouse and head of Yale's electrical-engineering department, Louis A. Ferguson of the Chicago Edison Company, and Charles L. Edgar of the Boston Edison Electric Illuminating Company. The composition changed now and then, but the general outlook remained consistent; later members included G. E. Tripp of Westinghouse, Theodore N. Vail, president of AT&T, Jewett, and Coolidge. With Jackson's arrival, and strong support from the advisory committee, the department embarked on a program of research for industry. "The [department's] present policies and ideals . . . were established in 1907," according to a later committee report. "In 1907 a definite effort in research and instruction of advanced students was entered upon."[65]

Jackson believed that research was invaluable both as a teaching device and as a service to industry. Early research for the Boston Chamber of Commerce and the Boston Edison Electric Illuminating Company was thus rapidly expanded, and the department quickly became a leading center of industrial service. Writing for financial support in 1910, Jackson assured Samuel Insull that although "this sort of research is of purely scientific character . . . I believe that the results of such research may have tremendous influence on the development of the electrical distribution of power." A department brochure prepared by Jackson that same year announced that "we are ready to undertake some of the more distinctively commercial investigations under the patronage or support of the great manufacturing or other commercial companies; but we hope to carry on the more important researches untrammeled by the limitations imposed by contributions of funds from commercial concerns."

The report of the advisory committee shortly thereafter declared its strong approval.[66] The committee fully realized the enormous savings of both time and money that the industries could enjoy if they could have their research done, at some public expense, by the universities.

> The expense of carrying on research work in a well-equipped laboratory in a technical school should be less than in the laboratory of a manufacturing company, inasmuch as the standard equipment in the laboratory of the technical school may be employed for some of such work. . . . Again, there is usually available in the technical school a staff of engineers of varied experience to act as experts in their particular specialties.[67]

The committee urged that the department conduct extensive research for industry on a contractual basis—like a private consulting firm—and suggested such possible areas of work as street-railway fares, energy consumption of motor vehicles, and the study of insulation.

By 1914 the report indicated that "the research work which was recommended . . . has progressed satisfactorily, and the field of work has been very greatly enlarged."[68] Studies were being made for the R. H. Macy Company, the Boston and Maine Railroad, and the New Haven and Hartford Railroads as well as the electrical companies. In 1913 AT&T stimulated such work with a grant of $50,000 for five years and the donation of the Vail (Dering Library) collection, potentially "one of our most valuable instruments of research."[69] Repeatedly the committee and Jackson emphasized the importance of research not only as a direct service to industry but as an aid in the training of future researchers. In a letter to President Vail summarizing the departmental use of the AT&T appropriations, Jackson described this dual function of research.

> As a consequence of the research contributions . . . [we are able to] put the more important problems . . . before graduate students and research assistants, who can prosecute their researches diligently and efficiently, under systematic and businesslike supervision, which not only enables these advanced students to be trained to higher effectiveness, but also enables their enthusiasm to be directed to definite ends. . . .[70]

Here, then, was an invaluable extension of the research labs of industry, and one with an important added feature: it could reproduce itself and provide manpower for industry at the same time. "We recognize it as our duty," Jackson wrote to Gerard Swope, "to contribute men to the industries and perhaps should be cautious about recruiting our staff through robbing the industries. . . . We need to not only train men for the industries but must train them for ourselves" as well. "We should 'feed' rather than 'feed on' the industries as far as men are concerned."[71]

Jackson built the MIT department into the best in the world. "The Department," the advisory committee reported in 1922, "is completely committed to a career of graduate instruction and research...there is now no other school in the civilized world where there is found equal activity in effectual advanced study and research in electrical engineering education." Although Jewett and his associates at AT&T decided to cut off the direct appropriations to the department in 1925—they had had difficulty coordinating the research with that of their own labs and feared "embarrassment" in the view of other universities—they declared their great interest in its development and urged that support "should be developed in other ways"; the next year the Bell System option of the VI-A Cooperative Course was begun with financial support from AT&T. That same year, GE, "on recognition of the service to the electrical industries of the graduates and staff of the Institute," voted an annual appropriation for continuance of the department's valuable activities, and for the support of additional research and teaching staff. Jackson's promotion of research for industry proved highly successful; moreover, he encouraged among his colleagues (including William Wickenden and the young Vannevar Bush) and students an avid interest and participation in the coordinated professional, industrial, and educational activities of such agencies as the AIEE, ASME, SPEE, NRC, National Electric Light Association (the utilities trade association), and National Industrial Conference Board.[72]

Outside of MIT's electrical-engineering department, similar efforts were made to establish industry-education cooperation in research. Most notable among these was the work of William Walker in the MIT chemical engineering department. Walker had left his teaching position at the institute to join Arthur Little in private contract research. In 1905, however, he returned as professor of chemical engineering and promptly expanded the industrial work of that department, establishing the Research Laboratory of Applied Chemistry for this purpose.[73] Walker shared the view of science held by his colleagues in the great industrial laboratories. His understanding of the scientist's role thus differed sharply from that of Michelson, who had lamented Jewett's industrial prostitution of his talents. "There is with scientific men," Walker wrote in 1911,

> a general awakening to the fact that the highest destiny of science is not to accumulate the truths of nature in a form no one but the select

few can utilize, but that the search for truth can be combined with a judicious attempt to make the truth serve the public good. Thus the distinction which has existed between the terms pure science and applied science is rapidly falling away.[74]

Walker's redefinition of science reflected a significant shift among university scientists, a shift that led Jewett himself to conclude twenty years later that "I think scientists in general have changed; I doubt whether the same atmosphere prevails now that did then."[75] In 1916 Walker joined Little in establishing the School of Chemical Engineering Practice, which further facilitated research for industry, on a contractual basis as in the electrical-engineering department, and also instituted cooperative courses of instruction whereby students received their training alternately in the classroom and the plants of industry.

In 1916, when MIT officially moved into its spacious new home across the Charles River (owing largely to the generous contributions of George Eastman, Charles Stone, Edwin Webster, and Pierre, T. Coleman, Irénée, and Lammot Du Pont), research for industry was the dominant theme of the day. Speakers for the occasion included Alexander Graham Bell, J. J. Carty, Michael Pupin, and, of course, Henry Pritchett. Shortly thereafter, however, MIT was plunged into a financial crisis, the resolution of which was to affect research operations at the Institute and elsewhere.

In 1911 the Massachusetts State Legislature had allotted MIT $100,000 per annum for ten years, thus continuing public support for education in the mechanic arts begun with the Morrill Act.* Another important source of income for MIT was the McKay endowment. In 1903 the executors of the estate of Gordon McKay, the shoe-machinery manufacturer (and employer of Bell's first backer, Gardiner Hubbard), granted to Harvard funds for the establishment of an engineering college. Instead of starting its own engineering school, however, Harvard attempted to absorb MIT for that purpose. After a decade of intense controversy over the matter at MIT, an agreement was finally reached between the two schools whereby MIT would remain independent of Harvard but would provide facilities and staff for the training of Harvard men in engineering subjects, in return for three fifths of the McKay endowment. The engineering faculties of the two schools were merged under MIT authority, and graduates received joint MIT-Harvard degrees. According to President Richard C. Maclaurin, "the agreement marked an epoch in the history of educational progress in

*From the outset MIT, as a "school of mechanic arts," had been the recipient of thirty percent of the land-grant-college appropriations for Massachusetts.

this country. The end sought was to build up an educational machine more useful to the community and to the nation than anything that could be maintained by either the Institute or the University acting independently."[76]

MIT depended heavily upon these two sources of income, and the electrical-engineering department received a considerable part of its operating budget from the McKay fund. In 1917, however, the Supreme Court of Massachusetts declared the agreement between MIT and Harvard to be in violation of the intentions of the McKay will, and ordered it dissolved; the state legislature shortly thereafter refused to renew its appropriation for MIT, which was to expire in 1921.

In search of ways to rescue the Institute from its financial woes, Maclaurin launched a major fund-raising drive and was again able to secure contributions from George Eastman and other science-minded industrial leaders. Maclaurin and his colleagues realized, however, that private donations would not suffice. Early in 1920 they formulated the so-called Technology Plan and took, in Maclaurin's view, "perhaps the most important step the Institute has taken since its organization."[77]

The Technology Plan, administered through the newly created Division of Industrial Cooperation and Research (under the direction of William H. Walker), was the means by which the various cooperative undertakings within the institute were systematized and coordinated as institutional policy. As Jackson explained the plan to skeptical engineering consultants, research for industry previously had been done "in a more or less haphazard manner; the Institute now undertakes . . . to put this line of information into a more definite order so that the industries may profit more fully from the relations which the Institute has, and the knowledge which it has, of its alumni and students, and in fact with the engineers in general throughout the land."[78] The essence of the plan was a standard contract made available by the institute whereby industry could take advantage of the resources of the Institute in exchange for a standard fee.* The Technology Plan did

---

*According to the contract, MIT agreed to make available to the companies "its library and files" and "to arrange for conferences with its technical staff on problems pertaining to the business of the company." MIT further agreed to "maintain a record of the qualifications and special knowledge of its alumni" and "to advise and assist [the companies] to obtain information regarding men for permanent employment." The Institute was also obliged to provide the companies with the records and qualifications of undergraduates and to arrange for employment interviews. Finally, MIT was committed to advise the companies about how they might best obtain long-term consultations, investigations, and tests or, for a fee, to provide such services itself. The full contract was published in William A. Walker, "The Technology Plan," *Chemical and Metallurgical Engineering,* XXII (March 10, 1920), 464.

more than merely facilitate the actual research work for industry; by the terms of the contract, MIT was obliged to function as a clearing-house of information for industry, providing ready access to both technical knowledge and the possessors of that knowledge. Moreover, through elaborate personnel procedures, as Walker put it, the Institute agreed "to maintain a steady stream of trained men constantly flowing into industry with the best preparation for scientific work which it is possible to give."[79] Aside from the obvious financial benefits accruing to the Institute, Jackson and his colleagues were certain that the close contact with industry would greatly "contribute to the fertility and effectiveness of the educational processes of the Institute." The plan was not therefore merely a money-raising gimmick.* It was the logical outgrowth of twenty years of effort toward industrial cooperation, and the realization on the part of its creators "that close cooperation between the industrial interests and the educational institutions of the country which in Germany was made so effective by the domination of both by the state can, in America, be brought about only by a voluntary personal relationship between the executives of the companies and the instructing staffs of the institutions."[80]

Not surprisingly, the Technology Plan was an immediate success; a prominent consultant himself, Jackson was well able to allay the fears of other private consultants by assuring them that the plan would stimulate industrial reliance upon scientific consulting work in general, and hence would be of help to them. The general response from industry was enthusiastic. Within a very short time over 150 companies had signed contracts with the Institute.† A writer in the *Bulletin* of the National Association of Corporation Schools (NACS), an organization concerned primarily with the training of manpower for industry, informed his readers that the plan would go a long way toward meeting the industrial requirements for technically competent personnel. "The Plan," he wrote, "is sure to have a far-reaching effect on educational and industrial institutions."[81] Indeed, MIT's plan served as the proto-type for undertakings at other institutions. Within a few months after

---

*While the industries and Technology Plan promoters were enthusiastic about MIT's new role, there was some opposition among alumni who accused these enthusiasts of "selling Tech." Editors of *Fortune, The Mighty Force of Research,* p. 16.

†Among those signing, the largest were GE, AT&T, United States Steel, United States Rubber, and the American International Corporation, which had been organized by Vanderlip in 1914 to promote industrial investment abroad. See William A. Walker, "The Division of Industrial Cooperation and Research at M.I.T.," *Journal of Industrial and Engineering Chemistry,* XII (April 1920), 394. See also "Dr. Walker Heads M.I.T. Industrial Division," *Electrical World,* LXXIV (January 24, 1920), 10.

it was officially launched, an editor of *Scientific American,* in an article aptly titled "The University in Industry," reported that a small number of schools had already adopted similar measures, and that others were preparing to do so.[82]

At the end of the first contract term at MIT, in 1924, President Stratton reaffirmed the Institute's commitment to the plan; he assured Everett Morss, treasurer of the MIT Corporation, that

> I have never questioned the importance of industrial research. I should like very much to place it on a more extensive basis than now exists. . . . I shall do all that I can to bring this about, as I think that, in addition to its value to the industries, it is of great importance to the Institute from the standpoint of training men.[83]

Stratton's successor, Karl Compton, reaffirmed the Technology Plan commitment when he became president in 1932. "The Division of Industrial Cooperation," he wrote, "is designed to make as effective as possible the assistance which the Institute renders to business and industry in solving their technical problems."[84] After World War II the Division of Industrial Cooperation was reorganized as the Division of Sponsored Research, expanded to administer research programs with the government and the military as well as with industry.

Cooperation between the universities and the industries provided the latter with an unequaled source of basic and applied scientific research, as well as a steady supply of technically trained manpower. And as the science-based industries expanded and other industries increasingly became science-based, industry reliance upon the contributions of the educational institutions grew accordingly. The universities, moreover, provided corporate investors with a large and ultimately profitable outlet for surplus capital. Thus, between 1900 and 1930, as enrollment multiplied fivefold (compared with a population rise of 62 percent), the dollar value of university properties rose from $2.5 million to almost $2 billion.[85] Much of this growth, especially after 1920, reflected the fact that "many business and industrial establishments collaborated with universities in the building up of research facilities, particularly in the engineering and the physical sciences."[86] In December 1916, for example, Nicholas Murray Butler, president of Columbia University and longtime advocate of industrial education, announced plans for the establishment of a great industrial-research center at Columbia. It was clear to Butler, and to those who heard him, that he was heralding a new day in university-industry relations, a revolution that would have lasting consequences for both parties and for society as a whole.

> Manufacturers throughout the country could bring to the university, with such a laboratory, their great chemical, mechanical, or other engineering problems for solution at the hands of experts who will devote their entire time to such work. Most of the large industrial plants have their own research laboratories, and it is proposed by Columbia to bring to the university the problems which are now submitted to private laboratories.
>
> The future of American industry is bound up with the future of American science. The schools of mines, engineering, and chemistry . . . are anxious and ready to undertake with great energy some of those specific tasks which will aid American industry to improve its products, to decrease its wastage, to coordinate its processes and to multiply its resources for dealing satisfactorily with the many-sided human problems which industrial relationships and industrial enterprise of necessity involve.[87]

Much of the new growth in university facilities, which better enabled the schools to investigate the technical, scientific, and "many-sided human problems" for industry, took place in the urban industrial areas, close to the industries to be served. Among the expanded educational service centers were the universities of Cincinnati, Dayton, Akron, and Pittsburgh, the Drexel Institute in Philadelphia, and Lehigh University in Bethlehem, Pennsylvania. Eastman Kodak and Bausch and Lomb cooperated with the Mechanics Institute and the University of Rochester; Westinghouse worked closely with the administrators and faculty of the Carnegie Institute and the University of Pittsburgh; GE at Schenectady teamed up with Union College, as GE at Lynn had done with MIT; and on the west coast, in Pasadena, George Ellery Hale, industry-minded head of the National Academy of Sciences, collaborated with lumber tycoon Arthur Fleming to transform the small Throop College "into an institution for the advancement of chemistry and other sciences as aids to industry"—the California Institute of Technology.[88]

"Destiny pointed to Cleveland as a center of chemical production and to Case as the source of its scientific manpower,"[89] William Wickenden, president of Case Institute, declared at the dedication of the new Chemistry Laboratory Building in 1929. Formerly a teacher at the Mechanics Institute in Rochester, the University of Wisconsin, and MIT, assistant vice-president for personnel at AT&T, and most recently director of the SPEE study of U.S. engineering education, Wickenden had brought to Case thirty years of dedication to industrial service and a long-standing commitment to giving "destiny" a helping hand whenever it appeared necessary.

Immediately upon his arrival he worked with Ambrose Swasey and other Cleveland industrialists to expand the facilities at Case and thereby improve its serviceability. He actively promoted the industrial use of university research facilities, staff, library, and graduates, and, after a Depression setback, succeeded in significantly enlarging the "industrial service work" of Case. "Most of the work done on the campus," he told the consultants of the Cleveland Engineering Society,

> is of a research of development character for which the parties contracting are unable to provide proper staff and equipment. . . . We have felt under obligation to render service of this type wherever possible . . .; without such contacts . . . our work would fall in time to a routine level which would not be satisfactory either to ourselves or to the professional and industrial groups whom we serve.[90]

Here, too, research cooperation was comprehended as a two-sided process. Not only would industry benefit from the research and information provided, but the educational part of the "process" would be "fertilized" and "stimulated" by industrial input, and this too would ultimately be of benefit to industry. "To be frank," the assistant manager of the International Nickel Company (Wickenden's brother), wrote to Case's president, "our motives are not entirely philanthropical as we feel that information planted in the minds of students today will in time bear fruit as they become factors in specifying materials in industry."[91]

The research-related cooperation between technical educators and their professional colleagues in the industries contributed greatly to the expansion of university facilities and to the transformation of the universities into a "functional unit" of a larger "industrial system." Indeed, the development of this new industrial resource became an industry in itself. "The people of the United States have a great national industry which is never mentioned in the summaries of the productive enterprises of the country," Samuel Capen told his audience at his inauguration as chancellor of the University of Buffalo in 1922.

> It is the industry of building universities. The industry has absorbed an extraordinary amount of creative energy. . . . It now represents an invested capital of $1,250,000,000. In cash it has never paid a penny on

the investment, which accounts for its common omission from the record of those productive undertakings that add visibly to the wealth of the nation. But indirectly what has been the return? Scientific discoveries and the application of scientific knowledge to manufacturing, to commerce, to agriculture, to engineering processes, to the prevention and cure of diseases which are responsible for a large part of the actual profits of the nation's business. Wipe out the contributions made by the universities during the last fifty years and the industrial life of the nation would shrivel to insignificant dimensions.[92]

While the primary mission of the university within the industrial system was the "efficient production of human material" according to "industrial specifications"—which made not only the building of universities, but education itself an industry—the role of universities as centers of research for industry was also a vital one. After its first survey of research activities in the United States, the National Research Council maintained that "the main sources of research in America have been, and must continue to be, the universities." And in 1957, in its evaluation of national research resources in the wake of the Russian launching of Sputnik, the National Science Foundation found that throughout the century American industry had supplied the largest percentage of financial support for research, while colleges and universities had been the principal performers.[93]

Industrial sponsorship and direction of university-based scientific research successfully shifted the burden of some significant costs, and risks, of modern industry from the private to the public sector. But this was not all. Perhaps more important, it redefined the form and content of scientific research itself. This involved more than the general shift away from the search for truth and toward utility which had already been well underway by the turn of the century. Now the shift toward utility assumed particular forms, molded by the specific, historical needs of private industry, by particular firms intent upon increasing their profit margins and their power. This reorientation affected not only what kinds of questions would be asked but also what particular questions would be asked, which problems would be investigated, what sorts of solutions would be sought, what conclusions would be drawn. Science had, indeed, been pressed into the service of capital.

Increasingly in the first two decades of the century, engineering educators and industry leaders cooperated to establish joint industry-university research programs throughout the country. Until the outbreak of World War I, however, there existed no central national agency to promote and coordinate these activities. During the two years of preparedness preceding U.S. entry into the war, three independent developments converged to correct this situation, giving rise to the National Research Council (NRC). These involved the establishment of the Naval Consulting Board (NCB), the Council of National Defense (CND), and the Research Council of the National Academy of Sciences.

In 1915 there existed no administrative mechanism, either within or without the government, adequate to the task of mobilizing the country's industrial and scientific resources for war. The Army and the Navy had anything but a dynamic approach to the military potentials of science, and those military men who did recognize that potential knew that the meager scientific resources of the government could not meet potential military problems; they realized that any mobilization of science for war would have to involve the coordination of the nation's scientific resources outside of the government. Such coordination, an exciting new prospect for military men, was no novelty to the engineers in industry.[94]

On May 30, 1915, *The New York Times Magazine* published an interview with Thomas A. Edison in which he stressed the importance of transportation and communication in wartime, and the corollary that technical men and inventive genius would be indispensable resources for effective military effort. "If any foreign power should seriously consider an attack upon this country," Edison boasted, "a hundred men of special training quickly would be at work here upon new means of repelling invaders. I would be at it myself." After the sinking of the *Lusitania* in July, Secretary of the Navy Josephus Daniels, having seen the interview, took Edison up on his proposal; he wrote to Edison about the possibility of creating "machinery and facilities for utilizing the natural inventive genius of America to meet the new conditions of warfare. . . ." The inventor responded positively and the two men met at Edison's home in Orange, New Jersey, to set the plan in motion. The membership of the new NCB, except for Edison and his chief engineer, M. R. Hutchison—who were appointed by Daniels— was to be composed of men selected by the eleven largest engineering societies in the country.* The NCB's official historian noted, "There was gathered on this board . . . a large number of business executives whose success had come about through their ability as engineers and

*Among these were Frank Sprague (AIEE), Baekeland and Whitney (ACS), and Howard Coffin (SAE).

inventors, but who, at the time they were selected for the board, were preeminently executives of large business enterprises, with an outlook on life which comes from the point of view obtained by the control over large forces of men."[95]

There were noticeably no men of pure science on the engineer-dominated board, and "only a few of the men chosen had any close connection with either university or government science." Quite early in the NCB's development Edison and other members of the board pushed for the creation of a special naval laboratory under the command of "civilian experimenters, chemists, physicists, etc.," and a considerable sum was ultimately appropriated for that purpose. Disputes erupted, however, over its allocation, and in April 1917, when the war broke out, the project was dropped. The major activity of the NCB throughout the war was a rather fruitless one: the screening of public suggestions and inventions for possible military value. (Of 110,000 suggestions, only 110 merited detailed examination, and only one actually went into production.) These results, however, did dramatize the need for coordinated, centrally directed, and problem-oriented scientific research. The most important contribution of the NCB was made by Howard Coffin, elected chairman of the Industrial Preparedness Committee of the board because of his achievements with standardization within the automobile industry. Coffin and his committee—which included future AT&T president Walter S. Gifford, then a company statistician, and Grosvenor Clarkson, a public-relations man from New York—contributed to the preparedness propaganda campaign, and provided an inventory of industrial resources which would serve as the basis for the mobilization work of the Council of National Defense and the War Industries Board.*[96]

With the sinking of the *Sussex* in April 1916, the preparedness campaign gained momentum. For Coffin and others who saw war with Germany as an immediate danger and inevitable, the pace was still too relaxed. Many Americans, they feared, still viewed war with Germany as a remote possibility and clung to the notions of neutrality. One of those who joined Coffin in his efforts was Hollis Godfrey, president of Drexel Institute and a management consulting engineer who had been trained in electrical engineering at MIT. As early as 1907 Godfrey had predicted war between England and Germany; by 1916 he was working

---

*Secretary of War Baker, in a speech before the Cleveland Chamber of Commerce, credited Coffin with having initiated and conducted the industrial-preparedness campaign of the NCB.

feverishly to arouse his countrymen to the need to prepare for war. He and Coffin together with Henry Crampton, a former MIT instructor and now a Columbia professor and curator of zoology at the American Museum of Natural History, lobbied tirelessly in Washington to drum up legislation for the establishment of a Council of National Defense. With the counsel of such men as Nicholas Murray Butler and Elihu Root, Godfrey formulated a plan for a council capable of marshaling the nation's industrial and scientific resources for war. On August 29, 1916, Godfrey's plan was adopted; six Cabinet members—the Secretaries of War, Navy, Interior, Commerce, Labor, and Agriculture—constituted the council, with Secretary of War Newton D. Baker as chairman. To handle the actual work of the council an advisory committee was established consisting of Daniel Willard, president of the Baltimore and Ohio Railroad; Julius Rosenwald, president of Sears, Roebuck; Franklin Martin of the American College of Surgeons; Samuel Gompers of the AFL; Bernard Baruch, a prominent Wall Street figure; Godfrey; and Coffin. Walter Gifford of AT&T was appointed administrative director of the council, and Clarkson, the public-relations executive, was made secretary.[97]

Coffin, as head of "Munitions and Manufacturing, Including Standardization," carried on the work he had begun on the Naval Consulting Board. Godfrey, chairman of the "Committee on Science and Research, Including Engineering and Education," was primarily interested in the coordination of technical education for the war. In addition, he pushed for the development of a Personnel Index, part of which was to be "the organization of leading men of pure and applied science in the country."[98]

The Naval Consulting Board and the Council of National Defense had both been the work of engineers; the nation's scientists, however, were themselves preparing for war. The National Academy of Sciences had been chartered in 1863 as a private organization for consultation by the government on scientific and military questions during the Civil War. It had been, in addition, an attempt by prominent American scientists—Louis Agassiz, Joseph Henry, and Alexander Dallas Bache—to centralize control over American science. The development of the various scientific bureaus of government, however, had usurped its advisory function, and the sponsorship of science in universities, institutes, foundations, and industry precluded its centralized control of science. At the turn of the century, the academy was moribund.[99]

The revitalization of the academy was due to the work of George Ellery Hale, through his association with former Secretary of State and United States Senator Elihu Root. Hale, an MIT graduate, had developed a new scientific field, astrophysics, in which the scientific principles of physics were wedded to traditional astronomical observation. He founded and became director of the Mount Wilson Observatory in California, and became closely acquainted in that capacity with Root, himself a student of mathematics and the son of a professor of astronomy, who was chairman of the board of the Carnegie Institution. Upon his election to the National Academy, Hale was informed by Root of its chartered role as a semigovernmental body and the official advisor of the government in scientific matters; Hale's dreams for the academy thus became linked with the promise of governmental cooperation.[100]

Hale sought to transform the academy into a powerful and prestigious force in American science, one which could cohere the diverse scientific resources of the country and bring them into sharper focus. Thus, Hale was, as historian Nathan Reingold has suggested, the J. P. Morgan of the scientific community.[101] In addition, he wanted to expand the scope of the academy so as to reflect the collapse of the rigid distinction between pure and applied science. As early as 1910 he succeeded in having the academy rules changed to allow for the development of a strong engineering section within it. He also began, along the same lines, to transform Throop College in Pasadena, of which he was a trustee, into a distinguished center of scientific education and one which symbolized the integration of pure and applied science. In both efforts he was joined by like-minded scientists with industrial experience, such as A. A. Noyes of MIT, Michael Pupin of Columbia, and Robert Millikan of Chicago. The majority of academy members, however, did not share Hale's dreams, and his other early efforts to secure funds for an academy building in Washington and expanded services were unsuccessful.

The creation of the Naval Consulting Board caused Hale to redouble his efforts to secure a powerful voice for American science. Hale saw in preparedness and the possibility of war a chance for reviving the governmental advisory function of the academy. Strongly anti-German himself, Hale tried unsuccessfully to have academy scientists appointed to the engineer-dominated board; the torpedoing of the *Essex* and the *Sussex* and the swell of preparedness fever, however, began to turn the tide in Hale's favor, and he was quick to capitalize upon it. In late April

1916, at the meeting of the academy's executive council, Hale introduced a resolution that "in the event of a break in diplomatic relations with any other country, the academy desires to place itself at the disposal of the government for any service within its scope." Hale was confident: "How Wilson can crawl out of a break with Germany," he remarked to a colleague, "is more than I can see."[102]

The resolution was adopted by the academy, and representatives were sent to the President to present the offer. Wilson agreed on the advisability of preparing the scientific resources of the country for war, but refused to allow his approval to be made public for fear that it might be construed as a step toward war. He did suggest, however, that the academy form a committee to "undertake such work as the Academy might propose." The stage was thus set for Hale and his supporters. Hale was made chairman of the committee, and Millikan, who had enthusiastically endorsed the resolution, became his most avid collaborator. Commencing to formulate plans for a "National Service Research Foundation," Hale wrote to a friend, "I really believe this is the greatest chance we ever had to advance research in America."[103]

As preparedness sentiment swept through the higher reaches of government, Root was able to use his influence to secure Wilson's formal order for creation of the National Research Council. Already by July 1916 Hale, Noyes, and Millikan had drawn up a comprehensive plan for the council. "The idea of keeping pure and applied science together in one all-inclusive organization," Millikan later wrote, "was one of the essential elements in Hale's thinking and one with which I myself was in thorough agreement."*[104] Hale, Noyes, and Millikan secured the cooperation of the American Association for the Advancement of Science's Committee of One Hundred, and through the efforts of Hale's close friends on the Engineering Foundation—Michael Pupin, Gano Dunn, and Ambrose Swasey—secured the financial support necessary for the work of the Research Council.

With its September organization meeting, dominated by Millikan, Dunn, Hale, Whitney, Carty, Stratton, and Pupin, the National Research Council officially began its career. Its stated objectives included

*Millikan was no stranger to coordinated applied scientific work; as early as 1910 he began to work with his former student Frank Jewett on the problems of wireless and the electron-tube repeater in the Bell Labs. He was keenly interested in the industrial application of work in physics and sent a flow of his students and assistants at Chicago to the Bell Labs. In addition, he was actively involved in the patent litigation between AT&T and General Electric which arose from wireless development. In Reingold's description, Millikan was "a Babbitt with a Nobel Prize."

the preparation of an inventory of scientific personnel, equipment, and current research work; cooperation with educational institutions and research foundations; the promotion of research relating to national defense; and the creation of a "clearinghouse" for the coordination of research projects and scientific information. With the help of Root, the council secured additional funds from the Carnegie and Rockefeller foundations (throughout the war, as after, private funds would outweigh governmental appropriations in the council's budget). Meanwhile, schools such as MIT and Throop College began expanding their facilities in anticipation of cooperation with the council.[105]

With the declaration of unrestricted submarine warfare and the subsequent severance of diplomatic relations with Germany in early February 1917, the Council of National Defense assumed responsibility for full-scale mobilization. The Naval Counsulting Board became the CND's board of inventions, and the National Research Council was charged with the organization of research. After some controversy, the National Research Council was recognized as the sole agency responsible for coordination of scientific resources for the war.*[106] Probably the most significant wartime work of the National Research Council involved antisubmarine research. Other tasks included the development of range-finders for naval gunnery, communication apparatus, and gas masks. Early in the summer of 1917 the national effort to coordinate scientific work received a great stimulus: the power of military authority. General George Squier, chief signal officer of the Navy, an electrical engineer with a Ph.D. from Johns Hopkins, a member of the AIEE, and a man who strongly believed in the potential of coordinated research, persuaded Millikan, by then the chief administrative officer of the National Research Council, to take a commission in the Army. Millikan did so, and soon many of the nation's leading scientists and engineers also donned uniforms, including the reluctant A. A. Michelson. Despite its military status, however, the wartime Research Council was hampered by a lack of funds and inadequate administrative staff; cut short by an unexpectedly early Armistice, it was unable to realize its objectives fully. Nevertheless, it did provide a clearinghouse of information and a centralized focus for scientific personnel. Most important, through its operations "American scientists became accustomed to working together for the quick solution of an immediate

*Hollis Godfrey had hoped to carry on such activities through his committee, but deferred to the National Research Council. On May 21 he formally had the name of his committee changed from "Science and Research, Including Engineering and Education" to simply "Engineering and Education."

problem," the coordinated efforts of the war having provided the "common experience of a whole generation of scientists."[107]

From the outset the NRC was an industrial as well as a military research agency. Its original purpose was "to bring into cooperation existing governmental, industrial, and other research organizations."[108] During the early part of the war such cooperation was sought expressly as a means toward military ends, the war thus providing the "opportunity" for the consolidation of effort along these lines dating back to the turn of the century. But as early as eight months before the Armistice, the leaders of the council actively undertook to shift the emphasis explicitly from military to industrial service. The council, formally created by executive order, enjoyed the same status as the Council of National Defense and the War Industries Board: it was an extrapolitical body, with official sanction, well suited to the management-minded engineers and business executives who were used to coordinating and directing the affairs of men from the top, without inefficient democratic mediation. With the creation of the National Research Council, the technical leaders of industry no longer had to rely upon periodic meetings in the faculty clubs of universities, the executive offices of industry, or their elite social clubs to achieve the necessary coordination of industrial research activities. The NRC provided them with an unprecedented vehicle for coordinating the resources of the nation to meet the needs of industry. In a confidential memorandum entitled "The Origin and Purpose of the NRC," which was circulated within the organization's executive council in May 1919, Hale made this quite clear:

> The Academy organized the National Research Council ... with a view to stimulating the growth of science and its application to industry and particularly with a view to the coordination of research agencies for the sake of enabling the United States, in spite of its democratic, individualistic organization, to bend its energies effectively toward a common porpose.[109]

One of the earliest and most far-reaching contributions of the permanent council was its support of pure research in physics and chemistry, through the National Research Fellowship program. The program had its origins in a letter sent by George Vincent, president of the Rockefeller Foundation, to Millikan, Michelson, and Steiglitz, in which he proposed that the wartime council be perpetuated beyond the war period and that a centralized research institute be created. "The industrial competition which will follow the war will test the scientific resources of the nation,"[110] Vincent wrote the scientists. Immediately a group within the council began to devote their attention to holding the council together and securing the scientific coordination necessary for American industrial supremacy.

Vincent and his collaborator at the Rockefeller Foundation, Edward C. Pickering, head of the Harvard Observatory and former Thayer Professor of Physics at MIT, had been pushing the idea of a centralized research institute since 1913, among the membership of the American Association for the Advancement of Science. Hale, Root, and Pritchett strongly endorsed the plan, but men such as Millikan and Whitney preferred a decentralized scheme, with the establishment of a number of research centers at different universities. Another controversy arose over whether the research was to be government-supported or strictly a private operation; most of the leadership of the council preferred private funding, without danger of government interference. After extended debate, both controversies were resolved in a compromise plan. Rather than a centralized research institute, as originally proposed by Vincent, there would be an NRC fellowship program for graduate study in physics and chemistry. The awards, while administered by the quasigovernmental council, would be funded entirely by the Rockefeller Foundation, and would be restricted to students at privately endowed institutions. The fellowship program, which was officially begun in 1919, became an important means of encouraging both basic research and the training of research scientists. According to Millikan, who was not given to understatement, the fellowship program was "the most effective agency in the scientific development of American life and civilization that has appeared on the American scene in my lifetime."[111]

Aside from the matter of support for basic research, the most pressing question was that of perpetuating the NRC. A committee composed of the wartime leaders of the NRC was able to secure a contribution of $5 million from the Carnegie Corporation, headed at the time by Elihu Root, which provided funds for a building and an operating endowment, and set the hopes for a peacetime council on firm footing. The Engineering Foundation pledged continued support two months later. On the advice of Root, the committee also submitted a proposal for a peacetime council to President Wilson, while Root himself pressured Colonel House into obtaining Wilson's approval. On May 11, 1918, the permanent NRC was created by executive order.[112]

Already by this time Hale had begun to define the "common purpose" which the NRC was to serve: "We have hitherto concentrated

AMERICA BY DESIGN / 156

most of our attention on the solution of military problems, but we must not neglect the equally important task of promoting research in the industries."[113] Accordingly, he informed his colleagues on the council, "we are forming a special section devoted to this purpose." The new section, the Industrial Relations Division, was composed of six men: J. J. Carty, chief engineer at AT&T; Raymond Bacon, director of the Mellon Institute; Frank Jewett, director of the Western Electric Laboratories; Arthur D. Little, president of the nation's largest engineering consulting firm; C. E. Skinner, director of research at Westinghouse; and Willis Whitney, director of research at GE. After a month of preliminary organization and the securing of President Wilson's executive order and the requisite financial support, Hale officially launched the new division on May 29, 1918, with a formal banquet at the University Club in New York.

"Hitherto, the National Research Council activities have been mostly devoted to war," he once again explained, "but plans have been under contemplation for industrial research and the time has arrived to put these plans forward."[114]

The themes of the banquet speeches were familiar to all assembled.* Pritchett and Maclaurin stressed the importance of university-industrial cooperation and the production of technical men for industry; Whitney "strenuously" called for the support of basic research in the universities; and Hale and Root emphasized the need for national coordination of industrial resources. Perhaps the atmosphere of excitement that pervaded the banquet hall was best reflected in the remarks of Swasey; "while deeply deploring the war," the minutes read, "Mr. Swasey directed attention to the marvelous advances it was bringing in the mental, moral, and spiritual realms, with consequent great benefits to mankind. . . ." "We who are living in these wonderful times," Swasey exclaimed, "have thrilling opportunities and correspondingly weighty responsibilities." On a somewhat less spiritual plane, Swasey pointed out, "as an instance of industrial advance, that whereas a year ago this country produced no optical glass, it is now manufacturing this

---

*Among those attending the banquet were Carty, Pupin, Gano Dunn, C. E. K. Mees, Millikan, Whitney and such university people as John Johnston of Yale, Fleming of Cal Tech, Maclaurin of MIT, and Rautenstrauch of Columbia. Those who formed the advisory committee included Theodore N. Vail, chairman of the committee and president of AT&T; Cleveland H. Dodge, vice-president of Phelps-Dodge mining interests; George Eastman, president of Eastman Kodak; Elbert H. Gary, president of United States Steel; Andrew Mellon, head of the Mellon banking interests; Pierre Du Pont; Elihu Root; Ambrose Swasey; E. W. Rice, president of GE; and Henry Pritchett, president of the Carnegie Foundation for the Advancement of Teaching.

material by the carload." In his tribute to the great progress made possible by the war—and the seizure of German patents—Swasey was not alone among his peers; while countless others in the general population might have been disillusioned by the war and its aftermath, these men were propelled to yet greater heights of industrial creativity.[115]

Aside from the naming of the division's advisory committee, the main outcome of the banquet was the proposal that some of the most prestigious of those assembled prepare statements heralding the new permanent Research Council, for publication in a pamphlet. Among those responding were Swasey, Mellon, Eastman, Vail (who was elected chairman of the advisory committee), Root, and Pritchett. Root, in "The Need for Organization in Scientific Research," argued that

> Science, like charity, should begin at home, and has done so very imperfectly. Science has been arranging, classifying, methodizing, simplifying everything except itself. . . . Scientific men are only recently realizing . . . that the effective power of a great number of scientific men may be increased by organization just as the effective power of a great number of laborers may be increased by military discipline.
>
> [After the war] the same power of science which has so amazingly increased the productive capacity of mankind . . . will be applied again and the prizes of industrial and commercial leadership will fall to the nation which organizes its scientific forces most effectively.[116]

Pritchett focused upon the problems of technical education; he realized, as had Jewett, that the United States, because of wartime draining of the universities, was approaching a "state where scientific man-producing machinery no longer existed."[117] His solution was cooperation between the industries and the schools, whereby the schools would be financially supported and replenished, and the industries would benefit from more research and a steady flow of technically trained manpower. His overriding theme, however, was that of effecting coordination of diverse research efforts for industrial progress.

> In the United States the relations between research men in the universities and institutes of research and those operating industrial plants have not yet come to a stage as intimate and fruitful as that which has existed for many years in Germany. It is today a part of our plan of progress for the future to establish such relations that the investigator and the manufacturer shall understand each other and shall cooperate for the promotion of science and industry.[118]

President Vail of AT&T elaborated on the same theme—that coordinated knowledge and organized knowledge-producers were the *sine qua non* of the knowledge-based industries.

> Organization and coordination of research for industrial purposes is urgently necessary. . . . Plans should be formulated at once. . . . Whatever is done should be national in its comprehensiveness. . . . Industry may be expected to support generously any organization which promises to effectively coordinate and correlate efforts for the increase of knowledge, since it is now generally recognized that industrial progress and success are chiefly dependent upon our knowledge.[119]

Decades later, when rebellious students and professors began to rail against the "knowledge factories"—machines which produced them and then employed them to produce others like them—few suspected how consciously those factories had been designed; because they could not share the larger corporate perspective which comes with being on the top of the process and looking down, few could perceive themselves as Pritchett and his colleagues perceived them: "The research men of a nation," Pritchett wrote, "are not isolated individuals but an organized and cooperating army."[120]

In 1918, of course, Pritchett was talking in terms of potential; the National Research Council was geared to realize that potential. The executive order which created the permanent council clearly indicated the scope of the undertaking, charging the council with six functions: to stimulate and promote scientific research; to conduct surveys of scientific and technical resources; to coordinate research efforts on a national and international scale; to bring scientists into active cooperation with the War and Navy departments and other government agencies; to direct research efforts toward the solution of military and industrial problems; and to collect scientific and technical information for "duly accredited persons." Written during the war, the order reflected the needs of the military as well as industry; at the close of the war the focus became exclusively industrial, and the "duly accredited persons" became the executives of industry.[121]

Early in 1919 the council was structurally reorganized on a permanent basis. Divisions were established to serve particular functions: Divisions of Military, Federal, Foreign, States, and Industrial Relations (to facilitate cooperation in each of these areas); Divisions of Physical Science, Chemistry and Chemical Technology, Geology and Geography, Medical Science, Biology and Agriculture, and Anthropology and Psychology (to promote cooperation in the sciences); an Educational Relations Division (to provide cooperation with educational institutions and associations); and a Research Extension Division (to promote research in industry). In addition, a Research Information Service, created during the war, was expanded to carry out the information-gathering function of the council.

The key to the council's effectiveness in each of these undertakings was its quasi-governmental status and intimate working relationship with the other bureaus of government. The council had official sanction to coordinate the efforts of scientists and technical men with the research facilities and informational resources of the country, and thereby to serve the nation's industry. As within the smaller domain of the industrial lab, the catchword of this new nationwide industrial laboratory was "teamwork." "Most of us are not geniuses," Vernon Kellogg declared to the scientists of the Entomological Society of America. "We are just capable, industrious, well-trained workers . . . able and willing to . . . work together." Isolated genius, rendered obsolete by the industrial research laboratory, was to be superseded by Pritchett's "organized and cooperating army" of researchers. "Let us not be afraid of organization," Kellogg continued.

> It means no real surrender of individual freedom or achievement. It only means that we direct our efforts more intelligently, to more important undertakings, with more material aid and more mutual encouragement. Organization lies in the very spirit of America. See what great things it has accomplished in American industry? . . . No one wants to organize the geniuses; no one proposes to; no one can. But I am no genius and most of you are no geniuses. Yet you and I counseling together, planning together, working together, can do something steadily to advance scientific knowledge.[122]

In addition to the "horizontal" organization of an army of competent researchers, the council was able to facilitate the "vertical" integration of science to engineering which was the essence of the science-based industries. "The gap between the engineers and the scientists is gradually closing, but is still wide in places," Comfort Adams, chairman of the Engineering Division of the council, reported in 1921. Adams, who had been the prime mover behind the establishment of the American Engineering Standards Committee, well understood that only the basic sciences, firmly wedded to engineering and industrial practice, could solve modern industrial problems. "Anything we can do to assist in this closing process," he urged, "will contribute largely to the progress of both groups."[123]

The industrial orientation of the council was more than rhetorical. Of its various divisions, four provided invaluable service to industry.

These were the Divisions of Anthropology and Psychology, and Educational Relations; the Research Information Service; and the Industrial Relations, Research Extension, and Engineering Divisions, which were formally combined in 1941.

The Divisions of Anthropology and Psychology and of Educational Relations brought to a focus the prewar industrial efforts to transform the educational institutions of the country into "man-producing machinery" capable of meeting changing industrial specifications. They also extended the field of management in both the industrial plant and the classroom. In the first division the work of Robert Yerkes and Walter Dill Scott in intelligence testing, begun during the war for the Army, was refined for industrial use. In addition, extensive work was carried on, under council auspices, in the development of student personnel techniques, college entrance testing, vocational guidance, and general personnel research. In the Educational Relations Division, coordination of educational institutions for the war was perpetuated for industrial purposes. The division conducted surveys of university research facilities and conditions and promoted research at educational institutions; it also pioneered, through the efforts of Dean Seashore of Iowa and Frank Aydelotte of MIT, in "sectionalizing" students according to aptitude and setting up honors courses for promising students as an aid to educational efficiency. Staffed by men such as H. W. Tyler of MIT, C. R. Mann, then professor of education at MIT, and Samuel P. Capen, who had become the first director of the recently established American Council on Education, the division cooperated extensively with other organizations in the fields of educational research and management—the ACE, the Society for the Promotion of Engineering Education, and other branches of the NRC itself.*[124]

"There is a common saying that 'Knowledge is Power,' " observed Charles L. Reese, research director of Du Pont, and founder of the Association of Directors of Industrial Research. "Information and knowledge are so closely related that it might be said that information is power and coordinated information is power plus."[125] The Research Information Service was the council's agency for providing "power plus" for industry. It was established in 1917 at the behest of Howard Coffin and Hollis Godfrey; as the Research Information Committee, under the direction of Samuel Stratton, its function was to effect informational cooperation between the United States and her European allies and to secure, classify, and disseminate scientific, technical, and

*The activities of these divisions are discussed more fully in Chapter 9.

industrial research information for the military and naval intelligence agencies. In January of 1919 it became the Research Information Service and began to reorient its activities for industry. For George Hale, the service constituted the very heart of the peacetime council.

> Properly regarded, the Information Service may be considered the pioneer corps of the Council, surveying the progress of research in various parts of the world, selecting and reporting on the many activities of interest and importance . . . and disseminating it to men and institutions which can use it to advantage.[126]

From the outset, the Information Service prepared compilations, source books, bibliographies, handbooks, and bibliographies of bibliographies in order to organize and systematize available information for those who could "use it to advantage." It provided a reference service and a photostat service, furnished information on specific subjects, and prepared indices along the lines of the *Engineering Index* (which had been established by MIT electrical engineer Calvin Rice of GE when he became secretary of ASME in 1906). In addition, the service prepared and continually updated scientific abstracts (along the lines of *Chemical Abstracts,* begun by W. A. Noyes in 1907), and lists of scientific and technical societies, industrial research laboratories, research personnel, doctorates in science, current investigation, and investment in scientific education. The service also developed a library of sources of research information and reported on scientific informational services throughout the world. Between 1923 and 1925, when the service became an administrative operation of the Executive Council of the NRC with restricted activities, it prepared an informational survey of scientific bureaus in Washington, D.C., a catalogue of available scientific apparatus throughout the country, lists of fellowships and scholarships supported by industry, a catalogue of graduate research in chemistry, and a census of graduate students in chemistry. The Research Information Service thus functioned as an active "intelligence agency," in Reese's phrase, which systematized scientific and technical information much as Magnus Alexander's "research arm of industry," the National Industrial Conference Board, systematized economic information. Both agencies were clearly directed toward the same end: the corporate comprehension of, and thus power over, the vicissitudes of a "knowledge-based" industrial society.[127]

Without question, the most active divisions of the council—and those with the most explicit industrial orientation—were the Industrial Extension Division and the Engineering Division. Together they were, in effect, the practical arm of the council, and had the most direct impact on industry. The Industrial Extension Division was organized as the Industrial Relations Division, and was officially launched with the banquet at the University Club. It subsequently became the Industrial Research Division, created "to consider the best methods of achieving [efficient] organization of research within an industry or groups of industries,"[128] and finally the Division of Industrial Extension. Although the prestigious advisory committee never did function as an active part of the division—owing largely to the death of Theodore Vail—its advice was never really necessary. The division was composed of men who fully appreciated the needs of industry and devoted their energies on the council toward meeting them.*

Their work in the early years of the council was impressive. They established cooperative research programs among manufacturers of enameled wares, refractories, glass, ceramics, and even macaroni, and initiated research projects with the Bureau of Standards in such areas as electroplating, metal cutting, and metal alloys. Their most enduring achievements, however, involved the establishment of industrial research institutes in cooperation with various trade associations.[129]

The Engineering Division was born in 1918, during the war, and from the outset was tied very closely to the industry-controlled Engineering Foundation. In May 1918 the Engineering Foundation formally became the research branch of the American Engineering Council (the unified association of the Founder Societies) and as such, cooperated with the National Research Council in the creation of an Engineering Division, with Ambrose Swasey of the foundation becoming a member of the new division. The foundation provided funds for the division and an office in the Engineering Societies Building in New York. The proper relationship between the private, industrially controlled foundation and the quasi-governmental council division was a

*The membership of the division included chairman John Johnston, of United States Steel, F. K. Richtmeyer of GE, Mees of Kodak, Carty of AT&T, F. G. Cottrell of the Bureau of Mines, C. P. Townsend, a patent attorney, Whitney, and Baekeland. Although Theodore Vail resigned as chairman of the advisory committee due to the pressures of his other corporate responsibilities, he nevertheless remained until his death one of the more ardent champions of coordinated research. He wrote to John Johnston that "nothing which has been done is comparable with what can and will be done by an effective organization of all independent efforts. . . . There is no question in my mind but that education and industry would be mutually benefitted, and through the voluntary contributions of the proceeds of industry, [the NRC] and educational institutions could be co-related so that the institutions could be better maintained and enlarged and brought within the reach of all desiring." See Gano Dunn, letter to Albert Barrows, June 12, 1922, Division of Industrial Relations Records, NRC Archives.

point of controversy for a few years. Some members of the division, such as Harvard's Comfort Adams, wanted the two entities to merge entirely, so that the engineering profession would have a unified voice in research matters and the research arm of the profession would have quasi-governmental status. Others, including Swasey, preferred close cooperation but separate identities so that the foundation would be free to pursue matters outside the council. Another group—some scientists within the council—wanted to abolish the Engineering Division altogether, relegate its activities to the foundation, and have the council become an association of scientists exclusively. The controversy eventually subsided by 1923 and the relationship between the division and the foundation remained essentially as it had been at the outset. It was this intimate relationship which strongly tied the National Research Council to the engineering profession and, through it, to the industries which dominated the profession.[130]

In 1918, under the chairmanship of Henry M. Howe, a prominent MIT-trained metallurgist, the division was preoccupied with war-related activity; at this point it was not yet clear where the emphasis within the division would be placed, whether on basic research or on applied industrial research. GE's Willis Whitney urged that the focus be upon basic research in the schools, thereby ensuring the cultivation of new areas for potential industrial research, and providing the training of research workers for the schools and the industries.* "Isn't it possible," he wrote to Howe, "to so steer your good ship that you can support scholastic, academic, or scientific research so that the schools will gain more than the industries, at least at first?" Whitney was wary of direct council aid to the industries. "I am not in sympathy with philanthropically supported industries ...; it is the risk of gaining something irregularly ... which I do not want to take, because if I do (or if anyone else does), there will ultimately be a just complaint that some of the industries have taken unfair advantage of [philanthropists and the government]." C. E. K. Mees of Kodak expressed similar fears; these men in the established industrial research laboratories were reluctant to endorse quasi-governmental support of fledgling industrial research, support which they had never had in the development of their

---

*Whitney was at this time trying to set up cooperative arrangements with Union College toward these same ends, and was a member of the committee on cooperation between the universities and the industries of the American Chemical Society, which maintained that "the most important contribution which the universities can make to the development of industry in this country is to supply the industries with sufficient numbers of men thoroughly and broadly trained. . . ."

own operations. Mees stressed that the council should concentrate on basic research in the universities rather than applied research in the industries—that the University of Illinois, for example, "would do much better to study the structure of the amino acids, than to study the best method of isolating them from water in which corn had been steeped." Mees argued that "assistance of specific industrial corporations is not in the best interests of the nation, the universities, and industry at large."*[131]

By the time Comfort Adams became chairman of the division in 1919, the matter had been settled: it would concentrate upon direct industrial research, while the promotion of basic research and research training would be left to the research fellowship program of the council and the Rockefeller Foundation. The division would devote its attention to the promotion of efficient research, and the coordination of the resources within the universities and governmental bureaus for specific industrial purposes. And it would strive to do so with as little cost to industry as possible.

The Division of Engineering, Comfort Adams wrote to the members of the AIEE,

> seeks to stimulate engineering research by industrial establishments, universities, governmental bureaus and other interested agencies; another rather general way of defining this objective is: to encourage and stimulate the application of scientific knowledge and scientific method to the solution of industrial problems.
>
> Funds for any considerable research work are supplied ... by the interested industry. ... However, much valuable work has been done by our committees without the collection or direct expenditure of any funds, by members of the committees in Government bureaus, university laboratories and industrial plants or laboratories, the Division acting as the stimulating, organizing and coordinating agency.[132]

To direct these activities, the division established two advisory committees, both of which were dominated by industry people.† The first

---

*The question of industrial support for academic research was raised again during the campaign drive for the unsuccessful National Research Endowment. Many industrialists were reluctant to fund research which promised little immediate return on the investment. See Tobey's discussion in his *Ideology of National Science,* pp. 200–25, and Lance E. Davis and Daniel J. Kevles, "The National Research Fund: A Case Study in the Industrial Support of Academic Science," *Minerva,* XII (April 1974), 207–20.

†The committee for electrical engineering was composed of Jewett, Whitney, Skinner, Craft, E.W. Rice, Dugald Jackson, Comfort Adams, and Elmer Sperry; the committee for mechanical engineering included A. A. Potter and Samuel Stratton, and was chaired by E. M. Herr, president of Westinghouse.

project undertaken by the division was a large-scale study of the fatigue phenomena of metals; financed by the Engineering Foundation and GE, it was conducted at the engineering experiment station of the University of Illinois. Other early work included a study of the heat treatment of carbon steel and a program of highway research; the expenses for the former were borne by the Bureaus of Standards and Mines, while those for the latter were covered by federal and state government appropriations. In addition to these, a major research study of electric insulation was begun, under division auspices, by Jackson and Whitney.[133]

In 1923 Frank Jewett, head of the Laboratories of the Western Electric Company (AT&T), became chairman of the division and his assistant at AT&T, E. B. Craft, became vice-chairman; Dugald Jackson was elected second vice-chairman. Jewett immediately saw the need for an administrative director for the division, a person who would provide continuity for the division's activities beyond the relatively short terms of the chairmen, and he appointed Maurice Holland, an MIT electrical engineer and former engineer with the Boston Edison Electric Illuminating Company; Holland had organized and directed the Industrial Engineering Bureau of the United States Air Service during the war and in that capacity had made a survey of the research operations of AT&T, Western Electric, Du Pont, and GE.

Under Jewett's leadership, the scope of the division's activities was expanded significantly; the separate divisions of Research Extension and Engineering were consolidated, and promotion of research became an official function of the new Division of Engineering and Industrial Research. Between 1923 and 1930 the many research projects sponsored by the division concerned the methods of locking screw threads; the welding of structural steel and steel tubing for aircraft; the fatigue phenomena of aluminum alloys; the strength of steel piers, brick walls, and columns; the waterproofing of concrete; the acoustic properties of building materials; the physics of plumbing systems; the efficiency of fan wheels in ventilators; the explosive properties of gaseous mixtures; the properties of oils used for insulating fluids; the gumming characteristics of motor fuels; heat transmission; and the thermal properties of liquids used as antifreezing compounds in automobiles. In addition, the division helped to establish the American Petroleum Institute and the American Bureau of Welding, which greatly stimulated the growth of the electric welding industry in the United States. Beyond the scope of industrial research proper, the division embarked upon an ambitious investigation of the "relation of quality and quantity of illumination to efficiency in industry." Conducted by Jackson, Vannevar Bush, and their colleagues from MIT at the Chicago plant of Jewett's company, the project evolved into the famous Hawthorne studies, which laid the groundwork for industrial psychology and sociology.[134]

Dugald Jackson, who initiated the industrial research activities of MIT's electrical-engineering department in 1907, became chairman of the division in 1930. He expanded upon Jewett's policy of inviting bankers and industrialists to division meetings by taking them on tours of the country's leading research laboratories, and published surveys and various promotional materials including the widely distributed *Research: A Paying Investment.* In addition, he broadened the division's scope by deliberately including within it such problems of research management as the training of research personnel, job analysis in the laboratory (a carry-over of industrial job analysis), the relations of the laboratory to the production and sales departments in industry, financial incentives for researchers, and patent policies for industrial research. (Much of this work eventually became the responsibility of the division's Industrial Research Institute, which was created in 1938 "to provide a forum for the study and discussion of problems of common interest affecting the utilization of science for industrial purposes.") This new emphasis upon research management reflected a conscious shift in the division's approach to research promotion from "why do research" to "how to do research." The cumulative efforts of three decades had awakened industry to the importance of research in modern profit-making enterprise; the new problem was to teach industrial leaders how best to conduct and manage it.[135]

Technology as People

The Industrial Process of
Higher Education—I

> The very word university comes from the Latin word for corporation
> and the college dormitory is simply a continuation of the plan of the
> guilds by which the master workmen not only trained their apprentices
> but took them into their households to live. That is where our circle
> began, but as it swung out on its wide arc, the world of education drew
> further and further away from the world of industry. . . . The Sorbonne
> and Oxford scarcely knew of the world of science and for the world of
> industry they had only disdain. But the two circles went swinging on,
> bringing industry and education ever closer and closer, until tonight
> they are closing back once more at the point of origin where industry
> and education are one; where corporation and university again mean
> the same thing.[1]
>
> —William E. Wickenden

Like every other social process, technology is alive. People—particular
people in particular places, times, and social contexts—are both the
creators of modern technology and the living material of which it is
made. Designers and builders of an ever more sophisticated productive
apparatus, they are at the same time the critical constituents of that
apparatus, without which it could not function. The corporate engineers of science-based industry, people very much aware of their role
in the technological enterprise, understood this fact. In their various
reform efforts, therefore, they strove to achieve the necessary production and organization of not merely the material elements of modern
technology but the human elements as well.

As they worked to standardize scientific and industrial processes and
secure corporate command over the patent system, the engineers strove

also to direct the human process of scientific research and to create an educational apparatus which could meet the demand for research man-power. Their concern with education, moreover, was not limited to the rarefied realm of scientific laboratories. In their view, education was the critical process through which the human parts of the industrial ap-paratus could be fashioned to specifications. These human parts fell into two general categories: the skilled and unskilled workers who executed the designs of the engineers, attended the machinery, and performed the human labor of production, and the engineer-managers who designed and supervised the capitalist production process. Accord-ingly, education was divided into two categories. "Industrial educa-tion" was the means for producing the former—a "new apprenticeship system," as it was called, to replace the moribund apprenticeship sys-tem of craft-based industry. Higher education, and especially engineer-ing education, was the means for producing the latter, the process through which the corporate engineers could reproduce themselves. Both forms of education were promoted, in the rhetoric of progressive educational reform, as "education for life." The one, however, was to prepare people for a life of labor; the other to prepare people for a life of managing labor.

The integration of formal education into the industrial structure weakened the traditional link between work experience and advance-ment, driving a wedge between managers and managed and separating the two by the college campus. Engineers, of course, were not alone in emphasizing formal education as the key determinant of occupational mobility. The legal and academic professions were growing in strength on the basis of strict educational requirements for membership, and leaders of the medical profession—the paradigm for all professionals—were upgrading their own educational standards in the wake of the Flexner Report of 1910. The efforts of the corporate engineers, how-ever, had the most far-reaching impact upon industrial society as a whole. Because of their unique social identity, they automatically inte-grated professional requirements with industrial and corporate require-ments. In emphasizing the role of formal education as a vital aspect of their professional identity, they at the same time laid the groundwork for the education-based occupational stratification of twentieth-century corporate America.

Representing those industries which were the first to employ college graduates in significantly large numbers, the corporate engineers also led industry and the schools in effecting close industry-university cooperation over matters of curriculum and recruitment. In their various managerial and executive capacities, moreover, and as educators, they promoted the industrial education of workers. In their own corporation training programs and through the reform of the public school system, they sought to habituate both the working population and potential workers to industrial discipline and to educate them to carry out the directives of management most efficiently. By 1920 the transformation brought about in large part through the efforts of these corporate engineers was becoming apparent. Commenting on the incorporation of the National Association of Corporation Training, an editor of the *New York Times* observed that "in the past we have cited it as a triumph of free institutions and a prime cause of our industrial efficiency that so many of our corporation presidents have risen from the ranks; but that past is closing behind us. Specialized science is yearly taking a larger part in industry. If advancement is to remain free, it can only be on the basis of liberal education for the deserving worker."[2]*

Engineering education was viewed by the corporate engineers in a rather special light; the recruiting mechanism of their profession, it was the source of their immediate subordinates as well as their potential successors. In their reform efforts, therefore, they sought to bring both the form and the content of that education into line with what they perceived to be the immediate manpower needs of industry and the long-range requirements of continued corporate development. The various schools throughout the country, for example, each operating according to its own unique requirements, had to be more closely coordinated with the industries and with each other. The procedures for rating and evaluating students had to be streamlined and standardized, and the information about the "educational products" had to be made accessible to the consumers of those products, the industries. The means through which the graduates were employed had also to be systematized as a cooperative operation of the industries and the schools. In short, the transformation of engineering education into a unit of the industrial system demanded the creation of an educational apparatus for the production, selection, and distribution of higher technical manpower, according to changing industrial specifications.

*Since corporate engineers perceived industrial education as an aspect of management, this subject will be discussed in the final chapter. Here we will focus upon developments in higher education, with particular emphasis upon engineering education.

The content of engineering education, like its institutional form, had to be more closely correlated with industrial requirements: the content of the education determined the kind and quality of product just as the form of the education facilitated the proper selection and distribution of the product. The content of the education had to provide the training necessary for technical work, especially for the early years of employment; it had to instill in the student a sense of corporate responsibility, teamwork, service, and loyalty; and it had to provide the fundamental training in the social sciences and humanities which was increasingly being perceived as the key to effective management.

During the first few decades of the twentieth century, engineering education in the United States was progressively geared to habituate engineering students to corporate life—to prepare them for, and to facilitate their fitting into, industrially defined "positions"; as such, it constituted the vanguard of reform in higher education as a whole. "With the growth of the technical industries," Frank Jewett recalled in 1924, "the engineering side of the business was the first to wake up to the necessity of taking college, university and technical school trained men into the business. The engineers were the first ones to organize college recruiting on a consistent basis, ... to create ... smooth working machinery for making contacts and getting in touch with the right type of men."[3]

Like the development of research, the transformation of higher education evolved at three levels. Corporate industry established in-house training programs, and coordinated their activities through the National Association of Corporation Schools; educational institutions formed cooperative programs with industry, independently and collectively through the SPEE; and, finally, new agencies were created during the war, such as the NRC and the American Council on Education (ACE), to coordinate these activities on a nationwide scale. At all three levels, the electrical industry, which had the greatest need for college graduates, provided the reform leadership.

Just as the larger corporations first instituted in-house research laboratories to provide the scientific advances which underlay industrial progress, so too did they first undertake the establishment of engineering-education programs within the plant. Corporation schools, as they were called, dated back to the early sales-training program of the R. Hoe Publishing Company and the apprenticeship schools of the Baltimore and Ohio Railroad, and were designed to meet the immediate manpower needs of the company. The majority of corporation schools were established by the electrical, railroad, gas, and machine industries, and were devoted to commercial, sales, office, and apprentice training.* Prominent among the innovators were the three major corporations of the electrical industry: AT&T (and Western Electric), GE, and Westinghouse. These companies, moreover, gave attention to another area important to science-based industry—graduate education for college-trained engineers.[4] The corporation graduate-training programs were designed to meet the needs of industry: to guarantee the technical proficiency of college-trained employees, to ensure their proper habituation to corporate life, and to prepare them for managerial responsibility.

In addition, the programs served as the institutional apparatus for the recruitment, selection, and distribution of graduates. In the early 1890s GE set up elaborate programs at both Schenectady and Lynn to meet these needs. Charles Steinmetz explained that the Test Course, as it was called, "originated from the experience that in the work of an electrical manufacturing company to secure efficiency to carry out operations, a theoretical knowledge is necessary."[5] The electrical-engineering training then offered in the colleges—with a few exceptions like MIT and Wisconsin—lagged seriously behind the industrial developments in the field; the expensive equipment necessary for "state of the art" instruction was available at only a few of the larger schools, and this situation restricted most instruction to blackboard fundamentals. The industries, rather than the schools, were at the forefront of discovery in the field, and the corporation schools thus served the purpose of updating theoretical training in addition to linking the fundamentals to the exigencies of engineering practice. For these reasons, the majority of electrical-engineering graduates flowed into corporation school programs to complete their professional training. Providing the crucial preparation for careers in designing, manufacturing, construction, consulting, research, education, and management, the corporation schools were a necessary part of the training of professional electrical engineers in the United States.

The Test Course at GE was, as Steinmetz reminded his colleagues, "not a philanthropic question . . .; it is merely a necessary part of the work of the corporation . . .; it is part of the corporation." The post-

*Among the earliest schools were those set up by the Burroughs Adding Machine Company, Yale and Towne Manufacturing Company, American Locomotive Company, New York Central Railroad, Carnegie Steel, Curtis Publishing Company, National Cash Register Company, International Harvester, Firestone Tire and Rubber Company, and Travelers Insurance Company.

graduate education of the college "raw material" was of critical importance to industrial development, and the Test Course at GE was
designed "to supply the demand, not only of the corporation, but also
of the industry at large." Of the testmen trained at GE before 1919, 54
percent stayed within GE while the remaining 46 percent assumed
prominent positions in the railways, government, utilities, communications, mining, and manufacturing. The former group did not limit their
activities to domestic enterprise, but played a role in American expansion abroad as well; a 1919 survey of ex-testmen indicated that

> the graduates of the Test Course . . . are scattered over the four quarters
> of the globe, doing their share in the fascinating work of electrifying
> China, harnessing waterfalls in India, installing electrical drive in the
> sugar mills in the West Indies, substituting electricity for steam or hand
> labor in the mines of Alaska and South Africa, building railways in
> Australia and refrigerating plants in the Philippine Islands.

The Test Course, therefore, was not merely a training program for GE
engineers and executives; it was part of the common experience of a
generation of corporate-minded engineers who devoted their energies
to the "modernization" of the United States and those parts of the
"open door" world which contained the resources upon which American corporate prosperity depended.[6]

Since the testing department of GE was scattered throughout the
various GE plants, the testmen were a floating population, "continually
shifted from one kind of work to another" according to their own career
options and because "the apparatus was tested where it was manufactured."[7] The Test Course was designed to give the graduates a broad
view of the operations of electrical manufacturing while at the same
time meeting the specific needs of the company. "The student course
at Lynn," course supervisor Magnus Alexander explained to his colleagues of the AIEE in 1908,

> is planned to meet the requirements of the GE Company, for designing,
> and estimating, construction and commercial engineers, and technical
> salesmen.
>   The company takes graduates of technical schools and trains them
> during a period of two years, giving them during this time practical
> experience in the handling and testing of apparatus, in order to fix in
> the students' minds the practical application of engineering theories, to
> enlarge their engineering knowledge in general, to acquaint them with
> the competitive value of the product of the factory, and to develop them
> along lines of their future usefulness to the company.[8]

In addition to the practical experience of testing equipment, the graduates heard lectures on theoretical subjects bearing upon the engineering side of the industry. Throughout the technical training program, emphasis was placed upon the business aspects of engineering, the relationship between the design and the dollar; Alexander observed that "consideration of the elements of time and money in carrying out practical work . . . although the important factor that makes for success in industrial life . . . is entirely neglected at college." The Test Course was thus the means by which the company instilled "the seriousness of business" in the college graduates, and gave them the "proper conception" of business values.[9]

Training in technical subjects in the Test Course was supplemented by instruction in the commercial and managerial aspects of the industry —preparation for future executives. In 1912 a "highly efficiently organized course for engineering salesmen" was introduced. "For a college graduate" the Test Course had become "the best if not the only route" into "responsible positions" in the industry.[10] By 1919 the number of former testmen in the commercial and management departments of GE exceeded the number in strictly engineering work. In addition to the organized instruction, the testmen received their management training in a more informal way. From the outset they were gradually initiated into the privileged world of the professional and industrial elite. They were encouraged to participate in the meetings of the ASME, AIEE, Society of Engineers of Eastern New York, NELA, and Illuminating Engineering Society (GE headquarters was the site of national as well as local meetings of these organizations), and enjoyed social contact with company leaders at the Edison Club in Schenectady and its counterpart, the Thomson Club, in Lynn.

The Edison Club was formed in 1904 and had over six hundred members by 1918. The members had "all pursued the same studies, . . . undergone the same training in the test course, and . . . lived the same life while being initiated into the electrical industry. The 'camaraderie' [existed] not only between the younger members, but the various social and athletic activities [offered] opportunities for the student engineers to be brought in contact with many of the officials and engineers of the company."[11] The facilities included a library, bowling, pool, movies, tennis, canoeing, music, golf, and restaurants and were the setting for AIEE meetings and company-sponsored lectures. As a writer for GE described it, the club had a "university spirit" and the alumni were bound together like those of any university; their common experience provided lifelong associations, institutional ties, and a shared social perspective: the corporate perspective of the leaders of the industry and the profession.

The Test Course at GE did not merely prepare technical and managerial manpower for the industry; it did so with unprecedented efficiency. Elaborate methods of recruiting and evaluating college graduates were devised, and, within the company, personnel files were kept on all testmen in order to chart their progress and determine their potential usefulness. Graduates were tested periodically and rated in terms of technical proficiency, willingness to learn, loyalty, dependability, appearance, tact, efficiency, cooperativeness, and ability to handle men. They were then classified according to job requirements within the organization and the industry. Such techniques of evaluation and selection were constantly refined as more effective means of fitting the individual to the job were developed. In effect, the programs like these within GE and the other large corporations were the pilot programs in personnel development and management which would transform American higher education in the decades to follow.

At Pittsburgh, also in the late 1890s, Westinghouse developed a Special Apprentice program for graduate engineers similar to the Test Course at Schenectady and Lynn. The educational work of Westinghouse was under the direction of two electrical engineers, Charles F. Scott, a Johns Hopkins graduate and inventor of the famous Scott connector for transformers, who became head of the electrical-engineering department at Yale in 1911, and Channing R. Dooley, a Purdue graduate who left Westinghouse after the war to direct educational and personnel activities for Standard Oil. "The notable point of contact between the engineering college and the electric company," Scott explained to engineering educators and other corporation educators in 1907, "is the engineering graduate. He is the product of the college and the raw material which is to enter into the human organization underlying the electric industry. . . ." Scott observed that "probably in no other field has there been such a growing demand for engineering graduates as in the electrical profession." (By the demand of the profession, of course, Scott meant the demand of the industry.) "This demand puts to a severe test the efficiency of the schools which are to furnish these men," and the manufacturing companies do not, therefore, "expect the graduate to be a ready-made engineer. They provide systematic courses for supplemental training."[12]

The supplemental training at Westinghouse, like that at GE, went far beyond technical instruction. Like Steinmetz, Scott explained that the "course . . . had not been established for sentimental reasons . . . but as a matter of necessity. The men need the experience and training and the new point of view which it gives before they are useful." Thus, the training at Westinghouse was also designed to adapt the graduate to corporate life. "The fundamental difficulty," Scott emphasized, "is lack of adaptation to new circumstances and conditions. We do not underrate knowledge and training, but we want [the graduates] to be of use . . . we want men who can see the situation and fit themselves to it. . . . The possibilities and the outcome depend . . . upon the ability of the man for harmonizing himself with his environment, and the more complete and efficient this adjustment . . . the more useful the life." Scott, Dooley, and most of the industrial participants in the discussion agreed that the college graduates did not have the proper "commercial or business point of view." They were too individualistic, unwilling to cooperate effectively in the "teamwork" of corporate enterprise. The major problem in the view of the representatives of industry, one educator observed, was that the graduates of the colleges "did not know how to adapt themselves to new conditions, . . . to adjust their personalities to the wishes and desires of their superiors. Preoccupied with the question 'How much can we get?' hardly once had the idea entered their minds, 'How useful can we make ourselves to somebody else? How can we be of more service?' " "They do not realize," the educator went on, "that until they have learned to work first for the success of the corporation, and only secondarily to consider themselves, and also have learned to subordinate their own ideas and beliefs to the wishes and desires of their superiors, that they can really be efficient [sic]." Another Westinghouse spokesman summed up this discussion. The graduate must "go in for teamwork and seek the place he is best fitted for," he argued. "Self-forgetfulness is what is required."[13]

Socialization for subordinate employment at Westinghouse was coupled, as elsewhere, with socialization for management. The graduate must learn "to work effectively with those about him . . . he must understand men," Scott argued. "His education should be one not of engineering subjects only, but should include the humanities."[14] In an article for the Society for the Promotion of Engineering Education, Scott and Dooley described the growing need for more broadly trained engineers in industry.

As the electrical industry increases and more exacting requirements are placed upon electrical apparatus . . . there is demand for greater ability on the part of the engineers and the managers who have to do with the

> production and operation of apparatus and the direction and manage-
> ment of manufacturing and operating companies. Many of those who
> now hold responsible positions in the various departments of such
> companies or as consulting engineers, have gained an important part
> of their experience in the training courses of manufacturing companies.
> It is the aim in the plans which we have outlined . . . to keep pace with
> the new conditions and to prepare men for the larger duties and respon-
> sibilities which they will face in the future as engineers.[15]

Westinghouse efforts to meet this need included intensive training programs. By 1910 training in purely technical subjects was formally
supplemented by courses in sales, management, economics, business
law, and the like. In a more informal way, the Westinghouse Club, like
GE's Edison Club, enabled students to rub elbows with corporate
management and thereby pick up the corporate point of view.

As it was at GE, the educational training for leadership at Westinghouse was conducted as an efficient operation of the corporation. The
problem, as Scott saw it, was simply "how many boys of different kinds
can be individually developed and fitted to varying needs." To that end,
the educational department devised personnel files, rating systems, job
specifications, and recruitment procedures. In developing the system at
Pittsburgh, Dooley, a Purdue graduate, worked closely with A. A.
Potter, the former GE testman who had set up the student personnel
system at both Kansas State and Purdue. Potter's system of personnel
cards and routine evaluation focused upon the "character, personality,
and physique" of the graduates, in addition to their technical competence; it screened graduates for "good traits" (such as persistence,
interest, loyalty, cleanliness, dependability, accommodation, tact, practicality, and efficiency), warned graduates of possible "deficiencies,"
and suggested ways of overcoming them to better fit the corporate
mold. "The new way," Dooley explained in 1913, "is to study each
student . . . and thus as far as possible scientifically to place each man
in that line of work for which he is best fitted."[16]

Those responsible for the educational programs in the Bell System
were concerned with the same problems. Albert C. Vinal of AT&T in
New York had by 1913 conducted "an extensive study of the whole
problem of selection," involving psychological testing and character
analysis; in addition, his department had developed a program of periodic testing of experienced employees at different ranks as a method of
determining job specifications, the requirements of each position, and
the basis upon which new employees were to be evaluated and
"guided." The most elaborate work in the Bell System was undertaken

by another Purdue electrical engineer, J. Walter Dietz, at the Hawthorne works of the Western Electric Company. Although a formal education department was not established at Western Electric until 1910, various educational programs dated back to 1898. "As we look back," Dietz reflected at a 1913 meeting of corporation educators, "the impression grows that we have been letting employees try to fit themselves to their work. Then came the period of training employees for classes of service . . . ; [now] we have arrived at a more encouraging period—that of the organized education of employees."[17]

Of primary concern in the educational work of Western Electric was the Graduate Apprentice Course for engineering graduates. This course was designed to "help college men to take up their chosen work intelligently, promptly, and in an organized way." In the words of another Western Electric official, Frank Jewett, the Bell System,

> in common with all other of the larger industries which are growing and which are developing new applications of science all the time, find it necessary, no matter how good and well-trained are the men who come to us, to ourselves put them through some sort of a course of training in order to fit them for the peculiar problems of their work . . . ; our specification for engineers is not essentially different from the specification which the GE or the Westinghouse or any big power or electrical development company would write.

Like the training programs of the other companies, the Bell educational departments were geared to prepare graduate engineers for managerial as well as technical positions. "The way things are developing in the industries based on science," Jewett observed,

> it seems to me . . . as though we must look in increasingly large measure to the group of technical graduates for the executives of the future. The problems of the industries which executives will be called upon to administer . . . are becoming increasingly involved in the just appreciation of fundamental physical science. . . . From the group of well-trained engineers should come many great executives.

The Bell training courses thus focused upon the same problems toward which the other industrial postgraduate training programs were directed: technical and management training, habituation to corporate employment and ideology, and efficient rating, selection, and distribution procedures. Dietz's summation of the "state of the art" of corporate graduate education for engineers, made in 1913, reflects the situation in the industry at large. "The nature of our engineering, manufacturing, and commercial work demands men of trained brains." Recruitment of the "raw material" from colleges, however, only begins to meet this demand; the graduates must then be

> broadly trained, and, if proper selection has been made, we have men of character, capable of independent thought, possessing adaptability, loyalty, capacity for growth, and a willingness to get along agreeably with their fellows, and physically sound.

"We have not been able, as yet, to prepare definite specifications on these points," Dietz reported, "so we are not surprised to find an occasional man with some 'human nature' in his make-up."[18]

The corporation schools put the corporations in the education business. In addition to the schools established by individual companies, trade associations began to create schools for trade and technical training: the National Foundrymen's Association conducted a foundry school at Winona Technical Institute in Indianapolis, and the National Metal Trades Association established a machinists' school and general apprentice school in Cincinnati.* Throughout the country, corporate educators experimented with different programs and published their results—evaluated in terms of labor quiescence, productivity, and efficiency—in the technical and trade journals. While there was thus some cooperation among them, however, they remained relatively isolated. In 1913 the National Association of Corporation Schools was created to correct this situation. It was this organization which first developed fully the concept of education as a vital function of corporate management; ten years later, with its activities significantly broadened, it became the American Management Association.

The NACS grew out of the experience of the New York Edison Company. As a means of developing its own educational programs for engineers and salesmen, New York Edison conducted a survey of corporation schools throughout the country and found that there was a need for some central agency through which they could all cooperate. President Arthur Williams of New York Edison, an electrical engineer who had worked with the early Edison power companies, therefore called together the nation's corporate educators, in January 1913, at New York University, to found the NACS. The meeting, like the new organization, was dominated by representatives from the electrical in-

---

*Others were instituted by the American Institute of Banking, Railway Education Bureau, Insurance Institute of America, American Institute of Laundering, and United Typothetae of America.

dustry: Williams and F. C. Henderschott, educational director at New York Edison; Alexander, Steinmetz, and others from GE; Dietz from Western Electric; and Dooley from Westinghouse. Other key figures included Lee Galloway, professor in the School of Finance and Commerce at NYU; E. St. Elmo Lewis of the Burroughs Adding Machine Company; and E. J. Mehren, managing editor of the *Engineering Record* and an associate in the efficiency-engineering firm of Harrington Emerson.*[19]

The concerns of the NACS educators did not differ substantially from those of GE, Westinghouse, and AT&T. "Man-stuff," in the view of Elmo Lewis, was the "most important thing" with which the companies had to deal; it was the substance "out of which they make their business." E. A. Deeds of the National Cash Register Company agreed; "I am most interested," he said, "in increasing the efficiency of the human machine." In addition to technical proficiency, these educators all stressed the need for training for management. "Electrical engineers," Arthur Williams observed, "are from the practical standpoint . . . men without a peer in running machines, in running plants, but not men trained, necessarily, in running human machines."[20]

Edison's F. C. Henderschott outlined the importance of educational and personnel "machinery" for the efficient development and utilization of manpower, and shared with his counterparts in the other companies New York Edison's three-pronged approach to the problem, focusing upon "positions to be filled" (job specifications, required qualifications); "the man" (investigation of types, tendencies, natural qualifications, adaptability); and "the means" (institutional training and guidance to fit the man to the position). "The field of activity opening for this body is unlimited," Henderschott declared. "Corporations are fast being converted to the theory of training their own men. They no longer expect to find satisfactory help ready-made, but are now applying themselves to the task of making men as well as commodities." Henderschott, who became the NACS executive secretary and editor of its publications, drove home the point without stating it: the corporations were turning their energies toward the production of men *as* commodities. Their approach to education, shared by all who gathered together in 1913, was perhaps most succinctly offered by C. D. Brackett

*Charter members of the NACS included GE, Western Electric, Westinghouse, Thomas A. Edison, Inc., Brooklyn Edison, Boston Edison Electric Illuminating Company, New York Edison, and Commonwealth Edison; others were Consolidated Gas Company of New York, Burroughs Adding Machine, National Cash Register, American Locomotive, Yale and Towne Manufacturing Company, Packard, Cadillac, Pennsylvania Railroad, Travelers Insurance, and Curtis Publishing Company.

of National Cash Register in his description of that company's Agents Training School: "Product: men and cash registers."[21]

The NACS was conceived as a clearinghouse for corporation-school education, with the stated object of "aiding corporations in the educational work of their employees by providing a forum for the interchange of ideas, and by collecting and making available data as to successful and unsuccessful plans in educating employees."[22] Its most immediate functions thus included the promotion of corporation schools throughout the various American industries, the development of educational methods and personnel procedures, the collection and dissemination of information, and the training of corporation educational directors. The visions of the NACS organizers, however, extended far beyond these modest tasks. They saw in education, properly guided according to corporate imperatives, the key to corporate prosperity and stability; by means of education they sought to eliminate the problems of "labor turnover," "labor troubles," and "lack of training," to bring about greater productivity and industrial efficiency. The temporary chairman of the organizing meeting, Lee Galloway, put their vision into words:

> We have associations formed for the consideration of various features of manufacturing, we have associations formed for technical work, we have efficiency societies, associations for the advancement of scientific management, etc., but in the last analysis . . . we find that the whole thing rests finally on some educational feature that must be evolved.
>
> I wonder how long we intend to leave the education of the workmen in the hands of the trades unions, in the hands of the I.W.W., and the Socialistic Party? . . . If a school is organized . . . within the corporation itself, to bring out the strong, practical purposes of the institution, and to show the art, ability and skill which is necessary to carry on a great industrial institution—if the dissemination of knowledge is carried on in an ordinary educational way within the corporation itself . . . it would tend greatly to change the attitude of the employees, and more than that, it would tend to change the attitude of the public, because it is the employee . . . who comes in contact with the public. . . .
>
> That seems to be the highest kind of insurance that any industrial corporation can have—to insure itself by creating a strong educational system among its own industrial forces, and if big industries are to assume the proportions of states, they must assume some of the responsibilities of states, and one of these responsibilities is to educate the people, and the welfare of these big corporations will be insured more safely by the education of their employees than in any other way. . . .[23]

The wide range of activity which Galloway and his corporate-minded colleagues envisioned for the NACS was implied in the three

"Functions of the Organization" formulated by the First Annual Convention: "to develop the efficiency of the individual employee"; "to increase efficiency in industry"; and "to influence courses of established educational institutions more favorably toward industry." Within the scope of the first two functions, the NACS, which changed its name in 1920 to the National Association of Corporation Training (NACT), undertook extensive projects to "fit the individual into his life's work." Dooley of Westinghouse chaired the committee which supervised the training of educational directors, first in cooperation with NYU and later at other schools. Vinal of AT&T led the Organization and Administration Committee, which "determined the best methods of organization of educational work as a function of management." H. A. Hopf of Du Pont headed the committee charged with the development of job analyses and job specifications, Kendall Weisiger of Southern Bell and W. M. Skill of GE directed the programs of technical and executive training, and Carl S. Coler of Westinghouse guided the work with regard to unskilled labor "to determine best methods of instruction to bring operators up to standard rates on specific tasks." By 1917, when J. W. Dietz of Western Electric became NACS president, the organization's activities included trade apprenticeship schools, accounting and office-work schools, advertising, selling, and distribution schools, codification, employment plans, safety and health programs, special training schools (engineering graduate training), unskilled-labor training, vocational-guidance projects, personnel-rating systems, psychological testing, and records systems. In addition to these education-oriented activities, the NACS studied and promoted profit-sharing and bonus plans, collective bargaining, workmen's compensation, welfare programs, life insurance for employees, employee representation in management, cafeterias for employees, works councils, the use of art, dancing, and music in personnel relations, medical departments, and Americanization programs for immigrant workers. In short, the NACS expanded its educational work to include much of modern corporate management. Underlying all of these efforts were the themes of corporate liberalism: cooperation rather than conflict, the natural harmony of interest between labor and capital, and effective management and administration as the means toward prosperity and general welfare. Lee Galloway expressed the faith that lay at the heart of the new management approach. "We must strive to keep the men in a pleasurable frame of mind," he said, "because we know that fear does not breed efficiency, and that pleasure does."[24]

The third function of the NACS, to "influence" the established educational institutions favorably toward industry, was essentially geared to put the corporation schools out of business by rendering the established educational institutions outside the corporations capable of, and disposed toward, providing the services for which the corporation schools had been created. Galloway put forth this purpose at the first organizational meeting: "It is time that our educational system was brought into some correlation with the business world."[25] M. S. Sloan, president of the Brooklyn Edison Company, later explained how the NACS might work to bring about such a correlation:

> No matter how willing educational institutions may be to train men better to fit the needs of industry and commerce, their training will be ineffective unless they know definitely what those needs are.
> [This] is the logical organization not only to promote the idea of practical experience on the part of teachers, but also to assist teachers in getting the industrial point of view. [This] is certainly the logical organization to help to formulate the training needs of industry, insofar as they can be definitely stated for application by the schools.[26]

The NACS thus undertook to act as the agency for industry-education cooperation. Through alliances with such organizations as the AIEE, which had its own "Cooperation with Educational Institutions" committee, the SPEE, the National Society for the Promotion of Industrial Education, the National Electric Light Association (the utilities trade association which carried on its own extensive propaganda campaign in the schools), the National Association of Manufacturers, and many schools, the NACS worked to integrate the vocational, public, and higher educational institutions within the industrial system. This work was greatly facilitated by the steady flow of high-ranking personnel back and forth between the industries and the schools, with the NACS serving as the primary medium for such interaction. Among the notable early NACS members were Ernest Hopkins, who represented New England Telephone in NACS before becoming president of Dartmouth in 1916; William Wickenden, who as an AT&T representative chaired the NACS committee on relations with collegiate institutions, headed the SPEE investigation of engineering education, and later became president of Case Institute; Charles Steinmetz, who directed GE's educational program for testmen, was an early NACS president, professor and curriculum reformer at Union College, and longtime president of the Schenectady Board of Education (in which capacity he reorganized the public school system); Frank Aydelotte, who taught English to AT&T employees while a professor at MIT and went on to become president of Swarthmore in 1921, organizer of the Guggenheim Foundation in 1924, and first director of the Institute for Advanced Study at Princeton in 1939; Walter Dill Scott, the industrial psychologist who became president of Northwestern University in 1920; and Herman Schneider, the father of "cooperative education" and president of the University of Cincinnati.

In 1913 F. C. Henderschott observed that "the universities and colleges are seeing in this new [corporation school] movement a link, long sought, between our institutions of learning and the business world, and are anxious to affiliate and push forward this new educational system." The corporate-minded educators in the engineering schools, who enjoyed the closest contact with the industries, were the first to attempt to close the education-industry gap from the educational side. E. J. Mehren, chairman of the NACS Committee on Allied Associations and Movements, reported in 1914 that his committee was "in very close touch . . . with two movements in particular—the vocational and continuation school movement and the cooperative technical methods being tried by the Universities of Pittsburgh and Cincinnati." "If by proper development of these systems, employees will be better prepared for the activities of life than now," Mehren observed, "it will mean a considerable curtailment of the educational work of the corporations themselves."*[27]

*This did not mean, of course, the elimination of corporation schools altogether. Although the movement, which had swelled its ranks considerably under NACS leadership, waned after the postwar economic slump, the science-based industries, while promoting cooperation with the schools, continued to develop their own in-house training operations. The electrical industry remained far ahead of the rest in this line of activity. Kodak, Dow, Du Pont, and the National Carbon Company had some of the earliest informal training programs for college graduates in the chemical industry, but it was not until 1934 that Dow set up the first comprehensive program similar to that established in the electrical industry over three decades earlier. Standard Oil of New Jersey led the petroleum industry in the education business and established a Student Engineer Training Program in the early 1920s. Goodyear instituted its famous Flying Squadron training program in 1913, which, under the direction of MIT engineer Paul Litchfield, became the core of its Industrial University. In the automotive industry, Packard created its technical graduate courses and Ford established the Technical Institute, both in 1919. General Motors conceived the General Motors Institute in 1926.

For further discussion of early corporation-school programs, see John Van Liew Morris, *Employee Training: A Study of Education and Training Departments in Various Corporations* (New York: McGraw-Hill Book Co., 1921); John H. Greene, *Organized Training in Business* (New York: Harper and Brothers, 1937), p. 180; Nathaniel Peffer, *Educational Experiments in Industry* (New York: The MacMillan Co., 1932); Don Whitehead, *The Dow Story* (New York: McGraw-Hill, 1968), pp. 85, 1218; William Cabler Moore, "Industry's Interest in the Professional Training of Chemists," *Journal of Chemical Education,* XVIII (1941), 576; S. L. Starks, "Training in Industry," *Journal of Chemical Education,* XXI (1944), 285.

The gap between the engineering schools and the industries had resulted from the historical fact that the majority of engineering schools had been created as extensions not of the industries, but of the established schools of science in the state and private universities. The college setting demanded that the engineering schools adopt an academically respectable approach to engineering, with an emphasis upon scientific theory rather than industrial practice. As a consequence, the schools remained relatively independent of industry and produced graduates who might be temperamentally ill-suited for disciplined industrial work and poorly trained in the practical application of their theories.

Those who called most strongly for some bridging of this gap were the "practicing engineers" in the industries and their like-minded professional colleagues in the schools. Increasingly, by the turn of the century, engineering-college alumni who had become leaders in industry put pressure upon their colleges to update instruction and bring it into line with the requirements of postgraduate professional practice. "In the college, the product is the workman himself," the president of Tabor Manufacturing Company of Philadelphia explained to SPEE members in 1912. "The molding of this material for the market which awaits it is surely an industry of preeminent importance."[28]

Frederick L. Bishop, SPEE secretary and dean of the College of Engineering at the University of Pittsburgh, agreed wholeheartedly:

> An educational institution resembles, in some respects, a manufacturing concern. . . . The goods produced must be of such design, finish, material, etc. as to satisfy its patrons; likewise, the graduates of educational institutions must meet the requirements of the concerns which are to employ them. . . . The inefficiency of a graduate (and I apply the word "inefficiency" to include all the undesirable characteristics) may be caused by poor material, due largely to improper preparation of the student at entrance to the college or to poor instruction, poor teachers or antiquated and improperly correlated courses.[29]

In Europe various means had been devised to bridge the same gap. In Germany, for example, young graduates of the *Gymnasium* were required to work in mechanical shops for at least a year before gaining entrance to the *Polytechnikum,* or scientific school; a similar scheme was adopted in the Scandinavian countries. Glasgow University in Scotland evolved the "sandwich system" whereby a six-month academic year enabled students to spend the other six months in the workshops of industry, and thus earn money for their education. In the United States the technical schools conducted inspection trips to the industries as a way of introducing their students to "real world" conditions.[30] The so-called shop movement, which began at Worcester in 1868 and MIT a decade later, brought the shop into the college for instructional purposes and to simulate industrial conditions. The shop movement, however, which spread to most engineering schools by the first decade of the new century, was largely unsuccessful and came under attack on two fronts. The industries maintained that the school shop could in no way approximate the actual situation in industry, and only a few of the best schools could even begin to afford up-to-date equipment such as existed in industry. The academics outside the engineering schools, moreover, had nothing but disdain for lowly shopwork and were loath to permit it on the campus, much less grant university credit for it.

The most promising resolution of the problem, one which intrigued engineering educators more than any other, was the cooperative course of the University of Cincinnati, set up in 1906 by Herman Schneider. As a young instructor of civil engineering at Lehigh University, Schneider had between 1899 and 1903 conducted what he called "pedagogic research into the problem of engineering education." In the process, he had visited "the largest manufacturing concerns in the Eastern and Middle States, in order to obtain from the employers of engineers their views on the subject. In a great many cases the men consulted were graduates of the best institutions in the country." In Schneider's view, and the view of those whom he consulted, the problem of bridging the gap between the schools and the industries boiled down to three questions: "What requirements should the finished product of an engineering school fulfill?" "Where and how shall we get the raw material to make the required finished product?" And "Through what processes shall we put the raw material in order to obtain the required finished product?"[31]

By 1905 Schneider, now a professor at the University of Cincinnati, had reached his "somewhat radical and revolutionary" conclusions. He formulated a plan for a "cooperative course" whereby engineering students would be required to alternate between the college classroom and the industrial workplace in the process of earning their degree. In 1907, when he became dean of the College of Engineering at Cincinnati, Schneider launched his project. Initially the course was a six-year program of instruction in mechanical, electrical, and chemical engineering, and was carried on in cooperation with a number of Cincinnati's electrical and machinery companies. Students alternated week by week between the industrial shops and the college during the school year and were required to work full-time in the industries during the four-month "vacation." In total, they spent four years in the industries and two years in the college classroom. Students were required to sign a contract whereby they were bound to meet the requirements of the school when at school and of the company when in the industries, and were paid at a starting rate of ten cents an hour for time spent at the workplace.

"The aim of the course is not to make a so-called pure engineer," Schneider explained, "it is frankly intended to make an engineer for commercial production. . . ." This approach, he boasted, "will furnish to the manufacturer a man skilled both in theory and practice, and free from the defects concerning which so much complaint is made." The Cooperative Course at the University of Cincinnati—the first of its kind anywhere in the world—focused on the same problems which concerned the corporation-school educators. "The shop discipline has had a marked effect upon the character of the work that they have done in the university," Schneider observed, and, in addition, "the students have gained valuable knowledge of the labor problem, and of time as being the very essence of commercial production."

> This course resolves itself down to a training in commercial production with a university preparation in the underlying science . . . ; we are operating our engineering college at the highest efficiency . . . ; we are educating only those who by mental, physical, and temperamental adaptability are worthy of the expenditures made.[32]

The educational director of one of the cooperating companies attested to the merits of the Cooperative Course from the manufacturer's standpoint. "The chief criticisms of modern technical education result from the fact that we try to take the shop into the school, whereas we should bring the school into the shop. The cooperative engineering course plan practically brings the school into the shop." The plan also brings the students "in close touch with the men," he explained, giving them an intimate knowledge of the condition and attitude of labor, which will be of value to them later as managers of men; "and it instills in them . . . the commercial sense of time." The cooperating company meanwhile is given the "opportunity to know the boys and form an exact knowledge of the abilities and possibilities of each individual. . . . Wherever [it] will use them they will, each one of them, be a known quantity."[33]

The Cincinnati plan was relatively well received by engineering educators and industrialists despite its "radical and revolutionary" nature. The poor economic conditions after the panic of 1907 hindered its rapid expansion, but by 1919, seventy-five companies—some, like the Boston Edison Company, quite a distance from Cincinnati—had joined the Cincinnati program, and by 1929 some twenty schools had adopted the Cooperative Course plan with only minor revisions, including Northeastern University, the Universities of Akron, Georgia, and Louisville, Georgia Tech, and Antioch.[34]

The Cincinnati Cooperative Course grew into an important laboratory experiment for industrial and college educators, and Schneider himself became a leading spokesman in educational and industrial circles. In 1914 he addressed the NACS on the problem of "selecting young men for particular jobs," and the editor of the NACS *Bulletin* promptly declared that he had "given to industry the most authentic knowledge available at this time" pertaining to this critical problem. Drawing upon his eight-year experience with the Cooperative Course and the production of over five hundred students, Schneider outlined a system of vocational guidance and personnel procedures based upon the "classification of marked characteristics which furnishes a rational basis" for manpower selection.

> Under present conditions our youth blunder into jobs . . . ; there is no method or agency to determine the general type of work for which a youth is talented and to classify the various jobs which fall under this type.
> Every individual has certain broad characteristics and every type of work requires broad characteristics. The problem then is to state the broad characteristics, to devise a rational method to discover these characteristics in individuals, to classify the types of jobs by the talents they require and to guide the youth with certain talents into the type of job which requires those talents.[35]

Among the various characteristics Schneider listed physical strength, music sense, color sense, manual accuracy, mental accuracy, and concentration. In addition, he cited certain polarities between which students might be rated: mental/manual; settled/roving; indoor/outdoor; directive/dependent; creative/imitative; small scope/large scope; adaptable/self-centered; deliberate/impulsive; and dynamic/static. Besides these characteristics of individuals, Schneider outlined some general racial guidelines which might help employers to select their personnel. The Chinese, he found, were "settled," Arabs were "roving," Sicilians were "impulsive," Hindus were "deliberate," Japanese were "manually accurate," and Persians had a "refined color sense." Whether he drew upon the empirical evidence from the experience at Cincinnati or fell back upon the racial stereotypes of his class, Schneider sought ways for industry to meet the new corporate challenge of "fitting the individual to his life's work."[36] His contribution was well received.

One of the first to recognize the potentials of Schneider's plan was Magnus Alexander, then director of the educational programs at the Lynn works of GE. In June 1907 he wrote to the president of MIT, Henry Pritchett, about a visit he had made to Cincinnati in order to observe the plan in operation. "The arrangement impressed me as a very practical one," he wrote Pritchett, "and I wish to suggest a similar cooperation between the MIT and the Lynn Works of GE." Alexander's proposal involved a six-year course of study leading to the Bachelor of Science degree in electrical engineering and consisted of "alternating fortnights" of classroom study and industrial work, and full-time work at Lynn during the summers. The cooperative program, he explained, would be supervised through the "joint-trusteeship" of MIT and GE officials. Attached to Alexander's letter was a supporting note from Elihu Thomson of GE: "I cheerfully endorse the proposal of Mr. Alexander," Thomson wrote. A short while later GE Lab Director Willis Whitney expressed similar enthusiasm.[37]

While Alexander rallied support behind the plan at GE, Dugald Jackson did the same at MIT. In an article in the AIEE *Transactions,* Alexander outlined the advantages of "The New Method of Training Engineers." He pointed out the great value of the cooperative program as a means of "bridging the gap between academic and practical training" and adapting engineering students to industrial and corporate conditions. The Cooperative Plan would "bring the regular engineering courses more fully into synchronism with the demands of modern industrial life," Alexander argued, whereby "the ideal of the college" could be "linked to the real of the factory." It would "well enable the student to adjust himself quickly to the interacting influence of the college and the factory," to make of him an energetic "industrial worker" in the classroom as well as the factory. "Under such a plan," Alexander contended,

> the freedom enjoyed by students during the college career is happily interrupted by the stern discipline that must prevail in a business organization; the advantage of this college freedom in the development

of the young man's character . . . is not lost, but his freedom is regulated by frequently recurring intervals of discipline in the factory, so that he may be prevented from soaring to the skies in his fanciful ideas engendered by his personal irresponsibility and after four years find himself all too rudely pulled back to earth by the stern call of practical life with its demand for cooperation of all forces.[38]

As far as expense to the company was concerned, Alexander pointed out that "all engineering apprentices will be utilized by the factory" and that "if the factory end of the cooperative course is well organized and efficiently conducted, an astonishing amount of good commercial work can be turned out by engineering apprentices." Unlike Schneider, Alexander emphasized the plan's value for the production of executive talent for business leadership. "The underlying thought of the proposed plan," he explained, "is to provide an education especially adapted to the needs of prospective engineers who are to specialize in life along administrative and executive rather than purely engineering lines." In the final draft of his proposal, Alexander indicated that the Cooperative Course students would be required to spend their entire last year at MIT studying, along with the scientific curriculum, such subjects as salesmanship, cost-keeping, factory management, and "the relation of employer to employee."[39]

Jackson conducted a quick survey of opinion on Alexander's proposal among his colleagues at MIT and the industries. Most of the responses were favorable if not enthusiastic, and the majority of those who opposed the plan did so merely on the grounds that it appeared inadequate to achieve the ends outlined, ends which they themselves strongly endorsed. A. M. Basford of the American Locomotive Company, GE's neighbor in Schenectady and one of its leading customers, urged acceptance of the scheme in light of its "real advantages";[40] Herman Schneider was of course most enthusiastic,[41] and Charles F. Scott of Westinghouse agreed that "on the whole . . . the proposed plan would accomplish your purpose of producing young men of better training and larger vision than are now being produced by the usual processes." Scott did offer, however, one note of "precaution" drawn from his own experience with the Special Apprentice Course at Pittsburgh.

> One of the objects of the present apprenticeship course is to get the young man away from his student point of view. It is to give him practical and commercial ideals and incentives. . . . If the short periods of shopwork . . . are so fully imbued with the school idea that the young

man, consciously or unconsciously, regards the whole course as a part of the college to such an extent that his ideas are academic rather than commercial . . . then the proposed course fails in an important particular.[42]

Jackson received only one response in his survey which did not reflect keen interest in this "important particular"; it was from Arlo Bates of MIT's English department. "The scheme seems likely to produce a state of things directly unfavorable to instruction in English composition and English literature," Bates complained, and, furthermore, the "spasmodic periodicity of the proposed plan will probably be a disadvantage to all Institute courses except those in the direct line of the work done in the factory." Bates' criticism, however, while incisive, was dulled by the timid way in which it was offered. Like many of his academic contemporaries (and successors), Bates was awed by the obvious power of the industries and the "real world" savvy of the engineers. He was thus caught between his own convictions and the fear of appearing stubborn, perhaps reactionary, and, at worst, out of touch with the times. "The Department of English has certainly no wish to appear as opposing [the plan]," he assured Jackson. It is "not so inflexible or so wedded to its present methods as to be unable to adapt itself to fresh conditions." Recognizing that the cooperative plan was an assault on the supposed autonomy of intellectual life, Bates was neither prepared nor willing to combat it. Consequently he took the most convenient way out. "In making this report," he wrote Jackson, "I wish it to be clearly understood that I am offering no opinion on the scheme proposed, since I do not regard myself as able intelligently to form one." Jackson did take Bates' criticism to heart, however, proposing that students in the Cooperative Course be required to write a theme each week while at work in GE's factory at Lynn.[43]

Despite these promising beginnings, Alexander's proposal was shelved, in the wake of the economic downturn following the 1907 panic. With the return of war-stimulated industrial prosperity, Alexander reintroduced his plan, emphasizing even more strongly its primary purpose: preparation for management. "The object of the proposal," he explained, "is the selection of a limited number of especially qualified college students and their subsequent training . . . in order to develop junior engineers of high capacity."

The new plan differed even more from Schneider's Cincinnati program: at MIT, students in the Cooperative Course would be selected rather than having all students participate; they would begin cooperation in their junior year rather than as freshmen; they would receive a Master's degree rather than a B.S. at the completion of the course; they would each be trained by only one "comprehensive company" throughout the period of cooperation; and they would receive formal instruction at the company as well as the institute. In addition, the alternate periods of training would be longer—thirteen weeks—to allow the lessons learned to "sink in," and the students would spend their last half-year of the five-year program at MIT studying "business law and organization" in courses "given by practical men of affairs."[44]

In contrast to their hesitancy in 1908, GE officials were now pushing for adoption of the proposal. Although "the proposed plan would cost the GE Company a large amount per annum," GE president E. W. Rice intimated to MIT president Maclaurin, "we are willing . . . for reasons more or less selfish, to try the experiment. . . . We believe [it] will mark the successful introduction of a greatly needed improvement in Engineering Education." Maclaurin agreed. "The experiment, if successful," he replied to Rice, "is likely to have far-reaching consequences in higher technical education." (Maclaurin complained, however, that MIT was in bad financial straits and that GE would have to bear the larger burden of cost. GE thus agreed to provide two thirds of the expenses.) Soon after the course was formally established, an Oversight Committee was formed to facilitate the "joint trusteeship" of the experiment, with Frederick P. Fish of the MIT Executive Committee as chairman. Representing GE were Magnus Alexander, Elihu Thomson, and Charles K. Tripp, supervisor of apprentices at Lynn. From MIT were Dugald Jackson, Comfort A. Adams, William Wickenden, and later William H. Timbie. Jackson immediately sent out letters instructing all juniors to report either to the office of Wickenden, who shared responsibility for the MIT side of the course with Timbie, or to the office of Alexander at Lynn. In the case of the latter, he advised them to "take overalls along."[45]

After another brief delay caused by the war, the MIT Cooperative Course in electrical engineering was finally launched in the fall of 1919. "It is my opinion," Jackson wrote, "that this is the most promising line of endeavor for the education of the future leading men in manufacturing lines, particularly those in the electrical branches, which has yet been undertaken."[46] At GE, where the course was "as carefully laid out as the work at the Institute," students took formal courses in corporation accounting, business psychology, contemporary English and American literature, and "human relations." Courses at the institute covered such subjects as contracts, purchasing, organization, production, employment, sales, and market analysis. At both sites, as Timbie explained, students learned "to estimate the strength and characteristics of men as well as the strength and properties of materials." In addition, Timbie continued,

> there is afforded an excellent opportunity to do some real training in teamwork and in the development of a sense of loyalty to the job. . . . This is the reason why a student receives all of his practical experience with one company, where he can be given a unified conception of the organization and a real appreciation of the policies and spirit of the company. . . . If these qualities are not gained, the most rigorous intellectual training is wasted.[47]

The success of the course—one half of the students were hired by GE after graduation, and these increased their salaries (i.e., rose into management) at a faster rate than regular engineering graduates—led to an expansion of the program. After the publication of a widely read promotional article by Jackson and Alexander, other companies entered the program, including the Edison Electric Illuminating Company of Boston, the Boston Elevated Railway Company, and Stone and Webster. A few years later they were joined by AT&T, New York Telephone, the Bell Labs, Western Electric, and a number of railroad companies.[48]

While the Course VI-A cooperative program enhanced the institute's "serviceability" (Jackson's word) to the electrical industry, a similar program in chemical engineering did likewise with regard to the burgeoning chemical industry. This course in Chemical Engineering Practice, established at MIT with much fanfare in 1917, was the culmination of a decade of effort by chemical engineers to define their calling in terms of industrial and corporate needs.

"In the primitive stages of chemical manufacturing," Alfred H. White, head of the chemical-engineering department at the University of Michigan, recalled,

> the all-important aim was to produce the desired product, and efficiency and cost were secondary factors. As competition became more keen [however] . . . and greater emphasis was laid upon mechanical operation and more economical production, the demand arose for men

with special knowledge and the question of the proper training for such men aroused earnest discussion.[49]

The early industrial chemists, who had been trained in laboratory chemistry rather than in engineering or works management, recognized the inadequacy of their own education as preparation for industrial practice; they sought a new type of training, one which would more satisfactorily link the laboratory with the industrial plant.

In 1910, for example, C. F. Burgess, a prominent consultant and professor of chemical engineering at the University of Wisconsin, suggested to his colleagues some ways of enhancing the "efficiency of the college graduate in the chemical industry." Similarly reminiscent of the efforts of men like Charles Scott and Dugald Jackson to define the educational needs of the electrical industry was an article by J. H. James, chemical-engineering professor at Carnegie Tech, outlining the proper "chemical education for the industries" and stressing the need for practical and management training. M. C. Whitaker, who established the chemical-engineering course at Columbia in 1911, emphasized the importance of creating an atmosphere of manufacturing and business efficiency in the classroom wherein the students could learn the basics of plant organization and management in addition to the principles of physics and chemistry and applied mechanics. In response to these challenges, the AIChE at its first meeting formed a standing committee on chemical-engineering education to serve as a clearinghouse and advance guard for educational-reform activities. Under the dynamic leadership of Arthur D. Little, this committee conducted extensive educational surveys (at first independently and later in cooperation with the SPEE), formulated guidelines for curricula, and eventually issued the report which set the criteria for accreditation of all courses in chemical engineering.[50]

The real beginnings of chemical-engineering education in this country, however, and of modern chemical engineering itself, centered at MIT. The chemical-engineering department was established in 1888 by chemist Warren K. Lewis, and in 1904 cooperative industrial research was made a routine part of the department's activities by William Walker. The inspiration for the cooperative School of Chemical Engineering Practice came a decade later, from Walker's former partner, Arthur D. Little. As chairman of the visiting committee of the departments of chemistry and chemical engineering, Little proposed the creation of the new course, one which would integrate training in chemistry, engineering, and management, and would be based upon

Little's key concept, unit operations.* "Chemical engineering," Little wrote, "is not a composite of chemistry and mechanical and civil engineering, but is itself a branch of engineering, the basis of which is those unit operations which in their proper sequence and coordination constitute a chemical process as conducted on the industrial scale."[51]

Two years after Little's report to the MIT Corporation, the School of Engineering Practice was established, although because of the wartime activities of its staff (Walker was head of the Chemical Warfare Service) the course didn't really start until 1920. The brainchild of Little, it was funded by Kodak's George Eastman and directed by Walker (who had also become director of the Institute's new Division of Industrial Cooperation and Research). The course was open only to select graduate engineers and involved eight-week shifts at three field stations plus two semesters of specialized study at the institute. The plants at the field stations served as learning laboratories, and all instruction was grounded upon the concept of unit operations.[52] Throughout, theory was integrated with practice, the physics of the laboratory with the economics of production and management. Scientifically trained chemists were geared to function as engineers, to incorporate automatically the imperatives of the market (and their superiors) into their scientific and technical work, and to organize and manage the activities of corporate employees accordingly. "Chemical engineering has enabled the businessman and the investor to view the chemical reaction from the standpoint of the efficient and economical manufacture of a product and to accurately forecast the profits to be expected from the application of this chemical reaction," Charles M. A. Stine, Du Pont's chemical director, explained. "Chemistry views the

*Little's revolutionary concept of unit operations—the resolution of the chemical process into a coordinated series of unit actions (e.g., pulverizing, mixing, heating, etc.) upon which plant organization is based—laid the groundwork for streamlined mass production in the chemical industry. Little had, in effect, done for the chemical industry what F. W. Taylor had done for the machine industry. As Walker and his colleagues W. K. Lewis and W. H. McAdams explained in the first textbook on modern chemical engineering, "If the underlying principles upon which the rational design and operation of basic types of engineering equipment depend are understood, their successful adaptation to manufacturing processes becomes a matter of good management rather than of good fortune." Not surprisingly, then, Little emphasized that "to be successful and render his full measure of service, [the chemical engineer] must know men and be able to work with them effectively." W. H. Walker, W. K. Lewis, and W. H. McAdams, *Principles of Chemical Engineering* (New York: McGraw-Hill, 1923), Introduction; Arthur D. Little, "Chemical Engineering—What It Is and Is Not," *Transactions of the American Institute of Chemical Engineers,* XVII (1925), 172. For a discussion of unit operations, see W. A. Pardee and T. H. Chilton, "Industrial and Engineering Chemistry," *Journal of Industrial and Engineering Chemistry,* XLIII (1951), 295.

chemical reaction; chemical engineering views the pocketbook reaction."[53]

Unlike those at Cincinnati, the cooperative courses at MIT in electrical and chemical engineering were restricted to select engineering graduates, men who seemed capable of becoming the successors of the corporate engineers who created the programs. When Gerard Swope of GE and Frank Aydelotte later initiated the MIT Honors Plan for superior students, they recognized that the cooperative courses were already honors-type programs. The students enrolled in these courses, moreover, were very much aware of their distinction. "The social cohesiveness of the VI-A groups was phenomenal," Karl Wildes of the electrical engineering department remembered, "and probably unmatched in all MIT history." Here, if anywhere, was an important new breeding ground of America's corporate elite.[54]

Another of the earliest and most prominent cooperative plans of engineering education, one which combined aspects of the Cincinnati and MIT plans, was developed at the University of Pittsburgh in 1910 by Frederick L. Bishop, an influential leader of the SPEE. A boyhood friend of Calvin Coolidge, Bishop was an MIT-trained electrical engineer; like Frank Jewett, he had gone on to Chicago to earn his Ph.D. in physics with Michelson, Millikan, and Mann. Rather than entering industry as Jewett had done, however, Bishop embarked upon a teaching career, becoming a professor at the University of Pittsburgh, dean of the College of Engineering, and a dominant figure in American engineering education as national secretary of the SPEE from 1914 to 1947. As dean at Pittsburgh, Bishop's "first noteworthy work" was the introduction of the cooperative system with the Pittsburgh industries. The Pittsburgh plan, like that at MIT, was under the joint supervision of the college and the participating companies, but, like the Cincinnati plan, it required all students in civil, mechanical, electrical, chemical, railway, and sanitary engineering to enroll.[55]

"The adoption of a cooperative system," Bishop explained, "is a logical development of engineering education in the United States. When the engineering schools were first established they were really schools of pure science. Those proving unsatisfactory, as judged by the output, shops were taken into the schools and shop-work made a part of the curriculum." But this educational method also proved inadequate to meet industrial requirements, and the industries were thus forced to establish their own training programs, "an acknowledgement by these concerns that the output of our engineering schools [was] unsatisfactory." Bishop viewed the cooperative system as a necessary step further, a development which would enable educators to "break a man into his job" in the early stages of his education, to "make him play the game" as early as possible.[56]

The Pittsburgh cooperative system, which began regular operation in 1911, involved the participation of some fifty railroad, steel, petroleum, utility, and chemical and electrical manufacturing companies by 1916. From the industrial point of view, the cooperative system provided a valuable service. The supervisor of cooperative students at the Westinghouse Machine Company (the major participant in the program) noted that, in return for the practical training received, they "earn their keep" by producing for the company. Channing R. Dooley, Westinghouse's educational director, wrote Bishop that the system enabled his company "to try out students each year and thus ultimately obtain some excellent men for our organization."[57] The general manager of the National Electric Signaling Company told the membership of the SPEE that the cooperative plan "is a strictly business proposition" allowing the manufacturer both to advertise his product and to habituate students to his company.

> It puts him in touch with a number of young men who are just starting their training for lines of work in which he is interested as an employer. He is afforded an opportunity of looking over these men and picking out such of them as appear particularly good. He can then suggest certain lines of training that will fit them for his requirements, and he, in this way, has a chance of picking up trained men for his force without expense to his company.[58]

The Pittsburgh industries, in cooperation with the University of Pittsburgh, focused upon the production of industrially oriented teachers as well as students. William Wickenden, chairman of the NACT Committee on Relations with Collegiate Institutions, reported in 1922 that the local chapter of the organization in Pittsburgh "is considering the question of requiring college graduates to hold an actual job for a year or two before teaching." Many business executives, Wickenden noted, "feel quite strongly on this point," and the Westinghouse Electrical and Manufacturing Company was already conducting "a special course for college professors whereby they may obtain such training." In addition to teacher education, the local NACT chapter in Pittsburgh developed the "Pittsburgh Idea"—cooperative programs and vocational guidance within the local public schools—in cooperation with the Pittsburgh Board of Education and the national NACT Committee on Public School Relations, headed by C. S. Coler of Westinghouse. Pittsburgh was thus among the more "progressive" cities in adopting the new corporate-industrial approach to the problems of education.[59] "In dealing with these problems with a view to larger industrial efficiency," the Chancellor of the University of Pittsburgh declared,

> We need not fear that our young people will be commercialized and the quality of our citizenship impaired. . . . It is quite possible to direct our education so that it will result in the largest efficiency, and at the same time guard against any possible danger of lowering the standard and the high aim of all education. In so doing we shall, I am sure, ultimately arrive at a much more sensible, more rational, more effective education in our entire system of instruction. The continual interchange of opinion, of bringing together the various elements of our social, educational, and industrial units, cannot but result in great advantage, material and educational, to every part of our country. I fail utterly as a prophet unless, as a result of the Corporation Schools movement and the calling in of these other institutions to consult and cooperate, there will be finally a very much more rational system of instruction throughout the country.[60]

Cincinnati, MIT, and Pittsburgh were not alone in their pursuit of a "rational" and industrially responsive educational system. Parke Kolbe, who founded and directed the new Municipal University of Akron in 1913, promptly set up a cooperative plan for engineering students, a "second-shift plan," and a "night college plan," all "designed to be of the greatest benefit to the rubber industry and for training those who work in the rubber industry." Similar arrangements were established at other universities, among them Case in Cleveland, Drexel in Philadelphia, Union in Schenectady, Marquette in Milwaukee, Harvard in Cambridge, and NYU in New York City.*[61] Elsewhere, valuable services were rendered to the industries without the establishment of formal cooperative courses. "The engineering schools can serve the industries," MIT's Dugald Jackson observed, "by keeping more studious and accurate records of the tastes, ambitions, and abili-

---

*Although not a school of college grade, the Rochester Mechanics Institute served as something of a model of industry-school cooperation. Kodak and Bausch and Lomb, two leaders in the cooperative and corporation-school movement, farmed out their educational work to the institute. Thus, as one student of industrial education explained, "the Institute [was] not a school for individuals but for corporations." In 1921 the cooperative system was formally established at the institute. It is interesting to note that William Wickenden, who did so much to transform engineering education in the U.S., first taught at this institute. See Nathaniel Peffer, *Educational Experiments in Industry* (New York: The MacMillan Co., 1932), p. 176.

ties of their students and accomplishments of their alumni, thus establishing an expanding store of information regarding the engineering intellect of the nation—a record invaluable to our growing industries."[62] As early as 1902, a committee at Cornell's Sibley College—headed by an assistant professor of electrical engineering—began to systematize that school's personnel records "in response to increasing inquiries from companies for technical graduates." The Cornell Faculty Committee on Employment within the next decade adopted a card and file system of evaluating qualifications of students and alumni and was able to match Cornell personnel with industrial job specifications "promptly and systematically." Likewise, the associate dean of Dartmouth College reported that "the information on file concerning a student has been found of value to companies seeking college trained men to enter their employ." The president of Dartmouth at the time had been drafted from the New England Telephone Company, where he had been company representative to the NACS.[63]

Probably the most elaborate personnel and placement system in these early days was that developed by A. A. Potter when be became dean at Purdue.* Designed in cooperation with the Indiana Manufacturers' Association, it involved the accumulation of extensive information about students and alumni: school grades, intelligence-test scores, aptitude-test scores, career aspirations, teacher evaluations, practical experience, hobbies, employer references, and "character profiles." Good traits of character were the focus of Potter's system, and those students who rated low and were thus deemed "deficient" in such areas as loyalty, efficiency, and adaptability were counseled about ways of improving their standing. "The personnel records and interviews," Potter explained, "have been helpful in pointing out to a student his deficiencies and in impressing him with the fact that good qualities of personality can be developed."[64]

Potter was quite explicit as to what constituted a good personality; in his letters to employers of Purdue graduates, he explained that

> Purdue University is interested in developing not only the mentality but also the character, personality, and physique of its students.
> We are trying to train men to be efficient workers in the engineering field. We are anxious to prepare them for conditions as they are, so that they will waste as little time as possible in adjusting themselves to the

---

*It should be noted that Potter got the job at Purdue in part as a result of strong support by two influential Purdue alumni, Dooley of Westinghouse and Dietz of Western Electric. See the biography of Potter by Robert Eckles, *The Dean* (West Lafayette: Purdue University Press, 1973).

needs of industry. Our whole course is being constantly studied and revised with this point in view.

The personnel system has been of considerable value in discovering the students' talents. ... During the past two years the employers of engineering graduates have been making much use of the Purdue personnel system in selecting engineers to meet certain specifications. Letters have been received by Purdue ... from many of the most prominent employers of engineers commending the system in bringing the man and the job together.[65]

Potter's system was widely copied. It served as the basis of Westinghouse's personnel system, which Potter and C. R. Dooley developed together, and became a model for other universities as well. The Dean of Engineering at the University of Minnesota, for example, copied Potter's system, noting that large employers of graduates were greatly aided "by the information included in this qualification record, especially that which shows the appraisal of the students' character ... as distinguished from his scholastic record." And at the Bell System Educational Conference for 1926, R. I. Rees, AT&T's vice-president in charge of education addressed educators and industry men on "The Selection and Development of Personnel," explaining how he and Dean Potter had "talked over the possibility of Purdue furnishing men on specification." AT&T had provided Potter with job descriptions and an outline of required qualifications and "we gave him a quota for Purdue." "That was very sketchy, of course," Rees noted, "but the time was limited." Upon receipt of the specifications, however, Rees went on, "Dean Potter agreed to turn out twice as many men as the list called for."[66]

The cooperative-education movement, while it began in the engineering schools, was not restricted to them; the need for trained managerial manpower turned the eyes of commerce and industry to the liberal-arts colleges as well. The National City Bank of New York, under the leadership of Frank Vanderlip, led the way. "I know the majority of businessmen trained in the school of routine work," Vanderlip conceded, "will doubt the feasibility of teaching in the classroom, in a scientific and orderly fashion, those principles which they have gained only through years of experience." But, he argued,

The engineers of an earlier day thought that blue overalls and not a doctor's gown formed the proper dress for the neophyte in engineering. ... We have come long ago to recognize that the road to success as an engineer is through a technical school. So, too, I believe, we will in time come to recognize ... that the road to commercial leadership will be through the doors of those colleges and universities which have developed courses especially adapted to the requirements of commercial life.[67]

To this end, the bank established a business-fellowship plan with various schools, involving one year of employment at the bank during the four-year college career. The "training of men for the management of extensive affairs," National City Bank Vice-President W. S. Kies wrote President Henry Suzzalo of the University of Washington, must "be made more systematic and effective" in order to increase executive efficiency and thereby increase industrial efficiency.[68]

Cooperative courses were established for similar purposes in the liberal-arts schools of the universities of Akron and Cincinnati and NYU. The most ambitious scheme for the production of executive talent was that devised by Arthur E. Morgan, the country's leading flood-control engineer and later the first director of the TVA. In 1920 when he became president of Antioch, a small Ohio college founded by Horace Mann, Morgan launched the famous Antioch Plan of cooperative education. Including such features as on-campus industries and very close correlation of technical and liberal-arts training, the plan was designed for training "primarily for proprietorship and management, not for subordinate employment." More than anything, it was geared to provide the country's top executives:

> The central core of all vocational and technical training at Antioch is to be the preparation for carrying the ultimate responsibility. . . . There exists in the United States a highly developed and enormously valuable technic of administration. This technic includes the ability and habit of analysis of a job into the factors that count for its success, such as cost analysis, plant operation, financial management, personnel organization, buying, salesmanship, development of morale, production methods, and analysis of supply and demand; while underlying all is the habit and ability of exercising responsibility and of being the final authority on matters of policy. This technic in its modern, highly developed form is in the possession of very few men and women. It is the business of vocational courses at Antioch to be the medium for transmitting this technic to a select group of students.[69]

Morgan fully understood the implications of the popular tenet that "knowledge is power," recognizing that while knowledge alone did not bring power, it was indispensable to those who had power. Antioch was thus designed not simply to educate its students but to groom them for power, while providing them with the habits and knowledge necessary for the effective use of that power. Morgan adopted an elaborate personnel system for screening applicants and charting student progress, including, along with high academic performance, rigorous standards of character and attitude. He observed that "an educational institution has certain points of resemblance to a factory" and that, although "the academic type of educator may object to this comparison, . . . a study of the points of similarity might profitably be made by many colleges." The small college, he found, was particularly well suited for the production of a well-defined product—such as top executives—the demand for which cannot be met by the standardized mass production of the great university:

> The small college, like the small factory, must select an output that the larger institutions either have neglected or cannot deal with efficiently, and should fortify its position by selection of its materials in a manner which the wholesale methods of its large competitors have made impracticable.[70]

"Violins can be made most cheaply in large factories," Morgan went on to say, "but only the small shop could produce the Stradivarius." Since the rare, finely crafted products of Antioch were to be the nation's top administrators, Morgan selected trustees who would most appreciate the importance of such a product, top administrators. They included Frank Vanderlip, president of the National City Bank of New York; Charles Kettering, chief engineer at General Motors; William Mayo, chief engineer at Ford; George Verity, president of the Amercian Rolling Mill Company; Henry Dennison, president of the Dennison Manufacturing Company and a prominent leader in the management movement; and Edward F. Gay, former dean of the Harvard School of Business Administration and an old hand at tutoring the power elite.

The Antioch Plan, then, was designed especially to meet private industry's demand for responsible executives. At least one observer, an engineering professor from the University of Toronto, noticed the emphasis at the outset. "While the idea of training to carry ultimate responsibility is stated and restated," he wrote to the editor of the *Engineering News-Record,* Morgan "makes no specific mention of training for administrative public service . . .; it would appear from his illustrations and from the outline of the proposed practical work, that it is the industrial field that the college will primarily serve." Morgan responded that Antioch would indeed provide training for responsible public administration as well as for corporate management; he failed to mention, however, that the growth of the corporate industrial system involved the transformation of political life no less than education; that the process—by that time well under way—signaled, above all, the collapse of any rigid distinction between "public" and "corporate."*[71]

On the industrial side of the gap between industry and the colleges, local efforts to correlate education with the needs of industry were promoted and coordinated through trade associations, the engineering societies, and, most importantly, the NACS. In the colleges the major national vehicle was the Society for the Promotion of Engineering Education.

The SPEE was organized in Chicago in 1893 by the World Engineering Congress. During its first decade the society was dominated by engineering educators from the state schools of the Midwest, and concerned itself primarily with the problems of an expanding curriculum, the proper relationship between science and shop work, and academic tensions between the engineering educators and their liberal-arts colleagues. At this stage, education and industry remained relatively isolated from each other; as A. A. Potter recalled, "cooperation with industry in engineering education was little before 1910."[72] Thereafter, however, largely in response to the demands of electrical engineers in the colleges and the industry, such cooperation became the society's major mission.

The growing importance of the electrical industry signaled a corresponding increase in the influence of electrical engineers in the engineering profession and among engineering educators. The emergence of an electrical-engineering curriculum distinct from that of mechanical engineering was of major concern to engineering educators at the turn of the century, and representatives from the industry were the most active participants in the discussions. Between 1900 and 1904 over twenty important articles appeared in technical journals on the proper method of training electrical engineers, many of them written by people

*For further discussion of the blurring of the distinction between the public and private spheres, and the politicization of corporate capitalism in general, see Gabriel Kolko, *The Triumph of Conservatism* (New York: Free Press, 1963); James Weinstein, *The Corporate Ideal in the Liberal State* (Boston: Beacon Press, 1968); Ronald Radosh and Murray Rothbard, eds., *A New History of Leviathan* (New York: E. P. Dutton & Co., 1972); and Jerry Israel, ed., *Building the Organizational Society* (New York: Free Press, 1972).

in the industry.* The theme was always the same: how to correlate the education of engineers with the industrial specifications for technical workers and effective managers.[73]

Efforts along these lines within the SPEE came to a focus in 1906, when Dugald Jackson became president. In July of the following year, at the fifteenth annual meeting in Cleveland, Jackson introduced a resolution calling for the creation of a Joint Committee on Engineering Education.† The committee, Jackson argued, would facilitate cooperation between educators and "practicing engineers" and, through them, with the industries as well. During the next four years the committee gathered material for a comprehensive study of American engineering education. Because of the many responsibilities of these prominent engineers, however, meetings were infrequent and progress on the work was slow. In 1911 the SPEE and the three technical societies acknowledged that the task was too great and their resources too limited, so they turned to the Carnegie Foundation for the Advancement of Teaching—and to its director, Henry Pritchett—for assistance. Pritchett agreed to provide the necessary funds for the project and appointed a Carnegie-sponsored "impartial observer" to direct the actual work: Charles R. Mann of the University of Chicago. Thereafter, according to one historian of the SPEE, "the joint committee, . . . established by a resolution in 1907, . . . [became] no doubt the most important SPEE committee in terms of achievement."[74]

Pritchett chose Mann to head this first major study of engineering education not simply because Mann was a prominent physicist. Mann, who had studied in Berlin, was familiar with the German form of technical education which Pritchett so admired, and he was already involved in the promotion of "education for life" in the United States. Besides being president of the prestigious American Physical Society, Mann was chairman of the Central Association of Science and Mathe-

*At the suggestion of the president of the AIEE, Charles F. Scott of Westinghouse, a joint meeting of the SPEE and the AIEE was held in 1903. The major address was delivered by Loyall Osborne, chief engineer of Westinghouse, and was entitled "The Proper Qualifications of Electrical Engineering School Graduates from the Manufacturers' Standpoint." During the next few years similar papers were presented by Scott himself, Dooley (his associate at Pittsburgh), the young A. A. Potter, and Magnus Alexander of GE.

†As originally established, the Joint Committee was composed of four men: Frederick W. Taylor, the father of scientific management, representing the ASME; John Hays Hammond, general manager of the Guggenheim mining interests, representing the AIME; Charles F. Scott, representing the AIEE; and Dugald Jackson, representing the SPEE.

matics Teachers, and of the North Central Association of Colleges and Secondary Schools, and was the leader of a national movement among physics teachers to correlate the teaching of science with the realities of industry. "There is nothing inherently incompatible between industrial education and the discipline of pure science," Mann wrote. "In fact . . . they are identical." Mann was enthusiastic about the influential Douglass Commission Report on Industrial Education and the various cooperative experiments undertaken in the high schools of Cleveland, Cincinnati, Chicago, and Fitchburg, Massachusetts; in his own work he promoted educational methods which were "significant both to the pupils and to the communities that support the school, methods rendering the industrial study of science, or the scientific study of industry . . . an effective weapon of genuine educational discipline." Mann practiced what he preached: he began his study of engineering education with a study of the needs of corporate industry.[75]

"A few months ago I knew nothing about engineers or their education," Mann confessed to the members of the NACS, "so I tried first to find out what it was that the professional engineers wanted of the engineering schools." Mann found, however, that he "succeeded in getting more specific information from the records that are kept at the General Electric Company, and Westinghouse" than he could obtain from "interviews and talks with engineers." Thus, during his visits to Pittsburgh, Schenectady, Lynn, and the New York offices of AT&T he readily adopted the industrial approach to the problems of engineering education and formulated the means for solving them.

> The principal point derived from these records—a point that was quite a shock to me as a schoolman—was that the efficiency of the students who went to those firms from the engineering schools was measured and estimated in terms of initiative, tact, honesty, accuracy, industry, personality, and other qualities of this kind. No schoolman ever thought of rating the students in this way.
>
> The question was raised how you are going to make the professional, industrial point of view clear to the schoolman. And, there is only one way to do it, and that is for the industrial professional class to define clearly what it means. This can best be done by means of tests which must be successfully passed by applicants for positions. The answer to your question . . . "How are the industrial men going to help the schools to understand what industry wants and needs?" is the same as the answer to your question "What sort of tests are we going to use for vocational guidance?" If you will devise and put into practice as a condition of admission to each occupation tests which really test the ability of the applicant for that occupation, the school will rapidly

modify its instruction so that the pupils will be able to meet those tests, and you will make progress in vocational guidance.

The one point that I want to bring out clearly to you is that definite objective tests which define the type of ability which you wish to have developed are most valuable, not only to yourselves as employers in selecting your help, but also as your most powerful means of controlling what is done in the school.[76]

As head of the SPEE Joint Committee investigation, Mann outlined its purpose in similar terms: "To give the schools a more definite conception of their purposes and ideals, and an increased appreciation of the importance of a continuous scientific study of their own educational methods, and a more positive and objective method for measuring their own results." Here, more comprehensively and deliberately than ever before, the ideals of the corporation schools became those of the colleges. In his *Study of Engineering Education,* the report of the investigation, Mann traced the history of technical education, stated the demands of industry, outlined the existing methods of educating engineers, and proposed the means by which they might be "improved." He recommended psychological testing, admission testing, rigorous standards of student evaluation, business and administrative instruction for management responsibility, job classifications, personnel systems, cooperative education programs, and a centralized agency for college accreditation. Although the impact of the report was somewhat overshadowed by the war, the response to it was favorable. "The Report represents to a considerable extent a composite idea of the present standing of engineering education together with an outline of the probable future development," SPEE secretary Frederick Bishop observed. It indicated "the most probable form of development which engineering schools would have to pursue to meet the requirements of the industries," and this, in Bishop's estimate, was the "fundamental conclusion" of the report.[77]

Immediately after the war the SPEE set up a committee to evaluate Mann's report. This committee recommended that the schools, above all, should adopt "objective tests" to "measure their results." They promoted "the establishment and adoption of standards or tests to determine the mental growth of the students and to evaluate these in terms recognizable both by educators and employers" and further urged the adoption of management training in the schools. In addition, they strongly proposed that "the Society should cooperate with the engineering societies to establish job classifications so that engineering students would have a clear understanding of the work for which they were preparing and the instructors would have before them a constant reminder of the main purpose of their teaching work." In short, the fundamental recommendations of Mann's report, which constituted nothing less than the program of educational reform of the electrical industry, were fully and enthusiastically adopted by the engineering educators. "Beginning with the Mann Investigation and Report," Harry P. Hammond, assistant director of the Wickenden study, later observed, "the engineering schools of the country have been engaged actively in a continual process of self-scrutiny and study under the leadership of this Society. . . . In no other division of higher education has more effort been expended to bring scholastic work into line with the changing needs of . . . industry." The Mann report, published in 1918, appeared at the close of the society's greatest period of membership growth. Whereas in 1907 there were 375 members, by 1917 there were nearly 1500. An editorial in the SPEE *Bulletin* for 1912 explained the reason for the marked jump after 1907, the year of Dugald Jackson's presidency. "Although the number of teachers in SPEE membership is not increasing greatly . . . there is a great increase of non-teaching engineers and businessmen."[78]

As it did in the other spheres of industrial consolidation—standardization, patent reform, the organization of research—the war provided the corporate educational reformers in the industries and the colleges with what Frederick Bishop called a "unique opportunity,"[79] the chance to extend rapidly the range of their activities. In the months of 1917 and 1918 the men who dominated the NACS and the SPEE were able to secure control over the entire higher-education structure of the country, to coordinate it as never before, and to infuse it with the imperatives of corporate industry. In a recent assessment of American higher education, historian William Appleman Williams has suggested that the "corporate elite"

> have seen to it that experts are a glut on the market. Their strategy has been awesome: they have created, in the disguise of what most citizens consider a college education, a vast system of unimaginative vocational training paid for by the very parents who consider it an escalator to power for their children and the key to the general welfare. It has been a covert coup d'état of almost classic proportions.[80]

The wartime activities of the corporate engineers possibly contributed more to this "coup" than anything done before or since. It is a period that merits close attention.

Perhaps no one better expressed the perception of the war as a "unique opportunity" than Charles Mann. Writing about the probable effects of the war on engineering education, he observed that

> Progress has always been hampered by the vested rights of individuals and of corporations so that none has yet dared to envision an entire community as a single working plant for the purpose of organizing it for the most intelligent production of human wealth. This can now be done. The war is opening many hitherto blind eyes to see that each gains more than he loses when he merges his strength with the might of all in an organization that is constructed for the purpose of releasing creative energy by giving each the work he is best qualified to do.
>
> The time has come for such an organization in every community and every state, because the Federal Government is struggling to shape the nation into an organization of this type. The responsibility for this work must finally be shouldered by engineers who are both masters of the mechanic arts and moulders of men.[81]

The engineers were among the first to recognize their responsibility. The earliest attempts to bring the colleges into this "single working plant" were made by the corporate engineers on the Naval Consulting Board, the Council of National Defense, and the National Research Council. By late 1916 they had begun to establish the means for locating and coordinating the scientific resources of the country, and had devised a personnel index of available scientific manpower. In February of 1917 a further step in the same direction was taken by Dean William McClellan of the Wharton School of the University of Pennsylvania; in cooperation with CND Director Walter S. Gifford (later AT&T president), Surgeon General William Gorgas, and the Secretaries of War and Navy, McClellan organized the Intercollegiate Intelligence Bureau "to facilitate the ready placement of college men (particularly graduates) in the government service." Two months later, in April 1917, the NRC responded to military requests by setting up a Psychology Committee under Robert M. Yerkes. Composed of men such as Lewis Terman, E. L. Thorndike, D. E. Seashore, L. L. Thurstone, G. Stanley Hall, John B. Watson, and Walter Dill Scott, the NRC committee represented the vanguard in the fields of behavioral and applied psychology and was charged with the development of psychological tests for the selection and placement of recruits.[82]

Another development along similar lines was the creation of the Committee on the Classification of Personnel in the Army (CCP). During the summer of 1917 Walter Dill Scott and his associates from the Carnegie Institute of Technology's Bureau of Salesmanship Research had joined Major Grenville Clark of the Adjutant General's staff at the Army training camps at Fort Myer and Plattsburgh to try to adapt their personnel "rating systems" for use by the Army. These systems, developed by Scott and Thorndike, had been used with some success to classify personnel in such corporations as the United States Rubber Company, the Winchester Repeating Arms Company, the Metropolitan Life Insurance Company, and Cheney Brothers Silk Mills; these experiments had been supervised by the Bureau of Salesmanship Research, in cooperation with the NACS.[83]

On the basis of the Plattsburgh and Fort Myer experience, Scott was able to convince Secretary of War Newton D. Baker that his system could be used to rate men "according to their industrial abilities" and identify those with officer potential; thus, at the end of the summer Baker created the CCP to carry on the work. Composed primarily of businessmen, psychologists, industrial employment managers, and "others who had specialized on the subject of personnel in industry," the CCP well represented the NACS in the Army. Nominally divided into two groups—the CCP proper, under the Adjutant General's staff, and the Psychology Division of the Medical Department, under the Surgeon General—the CCP membership included, in addition to psychologists Scott, Terman, Watson, Yerkes, and Walter Bingham, "civilian supervisors" of the personnel work in the various Army camps. Among these were the educational and personnel directors of Western Electric, Westinghouse, GE, Thomas A. Edison, Inc., Southern Bell, Pennsylvania Railroad, International Harvester, Winchester Repeating Arms Company, and Dennison Manufacturing Company—in short, the leaders of the NACS. The CCP developed a comprehensive and detailed description of the training needs of the military, specifications for eighty-four trades (a new art at this time), the means for rating personnel qualifications based upon testing (in cooperation with the NRC Psychology Committee), and the apparatus for matching the right man with a given position. "Under war conditions," the historian of the CCP wrote in 1919, "men were received by the hundred thousand . . . ; the new system had to be like a great factory where each process is separated and volume production is assured through rigid functionalization and organization. Men had to be sorted, recorded, and assigned as goods in some great warehouse and received, checked, sorted, and shipped on order." After the war the CCP was entrusted with a large share of responsibility in demobilization and devised the plans "for the reinfiltration of the soldiers into industry."[84]

Although the Psychology Committee and the CCP constructed the testing procedures and personnel systems which formed the heart of the educational work of the war, the major center of educational activity was the Council of National Defense, and particularly Hollis Godfrey's Advisory Committee on Engineering and Education. Godfrey, it will be recalled, was an electrical engineer, trained at Tufts and MIT, a disciple and associate of Frederick W. Taylor, and president of Drexel Institute in Philadelphia. His concerns included the enhancement of the prestige and role of engineers, the extension of management into education, and the development of Drexel Institute into a "demonstration plant" for educational reform (analogous to the "demonstration plants" used by the Taylorites to display the achievements of scientific management). In addition, as a somewhat erratic egomaniac, Godfrey was preoccupied with the furtherance of his own career as a distinguished "man of affairs."*

As president of Drexel and chairman of the SPEE committee on institutional membership, Godfrey was in intimate contact with the engineering educators of the country and understood the problems they faced because of the war. Above all, Godfrey wanted to prevent a recurrence of the British and Canadian disaster of having the pool of highly skilled technical manpower seriously depleted through the draft and enlistment. He lobbied to keep all technical students in school where they could continue their training and thus enhance their ultimate usefulness for the war effort; to provide military instruction for students in the colleges; and to coordinate the activities of the colleges with those of the military and industry to "reach the point of highest efficiency" in guaranteeing a steady supply of trained men. Early in May 1917, therefore, Godfrey called together leading educators "to formulate a comprehensive policy for cooperation between the higher institutions and the government which will make the most effective use of these institutions." In taking this step, Godfrey started what Parke Kolbe of the University of Akron called "a movement which has been of inestimable value to higher education." Seen in a different light, it constituted the first stage of the wartime *coup d'état* in American education.[85]

---

*Capen, Mann, and Bishop were not alone in considering Godfrey a bit "queer" and "superficial"; they admired him, however, for his ability to manipulate men. Capen wrote that "the man who accomplishes good is the man who starts something, no matter what his motive and no matter how crude his methods. That's why the Billy's [Godfrey] get on in this world." See Capen's letters to Mrs. Capen, September 2, 1917, and September 11, 1917; and to John A. Cousens (president of Tufts), July 23, 1920, in the Capen Papers in the Archives of the State University of New York at Buffalo.

The gathering in Continental Hall in Washington, D.C., included representatives of 187 educational institutions. Secretary of War Baker and United States Commissioner of Education P. P. Claxton both emphasized the important role that the educational institutions had to play during the war, particularly as the source of technically trained men. In response, the assembled educators resolved that they should modify their curricula to fulfill the need for technical and vocational training, coordinate their efforts so as to provide the "efficient use of institutional plant, force, and equipment," urge students below draft age—especially those in engineering and other technical fields—to stay in school to complete their training, and provide military training for all able-bodied college men. In addition, they proposed that an agency be established to serve as a link between the United States Bureau of Education, the CND, and the States Relations Service of the Department of Agriculture, and that it alone should serve as "the medium of communication between the Federal Government and higher education."

The Education Section of Godfrey's CND Advisory Committee was thus established "to advise all colleges and universities" and receive "all communications upon all questions relating to the present war emergency." The committee was headed by Godfrey and Henry Crampton of the CND and composed of educators such as Henry Suzzalo of the University of Washington, C. S. Howe of Case Institute, Frank McVey of the University of Kentucky, Winthrop Stone of Purdue, Frederick Ferry of Williams College, and A. Lawrence Lowell of Harvard. To direct the activity of the committee, Godfrey called upon his good friend and former Tufts classmate Samuel P. Capen. In this new capacity Capen became a key figure in the transformation of higher education in America.[86]

The first specialist in Higher Education of the United States Bureau of Education, Capen was already a knowledgeable spokesman in the field. His views, moreover, did not differ substantially from those of the engineers, and for good reason: he was—in the eyes of the engineers, at least—one of them.* The son of the president of Tufts, Capen had followed in his father's footsteps, preparing at Tufts, Harvard, Leipzig,

---

*This is as good a place as any to point out the class prejudices of men like Capen. They clearly viewed themselves as members of a superior breed and held their "lessers" in contempt, an important point to keep in mind when assessing the significance of their educational-reform efforts. Capen's class-consciousness is well illustrated by an anecdote he related to his wife in the summer of 1907. Standing on the upper deck of an ocean liner, Capen had knocked the ashes out of his pipe on the railing, only to have them land

and the University of Pennsylvania for a career as professor of modern languages. His experience in his first teaching job, however—at Clark University in Worcester—gave his career a new direction. Clark had been created as a graduate school, but in 1902 financial problems had prompted the establishment of an undergraduate college as well. Capen arrived at the start of this undertaking, accompanied by a new president of Clark, Carroll D. Wright, and was thus able to play a major administrative role in the development of the new college. Capen was strongly influenced in his view of education by President Wright, the first U.S. Commissioner of Labor and a leading advocate of industrial education. In 1908 the young professor married the president's daughter.

After taking courses in education at the graduate school of Clark, Capen taught courses in educational administration and became a leader of educational reform in Worcester. As president of Worcester's Public Education Association, he promoted efficient public-school administration and the destruction of the "political" ward system, and in this capacity made contact with the chairman of the Massachusetts Commission on Education, Frederick P. Fish, former president of AT&T. One of Capen's major projects was the development of methods for supervising college teaching, which, up to that time had been either nonexistent or, in his words, "of an irregular and desultory nature."[88] In this work he cooperated with two other educational reformers, Samuel Earle of the Tufts Engineering Department and H. H. Norris, head of electrical engineering at Cornell's Sibley College (and Bishop's predecessor as national secretary of the SPEE). In 1912, at the invitation of Norris and another former Tufts classmate, G. C. Anthony (then vice-president of the SPEE), Capen attended his first SPEE meeting in Boston. It was a joint conference with the AIEE and, in addition to addresses by leading electrical-industry spokesmen such as Steinmetz, included visits to the Edison Electric Illuminating Company, the Boston Elevated Railway Company, GE at Lynn, and the electrical-engineering department of MIT.[89]

In 1914, as a result of his educational work in Worcester and Clark —and the contacts he had made through his father-in-law—Capen was appointed Specialist in Higher Education by Bureau of Education Commissioner Claxton. In this capacity he was responsible for produc-

---

on the head of a man in steerage. Enraged, the latter responded with a few well-chosen words in his "foreign tongue." "I can't reproduce the tone of his voice or the indignation of his gesture," Capen wrote his wife, "but I got considerable amusement out of the thought of having figured for a moment in a little symbolical picture of class distinction." See Capen to Mrs. Capen, June 1, 1907, Capen Papers.

ing reports on all movements and conditions in American college education, preparing university and college statistics, and making special investigations of particular higher-education problems. Between 1915 and 1919 he personally conducted extensive surveys of the educational facilities of a dozen states and in the process came into contact with the country's leading educators. Among those whom he most admired were Henry Suzzalo, Parke Kolbe, and Herman Schneider, all of whom became his close friends; in this capacity also he learned of the personnel work of A. A. Potter, "a man out in Kansas." Another of the country's top educators with whom Capen worked, on problems of admissions standards and the classification of colleges and universities, was Henry Pritchett of the Carnegie Foundation. Such working relations with Pritchett were no doubt at least in part the reality behind the charge of one critic, a college educator in North Dakota, that the "Bureau of Education is working in cahoots with the Rockefeller and Carnegie Foundations in a deliberate attempt to control state universities and dictate their policies."[90]

The college surveys which Capen conducted required the development of techniques of inquiry which would reveal the "vital facts concerning the policies, administration, finances, and educational effectiveness of the colleges." Complaining to his wife that Commissioner Claxton was not interested "in the administrative side of educational work, the things that can be weighed and measured," Capen resolved to correct this situation; "I am anxious," he told her, "that whatever is done directly by my division shall be as coldly scientific as I can make it." Capen secured the assistance of his friend Godfrey for this survey work (Godfrey as a management expert had done similar surveys for the city of Philadelphia) and, because what he was doing "seemed to represent a kind of efficiency engineering," was invited to become a member of the SPEE and was listed in *Who's Who in Engineering*. In 1914, as a new member representing the Bureau of Education, Capen addressed the council of the SPEE and met Bishop and Mann. This was a turning point in Capen's life. Mann in particular overwhelmed him; to Capen, Mann's study of engineering education was a "revolutionary report," and he wrote to his wife of Mann that "his is a master mind . . . ; I've never seen so fine a mind as his, and he has learned the art of persuasive presentation as few men have." Capen was thoroughly awed, even intimidated, by the engineers. He described the conference as "brilliant, the best educational meetings I have ever attended," and confessed that "I don't grade up to them. I know it and I fear they do."

He excitedly confided to his wife, however, that he was "proud to be their humble associate."[91]

In late May 1917 Capen became the executive secretary of Godfrey's Educational Section and its link with the United States Bureau of Education. The "humble associate" of Mann, Bishop, and their colleagues joined enthusiastically in their collective efforts "to handle the educational situation in the United States." "I am getting to know a lot of engineers as well as educators,"[92] Capen wrote home. The Educational Section had the responsibility of monitoring the developments in the government and the colleges and keeping each posted on the work of the other. Its most immediate problem, however—as Godfrey and the engineers saw it—was to coordinate the technical schools with the government and to protect them from serious depletion of personnel through the draft and enlistment. To handle these matters the section created a Committee on Higher Education, headed by Capen, and a subcommittee On the Relation of the Engineering Schools to the Government. The latter was composed of Capen, Mann, C. S. Howe of Case Institute, and Milo Ketchum, president of the SPEE, and was chaired by F. L. Bishop.

The engineering-education subcommittee was the most active of the agencies under Godfrey's control. Within weeks, after meetings with representatives of the NACS and NRC, the committee succeeded in securing statements from P. P. Claxton, Newton Baker, and President Wilson urging students to stay in school, especially those in engineering and other technical fields. Through the mediation of Dean Frederick Keppel of Columbia (secretary to Newton Baker and later Third Assistant Secretary of War) they were able to have established the Student Officers Reserve Corps, which allowed students in engineering, medicine, and agriculture to continue their studies as officer training. In addition, they proposed to Secretary Baker that engineering educators be commissioned in the Army and "assigned the task of coordinating the needs of the Army for technically trained men with existing educational facilities."[93]

The War Department was not prepared to take such a step, however. As Baker saw it, the most pressing matter facing the Army was the demand for lower grades of technical skill—electricians, carpenters, mechanics, radio technicians—rather than engineers. Already men were being trained for the Signal Reserve Corps under the direction of the CND Committee on Telegraphs and Telephones (representatives of AT&T, Western Union, and the Postal Telegraph Cable Company), and "plans were being worked out to utilize various educational institutions for this work."*[94]

Bishop's committee on engineering education saw in the military need for vocational training a "unique opportunity" to coordinate and utilize the colleges. It promptly charged that the Federal Board for Vocational Education, set up to administer the new Smith-Hughes Act,† was inadequate to the task, and strongly lobbied for the creation of a special War Department committee. Largely through the lobbying efforts of Mann and Keppel, the committee succeeded in creating the War Department Committee on Education and Special Training (CEST) in February 1918; the CEST formally took charge of American vocational and higher education during the war and essentially rendered the Education Section of Godfrey's committee obsolete.[95]

The actual CEST was composed of three Army officers: Grenville Clark, Robert I. Rees, and Hugh Johnson. Clark, later a distinguished lawyer, had been the officer in charge of the personnel-rating work of Scott and his associates at Fort Myer and Plattsburgh; Rees, an electrical engineer, joined AT&T after the war and eventually succeeded Wickenden as vice-president in charge of personnel; Johnson became head of the NRA during the New Deal. The active part of the CEST, however, was the Advisory Committee, composed of Mann, Capen, Herman Schneider, Raymond Pearson, president of the Iowa State College of Agriculture and Mechanic Arts, and J. W. Dietz, educational director of Western Electric and president of the NACS.

The CEST was divided into four general divisions. The Vocational Division, in charge of all lower-grade trade and technical training, was headed by Channing R. Dooley, educational director of Westinghouse; the War Issues Division, responsible for "correctly interpreting the

*In October 1917, on the occasion of the twenty-fifth anniversary of Drexel Institute, Godfrey demonstrated how this might be done. Using Drexel as his "demonstration plant," Godfrey addressed the celebrants on the theme "The Service of the College to the State": "We have heard so much concerning the mobilization of industry. This I have translated to mean the mobilization of civil powers in which education will be the great part. . . . Education and industry will finally come together. We now are of one idea and purpose, and that is to serve." Ignoring for the moment the opposition of what Capen called "an organized faculty cabal," Godfrey pointed out admiringly how the curriculum had been expanded along vocational and military lines. Other speakers for the occasion —Capen, Bishop, Suzzalo, and Vanderlip—agreed that the new feature was an important step in educational reform, that it would serve to meet not only the needs of war but those of industry after the war as well. Edward D. MacDonald and Edward M. Hinton, *Drexel Institute of Technology, 1891–1941* (Philadelphia: Drexel Institute, 1942), pp. 59–65. Capen to Mrs. Capen, July 15, 1918, Capen Papers.

†See page 310 below.

issues of the war," was conceived and directed by Frank Aydelotte of MIT.* The Division of Educational Standards and Tests was divided into two committees, both of which cooperated extensively with the CCP and the Psychology Division of the Medical Department and the NRC; the committee on classification of personnel was headed by A. C. Vinal of AT&T, and the committee on coordination and needs was chaired by Dietz, Schneider, and Rees. Finally, the Division of College Training was led by Richard Maclaurin, president of MIT. In addition, there was a publications section, headed by W. H. Timbie, director of the Cooperative Course in Electrical Engineering at MIT, and district-level organizations around the country. A. A. Potter, was district director of the educational institutions in fifteen states of the Midwest; other district directors included Henry Suzzalo and William Wickenden.[96]

In his announcement of the creation of the CEST, Secretary Baker explained that it was established "for the purpose of organizing and coordinating all of the educational resources of the country with relation to the needs of the Army, . . . to represent the War Department in its relations with the educational institutions of the country and to develop and standardize policies as between the schools and colleges and the War Department."[97] In reality, the educational work of the military during the war was placed in the hands of the educational directors of AT&T, Western Electric, and Westinghouse and the leading advocates of the corporate reform of engineering education.

The early work of the CEST included promotion of the war-issues course in the colleges and the sponsorship, by Aydelotte, Capen, and Ralph Barton Perry, of national war-issues educational rallies to arouse support for the war effort. The first major project was the vocational-training program under the direction of Dooley's division and the Division of Educational Standards and Tests. During the summer of 1918 the Vocational Division trained 38,000 draftees in twenty basic trades. The work was based upon the comprehensive job specifications prepared by the CCP and was carried out at 140 National Army Training Detachments, educational institutions selected on the basis of Capen's surveys of educational facilities. The division worked closely with the CCP and the Psychology Divisions of the Medical Department and the NRC (in particular with Walter Dill Scott, L. L. Thurstone, and Lewis Terman) in the use of intelligence and aptitude tests and

*"Professor Aydelotte of MIT is the father and director of the war aims course." Samuel P. Capen, letter to George Zook, March 16, 1920, American Council on Education Archives.

Scott's rating system. Under Vinal's direction, they developed the most refined system yet available "for the distribution of trained material [men] according to qualifications so as to meet the requirements of different branches of service" and introduced psychological testing procedures and personnel-classification techniques into the colleges. In addition they devised and published, under Timbie's supervision, unprecedented short courses for intensive training in technical, vocational, and military subjects.[98]

The Committee on Coordination and Needs served as a clearinghouse in all of this work, obtaining and furnishing information on "army needs in terms of jobs, number and time," "the supply available in the educational institutions in terms of colleges, courses, ages, and classes," and "the success or failure of men supplied." Besides performing a great industrial service by introducing the most advanced personnel methods into the colleges, the CEST thus also provided the industries with a valuable pool of information from which to draw at the war's end. In January 1919, for example, Dooley furnished the National Radio Institute with CEST data on the men who had taken training courses as radio operators or radio electricians. The most important contribution of the Vocational Division of the CEST to the industries, however, was the large-scale experimentation which it provided for the development of new and better methods of personnel management. Indeed, the bugles had hardly ceased to sound when Dooley, acting under the auspices of the United States Bureau of Education, called a conference "to consider the permanent effect on industrial education of the plans and methods which were developed under the stress of the war emergency, . . . to formulate plans for utilizing the best of the experience thus gained."*[99]

Despite the impressive accomplishments of the other divisions of the CEST, none was more significant for the future of American higher

*In 1940 Dietz and Dooley again put their experience to use for the military. Together with Michael Kane of GE and William Conover of United States Steel they set up and ran the emergency training program during World War II. They trained some two million supervisory personnel in job instruction methods and developed the formalized scheme known as Training Within Industry (TWI). After the war they created the TWI Foundation, headed by Dooley, which was responsible for the widespread adoption of TWI techniques, as a means of increasing productivity, throughout industry in the U.S. and western Europe. Einar Wilhelm Sissener, "The Training Within Industry Movement and its Spread to Western European Countries," MS thesis, 1954, School of Industrial Management, MIT, MIT Archives. See also Fred Tickner, *Training in Modern Society* (Albany: Public Affairs Monograph Series, State University of New York at Albany, 1966), p. 112.

education than the work of the College Division. Essentially, the work focused upon the problem of supplying officers for the war effort through the training and recruitment of college students. In the summer of 1918 Dooley devised a plan which became known as the Student Army Training Corps (SATC).* Designed to prevent an unnecessary depletion of the colleges, through either indiscriminate volunteering or the draft, it offered students a "definite and immediate military status" while still in school, and otherwise aimed at developing into a "great military asset the large body of men in the colleges." Since the draft age was twenty-one, students between eighteen and twenty-one would be "encouraged to enlist" in the SATC program in their school; they would thus gain military status and, at the same time, come under military authority—the authority of the CEST.

Throughout the summer of 1918 Mann lobbied for the plan, talking repeatedly with Baker and more than once with President Wilson. When Maclaurin was appointed head of the College Section after some delay, Capen wrote to his wife that "the die is cast." He reported that Maclaurin, upon hearing of the SATC plan, exclaimed, "Why, this is a Department of Education you are building and more . . . ; it leads anywhere. It has the biggest kind of significance." Capen agreed. He and Mann went to the Navy "to try to get it to play with us too," and secured the support of the NRC as well. "The scientists see what we are up to in the War Department CEST without diagrams," Capen wrote. "Their enthusiastic approval and desire to play our game are doubly consoling." John Merriam of the NRC confided to Capen that "this is the most portentous thing for universities that has ever happened." While Mann, Maclaurin, and Godfrey—who, according to Capen, stood "ace-high with Woodrow"—lobbied for the SATC plan, Capen set about "to devise and launch a campaign to push our plan with the country." To do this he had to convince the college educators themselves.[101]

On their own, the educators were already organizing for the war, prompted largely by the May conferences held by Godfrey in Washington. In January 1918 Henry Pratt Judson, president of the University of Chicago and the person who originally pushed Mann into the physics-teachers movement, called a meeting of the executive committee of the nation's principal college associations. The assembled educators

---

*"Mr. C. R. Dooley, educational director of the Standard Oil Company of New York, was really the father and director of the educational work of the SATC." Capen, letter to George Zook, March 16, 1920, American Council on Education Archives.

asked Woodrow Wilson "to take steps looking toward the immediate comprehensive mobilization of the educational forces of the nation for war purposes under centralized administration, which would coordinate effort and stimulate defensive activities," and later that month met in Washington with Capen, Claxton, and Godfrey to coordinate their efforts with the CND, the United States Bureau of Education, and the War Department. Capen took the lead quickly, as chairman of an organizing committee which also included Bishop, Mann, and Rees, and recommended the creation of an Emergency Council on Education to centralize college activities. "The object of the Council," Capen's committee proposed, would be

> "to place the resources of the educational institutions of our country more completely at the disposal of the national government and its departments to the end that through an understanding cooperation: The patriotic services of the public schools, professional schools, the colleges and universities may be augmented; a continuous supply of educated men may be maintained; and greater effectiveness in meeting the educational problems arising during and following the war may be secured."

In Capen's view, the ECE would provide the means through which he and the CEST could "push our plan [the SATC] with the country."[102]

A recent history of the role of the colleges in World War I suggests that the country's educators readily allowed and even promoted the militarization of the campus. Carol Gruber has written that

> the absence of a *principled* objection to the militarization of the campus, particularly on the part of the professors, is startling. One looks in vain for organized appraisals by American academics of the implications of turning the colleges and universities into military training camps; one is struck instead by the alacrity with which they embraced the concept of an "essential industry" to describe the role of the university in war time. It is true that academics typically denounced the SATC as a failure; but ... they complained about the malfunctioning of the program and not about its purpose.[103]

Capen's experience with the ECE, as reflected in his letters to his wife, bears out this evaluation, and adds another dimension: the educators were clearly awed and intimidated by the engineers who conceived the SATC, much as Capen had initially been awed and intimidated by Mann. If they had "principled objections," they kept them to themselves for fear of appearing unpatriotic or out of the suspicion that they did not "grade up" to the forceful engineers in either mental or practical ability. With the authority of the Army behind them, the CEST leaders certainly displayed impressive power, much more than the academics could counter. Capen thus described the ECE as "the most docile organization we have yet encountered. It does just as it is told, no matter how disappointed it may be," and a short time later he boasted that "Mann and I succeeded in convincing the Council bunch to play our game, to campaign for college students not for the sake of the colleges but as a military measure initiated by the Army.... They saw the point and agreed to eat out of our hand again."[104] One would never guess that Capen was referring to the representatives of the twelve largest educational organizations in the country.*

The last stumbling block to the creation of the SATC was the United States Bureau of Education, and in particular its director. Commissioner Claxton, wrote Capen, "could not perceive its [the SATC's] significance.... He's still living in 1917—much water has passed under the bridge." Capen plainly saw himself in the midst of an educational revolution and in his tremendous excitement, his criticism of his slow-moving superior at the Bureau turned to outrage. "He's an irresponsible freak," Capen exclaimed; "he makes things unbearable; he's still about a year behind the rest of us. He is still talking in terms of surveys and keeping the schools going on exactly the old basis. He still thinks that Congress may provide specialists, and more specialists, to study interesting phases of American education—when American education is changing daily under his feet." Unable to educate Claxton as to the significance of the SATC, Capen decided to just leave him out of it entirely. "The chief consolation," he wrote his wife, "is that he can do no harm and the machine goes forward in spite of him."[105]

Once the SATC had been approved, Capen reported happily that "the Institutions are all falling in line and the majority of them are hailing us as the saviours both of education and of the trained personnel of the country." "With all due modesty," he confided to his wife, "I think their hail is correct."[106] With the creation of the SATC in the

---

*American Association for the Advancement of Science, American Association of University Professors, Association of American Colleges, Association of American Universities, Association of Urban Universities, Association of American Agricultural Colleges and Experiment Stations, American Association of Land-Grant Colleges and State Universities, Association of American Medical Colleges, Catholic Educational Association, National Association of State Universities, National Council of Normal School Presidents and Principals, and National Education Association.

summer of 1918, the Surgeon General, the Chief of Engineers, the Chief
Ordnance Officer, the Chief Signal Officer, and the Quartermaster General all turned over to the CEST, in Capen's words,

> the destinies of all men of draft age connected with schools or colleges
> as students or teachers who are or who might be candidates for the
> Enlisted Reserve Corps in any of these five departments. In other words
> we now have full control of the college situation in its relation to the
> draft. We can say which teachers and which students shall stay and
> which shall go . . .; through the SATC and the Enlisted Reserve Corps
> we can keep the men in college that ought to stay.[107]

The SATC was certainly an impressive achievement. By becoming in
fact the military staff charged with overseeing the college situation, the
engineering educators and corporate personnel officers had dramatically realized the original aims of Godfrey's first committee.

With their voluntary SATC system in full operation, the members
of the CEST watched with pride and excitement as the colleges became
military cantonments and the students became soldiers. Capen's letters
to his wife during this period reflect the unreality of the war which these
men fought and the strange lighthearted spirit in which they perceived
its horrors. "Bishop took me in hand at once on my arrival in Pittsburgh," Capen wrote cheerfully, "and showed me the soldier boys and
their quarters" at the University of Pittsburgh. "He is a wonder at a
thing of this kind, and still he dreams it bigger and bigger." A month
later Capen attended a regional SATC conference in San Francisco.
"The last two days I've been living at the Presidio in the officers'
quarters—great fun," he wrote his wife.

> They staged a drill, field manoevers [sic], and a review of the Officers
> Training Corps for us. They were bully, especially the field manoevers. . . . They had bayonet practice, which I hadn't seen before—right
> gruesome—and entrenching and going over the top and all the rest. It
> is a beautiful place, right on the Golden Gate.[108]

In August 1918 the CEST warriors were shaken out of their reverie;
due to unexpected losses overseas, the draft age was lowered to eighteen. The new situation posed a serious threat to the voluntary SATC.
Somehow the CEST had to figure out a way of saving and training
officer material in the colleges without appearing to favor the college
student over the noncollege man of the same age, both of whom were
now liable to the draft. The CEST innovators solved their problem with
a new plan, their boldest yet and one which was designed to place the

colleges literally in their hands: a compulsory SATC. Under the original SATC then in effect, the colleges remained independent of the War Department; only the actual SATC program, which was voluntary—although students were "strongly encouraged" to join—was under military authority. Again through the lobbying efforts of Mann, Maclaurin, and Keppel, the compulsory plan was pushed through to meet the new war emergency. Its stated object was "to utilize effectively the plant, equipment, and organization of the colleges for selecting and training officer-candidates [as] technical experts for service in the existing emergency." As Kolbe later described it, the new plan "amounted to the Government taking over the plants as officer training schools, to be carried on by the War Department." Since the CEST was the War Department agency for educational matters, it now actually took charge of all colleges of liberal arts, technology, business, agriculture, medicine, law, pharmacy, dentistry, veterinary medicine, all graduate schools, and all technical institutes in the United States. American education was placed under the military authority of the educational and personnel directors of corporate industry and the leaders of the new corporate brand of engineering education.[109]

The opportunity thus provided exceeded even the wildest fantasies of the educational reformers. Kolbe exclaimed delightedly that the SATC would allow them to break down "in a stroke" the academic barriers that had retarded their efforts for years, to "discard outworn or outgrown practices," reform curricula, and introduce "modern" personnel methods and testing procedures. Walter Dill Scott, who was charged with the "proper placement of the SATC product," called it the "most gigantic thing that has been imagined in the course of war operations." Perhaps no one was more enthused by the new prospects for reform than Frederick Bishop, the would-be "general" of the University of Pittsburgh and the national secretary of the SPEE. In the September SPEE *Bulletin* he proclaimed the new plan in capital letters: "A UNIQUE OPPORTUNITY."[110]

> To all intents and purposes the War Department through [CEST] takes charge of colleges on October 1, when all physically-fit male students eighteen years of age or over will be members of the [SATC].
> This means that the old courses of study are practically abolished and there are substituted short intensive courses which will fit men for specific duties with the Army.
> The change which this involves provides the engineering teacher with the greatest opportunity ever presented to teachers. Modifications in the curriculum, changes in methods of teaching, introduction of new

material can now be made at once while in times of peace it would require years of agitation to accomplish even a small fraction of these changes.

The teachers of engineering who have been in contact with the [CEST Vocational Division training centers] . . . have learned many things of value in training men for specific industries. The new course now being established by the War Department should be used by the teachers of engineering as a laboratory in which ideas can be worked out [along lines suggested by the Mann Report] . . . to meet the needs of industry for thoroughly trained men in normal times.

The [SPEE] is an organization with a unique opportunity to coordinate the best which can be developed and prepare the way for the adoption of these in the great reconstruction period which is to follow the war.[111]

On October 1, 1918, the new SATC was formally inaugurated at five hundred colleges throughout the country, and American higher education came officially under military command. As a display of the coordination they had thereby achieved, the CEST ordered that the ceremonies on all campuses be identical and take place simultaneously. "It is most fitting," R. I. Rees directed, "that this day, which will be remembered in American history, should be observed in a manner appropriate to its significance." Thus, at precisely noon Eastern Standard Time, on every campus in the land the national anthem was played, the pledge to the flag was made, a telegram from President Wilson was read, and the General Orders for the Day were issued. "At this moment," the telegram read, "over 150,000 of your comrades throughout the nation are standing at attention in recognition of their new duties as soldiers in the United States Army." President Kolbe of the University of Akron later recalled that "it was an impressive occasion, and one of deep moment for the higher educational system of the United States."[112]

Under the new plan the entire operations of the CEST were reorganized into two sections: Section A absorbed all work pertaining to the colleges and Section B absorbed all of that pertaining to vocational training, including the national training centers. Within five weeks under the SATC program, 527 colleges were "organized," 158,000 new "soldiers" were "processed," and over 15,000 new recruits were selected for officer training. Through the new compulsory program, the CEST was able to extend the use of psychological tests (notably the famous alpha and beta intelligence tests) and personnel systems in the colleges, and to break down existing barriers to the creation of the "efficient educational processing mechanism" required by corporate industry.[113]

The compulsory SATC provided the corporate reformers with an unprecedented chance to realize the objectives of their coup. That chance, however, was short-lived; the Armistice of November 11 caught the members of the CEST, like everyone else, by surprise. Most Americans, however, did not share their disappointment. The CEST final report expressed "regret that there was not more time to perfect the organization of the Corps and to overcome the obstacles that remained. This, it is believed, could have been entirely accomplished within another sixty days. . . . It is not felt that a final and conclusive experiment in the combination of military and academic training in the colleges has been made."[114] The war had allowed the corporate educational reformers to promote their own industrial objectives in the guise of military expediency; with the war over, and their military authority and "cover" gone, they were forced to carry on their "experiment" under different auspices. However, since they had been consciously preparing for it ever since the first days of the preparedness campaign, they wasted little time in making the shift.

Technology as People

The Industrial Process of
Higher Education—II

The American system of schools has a sanction in public efficiency as well as in equality of personal opportunity. It is a special system of getting brains for the public purpose. [University educators] have an immediate responsibility to make the prospect more effective. . . . Soon we must become as wise in pedagogical method as we have long been in scientific method. The processing of human beings through intellectual experiences is far more important socially than the processing of material things. Yet physical technology holds a place of respectability among us which human technology has not yet won.[1]

—Henry Suzzalo

The shift from war to peace no more disrupted the continuity of corporate educational-reform activities than had the shift from peace to war in 1916–17. If anything, the war had simply provided new vehicles for such continuity. Although, as Charles Mann reported, "the results achieved exceeded the fondest hopes of the committee [CEST]," its members realized that, in effect, their work had just begun. In carrying forth the activities begun in the corporation schools and the engineering colleges, the NACS, the SPEE, and, most recently, the CEST, the CCP in the Army, and the various psychology committees, the reformers proceeded under the auspices of four major national agencies: the United States Army, the permanent National Research Council, the permanent Emergency Council on Education—renamed the American Council on Education (ACE)—and the SPEE. The war had shown them the great potential for educational-reform work under "the influence of military training and discipline." Capen was not alone in noticing the "greater efficiency of a student body subject to a military regime" in terms of "physical fitness," the "development of courtesy," and the "spirit of service and self-sacrifice." Working in the postwar period through these new agencies, the reformers looked for ways "to retain these tangible advantages of the period of war training . . . the best fruits of the war."[2]

The Defense Act of 1916, which had allowed for the creation of the CEST, also enabled its members to retain their military status and authority in peacetime. In effect, it authorized the creation of an Army educational establishment along the lines drawn by the Morrill Act of 1861. One of the first effects of the act was the creation "in civil educational institutions" of a Reserve Officers Training Corps (ROTC).* During the war the ROTC units were merged with the Officers Reserve Corps, which had been established at the insistence of Bishop's subcommittee on engineering education to save engineering students under twenty-one from the draft and enlistment. Immediately after the Armistice, however, with the demobilization of the SATC the CEST adopted as its main function the revitalization of the ROTC. Early in December an ROTC Branch was created under the War Plans Division of the Army General Staff and placed under the direction of the CEST. "The administration of the ROTC has been committed to the CEST," the final report of the committee reads, "which thus continues as a living force in our national educational organization." In August and September 1919 the names changed, but the faces remained the same. The CEST was officially dissolved, but a new Education and Recreation Branch of the War Plans Division was created. The head of the branch, which would oversee all educational work of the Army, including ROTC, was Colonel R. I. Rees; the Civilian Advisory Board of the branch was chaired by Mann and included Samuel Capen, now head of the ACE, James Angell of Yale and the NRC, and Frederick P. Keppel, now Director of Foreign Relations of the American Red Cross and soon to become the new president of the Carnegie Corporation. The new E&R Branch would direct the military's educational activities throughout the 1920's.[3]

The enrollment in ROTC grew from 35,000 in 1916 to 120,000 in 1925, despite cutbacks in appropriations, and the program served to

---

*It is interesting to note that ROTC actually got its start at MIT. It was there that a professor of military science, Major Edwin T. Cole, initiated the plan in 1911. He thereafter urged President Maclaurin to bring it to the attention of the War Department and, after several discussions with the General Staff, saw his innovation adopted on a nationwide scale. See Samuel C. Prescott, *When MIT Was "Boston Tech"* (Cambridge: MIT Press, 1954), p. 295.

maintain a military presence on college campuses and thereby extend the benefits of "military discipline" beyond the active Army. Equally important from an industrial standpoint, the ROTC units on the campuses provided technical education for potential officers at government expense, in addition to purely military instruction. But the ROTC itself actually played a relatively minor role in the activities of the corporate educational reformers within the military. The E&R Branch was in effect the vanguard of the new "citizen army" called for in the amended Defense Act of 1920 and through its educational activities and industrial associations, as well as its ROTC program, it pushed for the extension of military discipline, and the "moral vigor" it generated, into all areas of national life. The notion of a "citizen army" greatly expanded the scope of military activities which were now aimed at the preparation of the entire citizenry for possible military service: the new military creed, which identified training for industry with military training, coincided nicely with the corporate need for an "industrial army" of properly adjusted and assembled "economic units."

The end of the war did not signal the end of "universal military training"; it meant only that such training would now be carried on outside of the military proper—out of uniform, as it were—in the country's various educational institutions. The E&R Branch was divided along the lines of its predecessor, the CEST, with units for vocational and technical training, general education, and testing. Under the heading of general education, the branch sought to promote physical fitness as well as "attitudes and dispositions toward loyalty and comprehension of American customs and ideals." Branch directors asked organizations like the National Amateur Athletic Federation, Boy Scouts, YMCA, and Camp Fire Girls to "undertake the job of defining the physical standards that young people should be able to measure up to at various stages," and worked with the Federal Council of Citizenship Training* to promote the proper "education for citizenship."[4]

The primary activity of the E&R Branch, however, centered on the committee on vocational and technical training; headed by Charles R. Mann, this unit had been set up expressly to carry on the personnel work begun by the CCP, the CEST, and the various psychology depart-

*Composed of the U.S. Bureau of Education, Federal Board for Vocational Education, Veterans' Bureau, Naturalization Bureau of the Department of Labor, and various agencies dealing with the problem of "Americanization."

ments during the war. "How to select, how to train, and how to assign men to the right jobs—these," wrote Mann, "are the basic problems of effective use of manpower, both in the military establishments and in civil life." The most pressing task of the E&R Branch was "to adapt practices that had proved effective for military mobilization to better organization of manpower for creative work," to consolidate "the gains made during the war by developing unified, reliable, and speedy methods of selection, assignment and training." At public expense, therefore, the E&R Branch sponsored research on the various aspects of the industrial personnel problem: job descriptions and specifications; techniques for the evaluation of human capabilities; and systems for rating, sorting, and classifying human material according to the occupational specifications. By the early 1920s, the branch had drawn up detailed specifications—so-called "job sheets" based upon analysis of "unit operations"—for 117 skilled occupations, thereby extending the "occupational index" prepared by the CCP during the war. It published twenty-three vocational manuals for short-course instruction patterned after the work of William Timbie during the war; it conducted, during the summer of 1920, intensive training in vocational guidance for officers and civilians in industry and education, in cooperation with the NACT, NRC, and ACE; and it tried "experiments" in testing "to perfect the methods used during the war and to coordinate progress in this field within and without the Army."[5]

The "experimental center" in which the E&R Branch carried on its early testing work was Camp Grant in Illinois, the so-called E&R Special School. With the passage of the amendment to the Defense Act, in June 1920, this one school was replaced by four permanent Special Service Schools (at Forts Hunt, Humphries, Holabird, and Vail) and the War Plans Division became G-3, Operations and Training; the Advisory Board, however, remained intact. The activities in the special schools, aside from recreation, included the experimental application—under the controlled conditions of military discipline—of the Army alpha and beta tests developed by Robert M. Yerkes' NRC committee during the war; this testing served as the basis for determining job specifications and complementary personnel training and classification. In June 1922, after further cutbacks in military appropriations, this experimental work was restricted to Camp Vail, the Signal School named after the President of AT&T. This school was selected because of its proximity to New York "and the headquarters of the great telegraph, telephone, and radio companies, . . . making possible close cooperation between the Army and civilian industry" in personnel-research matters.*

In addition to joint research projects, the E&R Branch cooperated with AT&T in the actual industrial application of the experimental personnel procedures developed at Camp Vail. The results of the first of these projects, at the Chesapeake and Potomac Company in Washington, prompted Mann to predict that "it will be possible . . . in two or three years . . . to establish in this field a standard terminology and training methods" which would both prepare men in industry for potential military service and train men in the military for industrial employment.[6] The E&R Branch of the Army thus played an important role in the development of personnel and training techniques for industry; in cooperation with the NRC, ACE, Engineering Foundation, and American Management Association (formerly the NACS), it later helped to establish the National Board of Personnel Classification and the Personnel Research Foundation, which became clearinghouses for such work.

The Army was only one agency which kept CEST activities alive in peacetime. The executive order authorizing the establishment of a permanent NRC provided another vehicle for their continuation. Within the council, the Psychology Committee under Yerkes served as the link with the Army. In his final report of the Psychology Committee at the close of the war Yerkes outlined the wartime achievements in testing and personnel classification which already had profoundly advanced "the status of the relations of psychology as a science and as a technology," and indicated that the results of this work would soon be published "in the interest of the Army and of other government agencies, as well as of education and industry."[†]

It is already evident that contributions to methods of practical mental measurement made by this committee of the NRC, and by the psychological personnel of the Army, are profoundly influencing not only psychologists, but educators, masters of industry and experts in diverse professions. The service of psychological examining in the Army has conspicuously advanced mental engineering, and has assured the im-

*It is hardly surprising, therefore, that experimental results coincided nicely with the requirements of the electrical and telephone industries; among the alpha-score/occupation correlations determined by the E&R Branch were: 103–120, electrical operating engineer; 87–102, power electrician, chief radio operator; 72–86, instrument repairman; 57–71, telephone installer; 46–56, telegraph lineman.
†For discussion of the impact of the war experience upon industrial personnel procedures, see Loren Baritz, *The Servants of Power: A History of the Use of Social Science in American Industry* (Middletown: Wesleyan University Press, 1960).

mediate application of methods of mental rating to the problems of classification and assignment in our educational institutions and our industries.[7]

As early as June 1919 the peacetime NRC began the task of coordinating the various industrial and educational applications of "fruits of the war." The Engineering Foundation, which was supporting the council financially, proposed that the NRC meet the need "for concurrent, coordinated research in industrial medicine, in psychology, in management, and in engineering ... a broad plan of investigation covering the whole question of personnel in industry—intellectual, moral, physical, psychological." After preliminary discussion among the various divisions of the NRC, a Committee on Industrial Personnel Research was created to develop the proposal; the committee, which was directed by Yerkes (now chairman of the NRC Anthropology and Psychology Division) and Capen (now head of the NRC Educational Relations Division), posed what it saw to be the critical areas for study and experimentation.[8] Among the former were listed the "causes of labor unrest and resultant excessive high turnover, low production, and high costs"; the "physiological and pathological aspects of labor problems relative to health, efficiency, and productiveness in industry"; and the ongoing "analysis, classification, and specification of industrial employments." It proposed experimentation with work-incentives schemes and the shortened work week, and sought ways to enlighten "the directive element in industry ... as to the results of scientific selection, assignment, and promotion of employees." Most importantly, it hoped to develop "methods for overcoming [worker] misapprehensions and prejudices" and "countering in the school sinister industrial tendencies and fallacies" while at the same time coming to grips with "the psychology and psychiatry of trouble-makers." Thus, at the outset, the NRC committee comprehended the whole range of capitalist social production in the new era, the exigencies of the postwar corporate order. They also began to ask the question which challenged many of management's prewar assumptions and opened the way for a consumption-based economy: "Is the only possible way of continued industrial prosperity and progress to be found in increased production?"[9]

In a series of conferences sponsored by the Commonwealth Fund between August and December 1919, the NRC committee discussed these problems with representatives from various organizations: the National Association of Employment Managers, which had been set up in Rochester, New York, during the war (represented by J. C. Bower of Westinghouse and Mark Jones of Thomas A. Edison, Inc.); the National Industrial Conference Board (represented by its director, Magnus W. Alexander); the NACS (represented by F. C. Henderschott, of New York Edison, and J. W. Dietz, now with Standard Oil of New York); and the United States Chamber of Commerce. While all present agreed that the American Federation of Labor should be represented in any central personnel agency, they decided that it would be "wiser" to defer their invitation to Samuel Gompers until they had "given their program preliminary formulation."* Taking advantage of the absence of a labor representative, Mark Jones suggested that the new organization should consider the question of "conversational hygiene." "Conversation among the rank and file of workers," Jones complained, "is too often of a filthy and vulgar nature, and the state of mind which it encourages is that which ultimately becomes a most fertile field for all morale-destroying influence."[10]

The outcome of this series of NRC-sponsored conferences was the establishment of the permanent Personnel Research Federation (PRF),† which became the country's central agency for personnel research. Among its various functions, the PRF was responsible for the "collection and dissemination of information," the "stimulation of research," the "coordination of research agencies," the "formulation of problems and allocation of study," and the "encouragement of training in personnel problems." Most important, it brought together in one agency the personnel activities of the AFL, Bureau of Industrial Research, Industrial Relations Association of America (formerly the National Association of Employment Managers), Taylor Society, National Bureau of

---

*By November 1920 Gompers had been invited to join in the project, but only after being assured that nothing had yet been done, that he was getting in at the start. The care with which the NRC committee approached Gompers—assuring him that the Engineering Foundation "could not by any possibility be considered as the representative of capital" and submitting nothing to him in writing—ignored the fact that Gompers was more than willing to share in the enterprise; he readily identified the progress of labor with industrial progress and even suggested that representatives from the National Civic Federation, an organization of the nation's most progressive corporate leaders, and the Taylor Society, the forum for scientific management, be invited to join the new enterprise. See Alfred Flinn, memo on Conference with Samuel Gompers, December 9, 1919; and Gompers to Flinn, July 14, 1920, NRC Executive Committee, NRC Archives.

†The leading advocates of the PRF, and the men who officially launched its operations, included Capen, from the NRC Educational Division and the ACE; Yerkes, Comfort Adams, and James Angell, from the NRC; Frank Jewett and Alfred Flinn, from the Engineering Foundation; Walter Dill Scott, Walter Bingham, and Beardsley Ruml, from the Scott Company and the Carnegie Institute Bureau of Personnel Research; and Robert I. Rees, from the War Department and AT&T. Rees became secretary of the PRF.

Economic Research, Carnegie Tech Bureau of Personnel Research, National Industrial Conference Board, U.S. Bureau of Labor Statistics, and other agencies of the federal government. Here, as never before, were concentrated the corporate efforts to standardize the "human material" that comprised American society for efficient, profit-making enterprise.[11]

In addition to coordinating personnel research for industry, the NRC contributed to the standardization of vocational terminology and the preparation of job specifications. Whereas personnel research focused upon the measurement of human usefulness, these efforts concentrated upon analysis of the slots into which the human pegs were to be fitted. The prime mover in this area was Charles Mann of the Army's E&R Branch. In November 1922, after two years of experimentation under the joint auspices of the War Department and AT&T, Mann urged the NRC to help create a central agency "for the purpose of coordinating all the activities in the country engaged in writing occupational specifications and defining terminology." Mann argued that since "the national defense is based upon the citizen army," the needs of industry and the Army coincided, and that "the handling of occupationally skilled workers" was one of the most important of them.

> What can we do to get all the activity . . . on this subject flowing in the same channel, directed toward the same common goal, so as to get us somewhere in regard to standard terminology of nation-wide origin? What can be done to stimulate technical training schools and industries to produce more accurate occupational specifications? What can be done to coordinate the terminology and specifications for occupations as drawn by governmental agencies and industrial organizations? What sort of a central coordinating agency can be created to standardize terminology and coordinate specifications?[12]

With the help of the new Secretary of War, John W. Weeks, Mann secured the cooperation of the NRC in sponsoring a conference on Standardization of Vocational Terminology, in January 1923. Out of this conference emerged the new National Board of Personnel Classification, representing all governmental, industrial, educational, and engineering agencies concerned with the standardization of job specifications and vocational terminology, the complement of standardized personnel measurement and classification. The movement to standardize measure and materials for industry was now coupled with the movement similarly to standardize human beings.[13]

In the discussions during the January conference the question was raised, "Who is going to take responsibility for achieving" the standardization of terminology and specifications? Mann pointed out that "if the work is undertaken by the military establishment, it is at once doomed to failure, for the reason that people do not care to be classified for military purposes"; while if, "on the other hand, it is undertaken in chambers of commerce and by employers only, then the work is much handicapped by the idea that this is a new dodge to put labor in its place and reduce wages. The converse is true if it is undertaken by labor organizations." The only viable solution to the question "of where to centralize or locate responsibility for the work," Mann argued, is "that the educational institutions should be in official charge of this operation."

> These specifications not only have an enormous value for industries and for employees, but they have a very fundamental significance in education, because it is well recognized that our training systems of the country do not turn out a product which is available for use in the industries of the country. One of the fundamental difficulties is lack of specifications of what the institutions are trying to do. Therefore, it has a very basic educational significance and all hands are willing to give the problem over to educational institutions. They get the specifications, cooperate with the industries in setting them, and take them back to the school and use them as definitions of objectives for training. The educational authorities decide how educational courses can be organized to train men so that they will be able to achieve the objective defined in the specifications.[14]

Mann's view that the educational institutions should bear the responsibility for national standardization of job specifications and vocational terminology—since after all, it was they who had actually to produce people according to them—was generally shared by the governmental, educational, and industrial representatives of the new board. While the board would serve as a clearinghouse for such activity, the educational institutions of the country would actually do most of it. To aid the educational institutions in this work, the NRC sponsored studies and conferences throughout the 1920's. In 1923 L. L. Thurstone, who was already conducting extensive experiments in vocational guidance in engineering colleges under SPEE auspices, became head of the Anthropology and Psychology Division of the NRC, which from then on became a "clearinghouse and service agency for the colleges" and sponsored such activities as the study of intelligence tests, admission tests, examinations, classification of students, personality traits, and vocational guidance. From 1923 to 1925, the division set up a series of conferences on the problems of "career counseling," focusing upon such problems as the "analysis of mental and physical traits required in each occupation"; the "evaluation or measurement of the mental and physical traits of individual students with regard to their occupational objectives"; and the "administration of vocational guidance service in colleges and high schools."

In 1925 the division cooperated with the ACE (of which Mann was now director), the American Association of College Registrars, the Personnel Research Federation, and the Educational Relations Division of the NRC to take "steps toward united and systematic study of college student personnel problems," a project begun by A. A. Potter at Kansas State two decades earlier. Among the most active conferees were leading educational reformers in the engineering colleges, men whose first experience with such work had been in the corporation schools of the science-based industries.[15]

The NRC's Educational Relations Division was funded by the General Education Board of the Rockefeller Foundation. Directing the work of this division were Capen, Rees, Yerkes, Mann, Harry Tyler, Angell, Frank Aydelotte, Dean Seashore, and Henry Pritchett. In addition to making surveys of American educational facilities with regard to research work and the production of research workers, its primary function, the Educational Relations Division undertook studies of "gifted students" which were supervised by Aydelotte and led to the introduction of "honors programs" in colleges throughout the country; "sectioning on the basis of ability"; "sifting out the university student"; entrance exams; and the development of "university services for occupational placement and vocational guidance." Capen, chairman of the ACE and now chancellor of the University of Buffalo, chaired the committee which dealt with the "possible reorganization of administrative units to save time and to increase efficiency in the whole educational process," thus promoting further the "coldly scientific" educational work which he had begun a decade before as Specialist in Higher Education for the United States Bureau of Education. The Division of Educational Relations of the NRC, however, never became a major center of reform activity. It suffered from lack of funds throughout the 1920's; more important, the major reform activities in which it participated were carried on by its own members elsewhere, in the SPEE and the new American Council on Education.[16]

As before the war, the dominant themes in postwar engineering-education reform were industry- education cooperation and education for management. The events of the war had overshadowed the publication of the seminal Mann Report, which had proclaimed these educational goals as never before, but the experiences of the war dramatized and drove home the message. In the fall of 1919, for example, Dean Walker of the University of Kansas observed that "in general, it seems as though there has never been a time when school men were more ready to respond to new ideas. . . . It is being realized that technical schools are a real part of the industrial system, and the more vividly that idea stands forth the more the industrial atmosphere will prevail." Officer training during the war had given Walker and his colleagues in the Engineer Reserve Corps a profound lesson in the need for management training, for educating men broadly for positions of responsibility in war and industry. In addition, as Walker noted,

> The war gave a great impetus to the idea of testing and grading men as a means for estimating their capabilities. The intelligence (or psychological) tests and the trade test were most in evidence, along with the system of grading on personality characteristics. . . . There is a strong tendency in the schools to take up the matter. From the beginning the weeding out of the unfit and the adjusting of the pegs . . . to fit the industrial openings, have been among the difficult problems. Educators are disposed to welcome any system which promises assistance.[17]

Mann himself, evaluating "the effect of the war on engineering education" for the SPEE *Bulletin,* reaffirmed his prewar conclusions. "There must be closer cooperation between school and industry," he urged, "and there must be more attention to the assessment of values and costs." Reasserting the major themes of his study, Mann observed that, whereas under the Morrill Act of 1862 "many new colleges had called themselves 'industrial universities,' they soon dropped the 'industrial' from their titles, fearing lest they lose caste in academic councils." But now, after the war, "if they grasp the opportunity opening before them, [they] will claim with pride their abandoned surname" and create a "true university, with its feet firmly planted in industry."[18]

In effect, the war experience had raised the prewar experiments in cooperation to a higher, and national, level. Fresh from their Army experience in personnel classification and officer training, the educational reformers in the engineering schools had begun to envision cooperation on a grand scale: industry as a whole would furnish the job specifications and employment requirements the schools demanded, and the schools would provide the complementary testing, training, selection, and distribution of manpower for industry. This mass-production approach to education was perhaps nowhere better articulated than in an article written by the treasurer of the Winchester Repeating Arms Company, in 1921:

> The employer of engineering graduates is probably better qualified to determine what the product of the engineering schools should be than is the teacher of engineering students. On the other hand, if the specifications for the product of the engineering schools are determined, the pedagogical expert is far better qualified to determine the educational processes which will produce the required result, than is the employer of engineering graduates.[19]

In the spring of 1920 the first postwar attempt to set up the national machinery for such cooperation was undertaken by Hollis Godfrey, now president of Drexel Institute. As a disciple of Frederick Taylor, Godfrey quite early had called for the teaching of scientific management in the engineering schools and just the previous year had formally introduced the cooperative plan of engineering education at Drexel. Godfrey's new enterprise, however, was much larger in scope than these. He was preoccupied—as were many leaders in industry—with the shortage of highly trained personnel available to industry, a shortage which had come about through war casualties, the draining of the colleges, and the decline in immigration. To correct this situation, he envisioned a grand scheme of education-industry cooperation which would involve the preparation of "joint specifications" for both technical and managerial manpower. Accordingly he called a conference at Drexel, attended by representatives from a large number of eastern colleges and some seventy-five corporate executives, "to work out a definite course of technical training for colleges to meet the specific needs of American industry." The plan Godfrey unveiled involved the creation of a joint committee representing education and industry which would "define the kinds of specifications of product that are useable and intelligible to educational institutions, . . . review the specifications submitted by industry and criticize them, . . . and circulate them with comment among the different types of educational institutions." To represent the industry side of this ambitious joint venture, Godfrey launched the new Council for Management Education. On the education side, Godfrey prevailed upon his friend Capen, now director of the ACE, to set up a special ACE committee on cooperation with the industries (composed of Capen, Bishop, and Mann).[20]

For a year after the Drexel conference Godfrey forcefully promoted the new CME, calling it "the only organization which has carried wartime operations into a time of peace." He wrote articles for trade journals, issued press releases, and used his wartime prestige to secure the support of many industrialists, among them Sam Lewisohn of the Miami Copper Company, the first president of the American Management Association. The project collapsed by the end of 1921, however, having succeeded merely in drawing up a sample set of joint educational specifications covering a single industry, prepared primarily by Mann. There were a number of reasons for the failure. First, the project was unwieldy, requiring a kind of patient, long-run investigation which Godfrey's promotional gimmickry defied. Second, there existed a general wariness, among industrialists and educators alike, of Godfrey's superficial, erratic, and egotistical behavior. In addition, the ACE was never enthusiastic about the project, and was drawn into it more by Capen's friendship with Godfrey than by anything else.* Finally, and most important, the industrial and educational leaders who would have been most likely to support Godfrey's program were already beginning to formulate a similar plan under the auspices of two established agencies, the SPEE and the National Industrial Conference Board.[21]

The National Industrial Conference Board was the brainchild of Magnus Alexander. As early as 1914, while director of personnel at GE, he had proposed the creation of an industrial research agency; his own research into the complicated problem of labor turnover had convinced him that there was a need for some machinery through which the industrial community could study pressing social and economic problems, devise and promote their own solutions, and thereby coopt potentially disruptive political reform. Such a board, Alexander argued, "would have realized," for example,

> the inevitability of workmen's compensation legislation in response to the economic belief of our people, and would therefore have carefully studied the subject . . . and prepared its case for presentation in the various legislatures. At the same time it would have educated the employers themselves to a proper understanding of the issue and would have worked out a practical, fair, and yet conservative proposal for introduction in any state in which public opinion demanded a work-

---

*From the beginning, Capen preferred that the educational side of the project be handled by the SPEE so that he could devote his time to building a broad and solid base of support for the ACE. Godfrey, on the other hand, wanted little to do with the SPEE—he had lost the election for its presidency in 1918 after a bitter contest—and demanded the support of the ACE because it represented liberal-arts as well as technical colleges.

men's compensation law. The result would have been greater uniformity of enactments and greater sanity and practicability of their provisions. While this particular opportunity has passed, similar opportunities for effective cooperative work are right now before us and we should not let these opportunities go by unheeded.[22]

To promote his scheme for a "research arm of industry," Alexander called a series of Yama Conferences on Industrial Efficiency at the Yama Farms Inn, a resort in the Catskill Mountains catering to the elite of United States business. Joining him in this venture were his friends Loyall Osborne, vice-president of Westinghouse, E. W. Rice, president of GE, Frederick P. Fish, fomer president of AT&T, and Frank Vanderlip, president of the National City Bank of New York. At the second of these conferences, in September 1915, they proposed that a National Industrial Conference Board be created to serve the needs outlined in Alexander's original proposal, and it was formally founded in May 1916.[23]

For the first four years of its existence the NICB was dominated by William Barr of the National Founders Association and leaders of the National Association of Manufacturers, men who did not share the corporate liberal approach to social problems. As a result, the board functioned as an open-shop propaganda platform, a focus for antiunion forces in the country. Representing twelve important employers' associations which in turn represented 15,000 employers (employing seventy-five percent of the nation's workforce), the NICB understandably was perceived by labor as a major management counteroffensive against unionism (in response to the formation of the board and the inflammatory antilabor pronouncements of Barr, the railroad brotherhoods were forced into a defensive alliance with the American Federation of Labor). By 1920, however, Alexander and the other corporate liberals who had originally conceived the board had regained control over it and begun to steer it in the directions for which it had been designed. To signal their benign intentions, they issued a public statement indicating that henceforth "the Board will refrain from all political activity."*[24]

One of the pressing problems facing the board, and industry, in the early 1920's was the shortage of technically trained manpower and especially the lean supply of men educated for business leadership. These, of course, were problems to which the leaders of the science-

---

*The NICB has continued to serve as the "research arm" of corporate industry, and is now called simply the Conference Board.

based industries (Alexander, Rice, Osborne) were particularly attuned. In the summer of 1921, therefore, Alexander initiated a cooperative venture with the SPEE to try to meet the problem on a nationwide scale.[25]

On the education side, meanwhile, the SPEE itself had begun to implement the "fruits of the war," to place the engineering schools in a better position to serve the industries. In 1919 the society significantly extended the heretofore limited psychological testing of engineering students, begun by L. L. Thurstone at Carnegie Tech in 1915.* Drawing upon his war experience with Yerkes' NRC committee and the CCP in the Army, Thurstone, who was now at the University of Chicago, directed extensive experiments with vocational-guidance tests in twenty-nine engineering colleges under SPEE auspices. By 1927 he had developed procedures for evaluating "the different mental traits that are basic for the engineering course," and had set up the Vocational Guidance Service to "guide" able high-school students into engineering careers. In addition to testing and other personnel procedures, the SPEE began by 1920 to focus upon national industry-education cooperation as never before. In his presidential address for 1920, Arthur M. Greene chose the theme "Requirements: Cooperation Between the Preparatory School Colleges and the Industries as Viewed from the Standpoint of the Educator." This address was followed by others conveying the same message: Roy D. Chapin, president of Hudson Motor Car Company, discussed the complementary "Cooperation Between Education and Industry from the Viewpoint of the Manufacturer"; Samuel P. Capen outlined the aims of the Council for Management Education; and William Wickenden, personnel director for AT&T, presented the results of an AT&T national survey of technical and managerial manpower in "The Engineer as a Leader in Business." Each speaker emphasized the same points: the pressing demand for nationwide industry-education cooperation and the need for broadening the engineering curriculum to prepare men to handle, as leaders in business, the complexities of industrial life highlighted by the war.[26]

At the next annual convention the SPEE executive council, under the direction of its new president, Charles F. Scott, took steps to move

*Along similar lines, Professor E. L. Thorndike of Columbia University before the war had devised tests for seniors in electrical engineering and first-year graduates working at Westinghouse. This work was done in cooperation with, among others, Charles Mann, Dean Keppel of Columbia, and Dooley of Westinghouse, all of whom later played major roles in the wartime CEST. See Mann's *Study of Engineering Education* (Boston: Merrymount Press, 1918), pp. 70–1.

engineering education along these lines. It established a committee to implement the Mann Report proposals, the most active members of which were Walter Dill Scott, Dugald Jackson, and Frederick Bishop. Not long after the committee was set up, Magnus Alexander contacted Scott and Jackson about the creation of a joint body which could coordinate the complementary activities of the two organizations, with the NICB approaching the problems of engineering education "from the standpoint of the man in industry who employs engineers and from the standpoint of the practicing engineer," and the SPEE approaching them "from an educational point of view." The upshot was the Advisory Joint Committee on Engineering Education. Among those representing the NICB besides Alexander were Howard Coffin of the Hudson Motor Car Company and E. M. Herr, president of Westinghouse. Representing the SPEE were some of the vanguard of the cooperative movement: Herman Schneider of the University of Cincinnati, Dugald Jackson of MIT, H. J. Hughes of the Harvard Engineering School, and J. W. Roe of NYU. The chairman of the joint committee was Frederick P. Fish. The committee thus, in effect, constituted the MIT-GE cooperative plan raised to a national plane since the same men were responsible for both.[27]

As its first project, the joint committee sponsored preliminary surveys of the technical and managerial manpower problem, which were published by the NICB. (The most comprehensive study was entitled *Engineering Education and American Industry;* * separate shorter studies were prepared for the electrical, chemical, and metal-trades industries.) Essentially, these reports presented variations on Alexander's theme that education was "one of the important arteries in the industrial system" responsible primarily for developing young men for "effective service in industry." In 1922 Alexander sent a telegram to the Urbana SPEE Convention, formally calling for a major cooperative study of American technical education. "We are keenly interested in engineering education as an essential part of the industrial system," Alexander wired. "We welcome heartily cooperation of engineering professors in working out fundamentals on which relationship of engineering education and industry must rest so that engineers may be trained efficiently to render effective constructive service to industry."[28] Not surprisingly, SPEE President Scott strongly endorsed the proposed

---

*Oliver S. Lyford, *Engineering Education and American Industry,* Special Report Number 25 (New York: NICB, 1923), published also as "The Engineer as a Leader in Industry," SPEE *Proceedings,* XXXI (1923), 135–265.

study. At a joint conference of the SPEE and the American Association for the Advancement of Science, he expressed his own view of education, one which did not depart significantly from that of Alexander.

> The industries constitute the organization through which engineering utilizes the forces and materials of nature and organizes and directs human activities for the benefit of man. Today the practicing engineers and representatives of industry meet with the association of engineering educators; this is a conference between users and producers of human material for leadership in industry.
>
> If producers and users of steel rails were in conference they would discuss the uses which rails are to serve, classifying the kinds of service, considering wherein past products had failed, inquiring as to chemical analysis and metallurgical treatment. They would see improvement in production and discrimination in use. But the more difficult problem of the human material for technical and administrative leadership has received less attention . . .; how seldom do representatives of engineering industry and of engineering education meet together for conference! Yet they are users and producers of a vital product. Let us try to agree on what we want and then determine how to get it and how to use it. How many boys of differing kinds can be individually developed and fitted to varying needs.[29]

While the NICB continued to survey the needs of industry for technical and managerial manpower, the SPEE proceeded to establish the apparatus for conducting a major study of engineering education, to survey the existing and potential means for meeting those needs. In the summer of 1922 the Developmental Committee officially became the Board of Investigation and Coordination, the SPEE agency charged with directing the study.* During the following year the board formulated an outline for the proposed study; it was aided in this task by Henry Pritchett, then temporary president of the Carnegie Corporation —and the man who had been president of MIT when the cooperative plan was hatched there. In 1923 the new permanent president of the Carnegie Corporation, Frederick P. Keppel, was elected. Keppel, it will be remembered, was the former Third Assistant Secretary of War who had been instrumental in the creation of the CEST and the SATC; soon after his appointment he notified the new SPEE president, A. A. Potter, that the Carnegie Corporation would gladly support the project. To direct the investigation, the SPEE Board of Investigation and Coordination selected the man who had already conducted an impressive

---

*Composed of men such as Scott, Jackson, Bishop, Potter, Rees, Kimball, Aydelotte, and Seashore—all leaders in cooperative education.

survey of technical education for AT&T, William Wickenden. Wickenden, of course, was no stranger to the members of the board; he had supervised the school side of the GE Cooperative Course back at MIT.[30]

Thus the famous SPEE Wickenden investigation was born, in cooperation with the NICB, the United States Bureau of Education, the national engineering societies, and Eta Kappa Nu, the electrical-engineering honor society.*

The Wickenden study of engineering education was the most comprehensive investigation of the technical-manpower needs of American industry, and the educational means of meeting them, ever undertaken. It involved the close cooperation of some 150 different schools over a six-year period, and included studies of such problems as the structure of undergraduate curricula, the social and economic content of curricula, the function and scope of the engineering college, the history of technical education in the United States, comparative development of technical education in the United States and Europe, the value of the cooperative method of engineering education, engineering degrees, graduate education, teaching personnel, quality of teaching, comparative success and economic status of graduates and nongraduates, admission requirements, testing, and the reasons for failure. The Wickenden staff conducted surveys of the opinions of graduates, nongraduates, teachers, industrial employers, and practicing engineers in an attempt to formulate scientifically the shortcomings of existing educational methods and the means of overcoming them. It sponsored summer courses for teachers as well as industrial employment for students and teachers, and produced the first major study of technical institutes, proposing that the two-year technical institute be geared to meet the industrial demands for technical manpower now that colleges were being geared to produce managers and executives. Not surprisingly, the major recommendations which emerged from the six-year study called

*In addition to the initial Carnegie Corporation grant, funds for the massive study were secured from Pritchett's Carnegie Foundation for the Advancement of Teaching, the Engineering Foundation, GE, Westinghouse, AT&T, Western Electric, Detroit Edison, Western Union Telegraph, and Stone and Webster. Individual contributors included Samuel Insull, czar of Commonwealth Edison; Frederick Pratt, son of John D. Rockefeller's Standard Oil partner and head of Pratt Institute; and James H. McGraw, founder of the McGraw-Hill publishing empire, who began his career with the ownership of the major electrical-industry journals—*Electric Railway Journal, Electric World,* etc.—and was a lifelong member of the AIEE and NELA, and an intimate of the elite of the industry. See *Report of the Investigation of Engineering Education* (Pittsburgh: Society for the Promotion of Engineering Education, 1930), I, 15, 16.

for closer cooperation between industry and education and for a broad-ened curriculum to provide both scientific training "in the fundamen-tals" and social-science training for future management respon-sibility.[31]

In addition to the Wickenden investigation, which was the focus of engineering-education attention throughout the 1920's, the NICB con-tinued its inquiries and produced more reports on industrial require-ments, under the auspices of the joint SPEE-NICB committee.[32] The themes of industry-education cooperation and management training for engineers were also promoted through the efforts of the educational committees of the NELA, and through a series of major Bell System Educational Conferences for educators from both engineering and lib-eral-arts colleges, supervised by Rees in 1924, 1926, and 1928. Sam Lewisohn, president of the American Management Association, also held a series of informal conferences of engineering educators at his New York City home between 1926 and 1930, aimed at promoting the teaching of "industrial relations" in the schools.[33] The most lasting result of all this effort, and particularly of the Wickenden investigation, was the establishment, in 1932, of the Engineers Council for Profes-sional Development (ECPD).

Under the umbrella of the SPEE Board of Investigation and Coordi-nation, a committee had been established to cooperate with the various engineering societies. This committee coordinated its efforts with a committee of the ASME which was conducting a study of the "eco-nomic status of the engineer."* In 1932 the ASME committee proposed the formation of a permanent agency to direct all activities relating to the education and professional practice of engineering; major propo-nents of the move included Wickenden, Bishop, Jackson, Calvin Rice, secretary of the ASME, Scott, C. F. Hirschfeld, Rees, and Potter. As originally constituted, the new ECPD represented the four founder societies (the ASME, AIEE, ASCE, and AIME), the SPEE, the Ameri-can Institute of Chemical Engineers, and the National Council of State Boards of Engineering Examiners. Its immediate objective was the formulation and enforcement of "minimum professional standards," guidelines for professional self-regulation which would serve as a defen-sive bulwark against compulsory state licensing of engineers; ulti-

*The ASME committee was composed of C. F. Hirschfeld, an electrical engineer and director of research at Detroit Edison; C. N. Lauer, president of the Philadelphia Gas Works and a noted proponent of scientific management; Dexter Kimball; and Wickenden himself.

mately, however, it was aimed at control of every aspect of professional engineering, from the cradle to the grave.

The ECPD focused upon the engineering student, the engineering college, and the professional practicing engineer, in its comprehensive, long-range program for "implementing" the Wickenden report; and it soon became the recognized central agency for all matters relating to the engineering profession, including college accreditation, professional standards of ethics, determination of competence for practice, professional recruitment from high schools, and the general definition of what it means to be a professional engineer in America. At the same time, it signaled the complete triumph of the corporate engineers and their particular brand of professionalism. Success in the profession now officially meant promotion up the corporate ladder, and education for the profession now officially meant education for both subordinate technical employment in and responsible management of corporate industry. The first chairman of the council was C. F. Hirschfeld, director of research at Detroit Edison; of the twenty-one men who sat on the first council, twelve were actively engaged in corporate industry, three were either acting or former vice-presidents of AT&T, one was a former educational director of Westinghouse, and another had done early educational service for GE.* As never before, technical education in the United States, training for the engineering profession, had become an integral part of the corporate industrial system.[34]

Although it began within the engineering schools and the science-based industrial corporations, the corporate reform movement in American education had, by the end of the war, moved far beyond both. A survey conducted by the NACT in 1921 revealed that within the industrial community there was now a "progressive dependence upon higher educational institutions as sources of employee supply," and that "the prejudice of many businessmen to higher education as a factor in employment is being rapidly overcome." The survey indicated also that there was a marked "readiness on the part of industry to receive

*The chairman of the ECPD Committee on Engineering Schools was President Karl Compton of MIT; the chairman of the ECPD Committee on Professional Recognition was C. N. Lauer, president of the Philadelphia Gas Works; and the chairman of the ECPD Committee on Professional Training was R. I. Rees, vice-president for personnel at AT&T. Directing the early educational activities of the ECPD were William Wickenden, now president of Case Institute, R. I. Rees, Charles F. Scott, Dugald C. Jackson, A. A. Potter, and Frederick Bishop.

into its ranks untried, inexperienced human material which must be moulded by training and discipline into the various forms required." Increasingly, companies were taking on students part-time and during the summers so as to "test out ... prospective employees" and "enable the employment manager to keep a waiting list of promising applicants." More and more concerns, the survey also found, were carrying on "negotiations" with school placement bureaus and school officers for specified training and needed personnel. The barrier between the schools and the industries was fading away, Charles R. Mann noted approvingly, as "industries are beginning to see that they are fundamentally educational institutions."[35]

The progressive convergence of the industries and the schools was reflected most dramatically in the schools, especially in the colleges. Already the application of management principles to educational institutions had begun to transform the shape of higher education. Thorstein Veblen subtitled his *Higher Learning in America,* which was written in 1908 and published a decade later, "A Memorandum on the Conduct of Universities by Business Men," observing that:

> Business principles take effect in academic affairs most simply, obviously and avowably in the way of a businesslike administration of the scholastic routine, where they lead immediately to a bureaucratic organization and a system of scholastic accountancy. ... The underlying businesslike presumption accordingly appears to be that learning is a merchantable commodity, to be produced on a piece-rate plan, rated, bought, and sold by standard units, measured, counted and reduced to staple equivalence by impersonal, mechanical tests. ... It appears, then, that the intrusion of business principles in the universities goes to weaken and retard the pursuit of learning, and therefore to defeat the ends for which a university is maintained.[36]

Veblen understood that the "free pursuit of knowledge," which he considered the "end for which a university is maintained," was being thwarted, and its radical potential checked, because it was being incorporated within the larger closed framework of big-business enterprise. The "pursuit of knowledge" would gain considerable social stature and financial support as a result of the new industry-education cooperation, but it would no longer be really "free." While freedom of inquiry would be maintained and guarded, it would now become, as Mann phrased it, a "controlled freedom," a "disciplined initiative," an energy directed into, rather than against, the capitalist industrial system which sustained it.[37] Not many of Veblen's contemporaries in academia shared his insight, and no wonder. All that was asked of the immediate benefi- ciaries of the new support was that their efforts be on the whole and ultimately useful and consonant with the demands of the corporate system. The corporate reformers never required that all who pursued higher learning in America be conscious of the utility of their work, nor even that all such work be of ultimate utility. Rather, they created an institutional apparatus which would correlate the activities of academ- ics "behind their backs," thereby rendering such consciousness of pur- pose unnecessary.

Mann was another "schoolman" who evaluated higher education as the war drew to a close. For him, the changes engendered by the wartime emergency meant the dawning of a new day for the colleges and a golden age for corporate industry. "There were embodied in the organization of the Student Army Training Corps," Mann proclaimed, "several large conceptions of fundamental importance in the develop- ment of a national system of education; [the SATC] united all the institutions of higher education in a single enterprise for training for national service, . . . into a single university of Uncle Sam." This experi- ence in educational coordination constituted, in Mann's view, "a model that may safely serve as a guide for the future." Beyond this, the war experience provided a new sense of purpose for American higher educa- tion, the spirit of service. With the war over, and the emergency which fostered that spirit gone, new means had to be established to perpetuate this creed of the new corporate order.

> The schools must recognize, as the Army has, that every citizen has abilities that render him capable of some useful service. It is one of the functions of the educational system to discover each individual's ability and develop it for useful service. The methods of rating, sorting, classi- fying, and placing men as developed by the Army are available for school use. . . . Hopefully, the schools will adopt a plan of training that promises to deliver goods on a similar scale.[38]

To guide them in their new mission, the universities and colleges re- quired some "centralized organization" which could "perpetuate the system established under military control." What is needed, Mann observed, "is a Federal education council or department of education or national university that would define the national problem of educa- tion. . . ."[39] The American Council on Education emerged from the war to meet this need.

From the outset the organization established to coordinate the war- time activities of the colleges was under the control of Mann and his

CEST colleagues. The first meetings of the Emergency Council on Education were held in Capen's office or at the Cosmos Club, an elite Washington social club. These meetings were chaired by Capen, in his capacity as Specialist in Higher Education for the Bureau of Education, and were dominated by the forceful spokesmen for the Army: Rees, Mann, and Bishop. "The Emergency Council needs some steering," Capen wrote his wife after the first meeting.[40]

The focus of ECE activities was not the war itself but rather the peacetime "reconstruction" to follow the war. According to the initial announcement, the ECE had been created so that "a continuous supply of educated men may be maintained; and preparation for the great responsibilities of the reconstruction period after the war may be anticipated." Indeed, barely two months had passed before the ECE was made a permanent organization, renamed the American Council on Education. Significantly, the meeting which became the founding convention of the ACE was held at MIT; H. W. Tyler of MIT's department of mathematics—a close associate of Dugald Jackson and head of the new American Association of University Professors (AAUP)—wrote the ACE constitution. The industrial orientation which would mark the postwar activities of the ACE was fully apparent at this first meeting: Mann moved to invite the United States Chamber of Commerce to join the council; Bishop moved to invite the NACS; and Tyler moved to invite the National Society for the Promotion of Industrial Education —a reform organization directed by Henry Pritchett and Magnus Alexander.[41]

The ACE emerged from the war as the central agency representing higher education in the United States, having brought together for the first time the nation's largest educational associations. "The most influential men in higher education believed in the Council and wanted to see it start," Capen wrote his wife; what is more, he added proudly, they are all "quite willing to follow my lead."[42] Thus Capen became at war's end the chief executive officer of the most important educational organization in America. He was hardly unaware of the possibilities.

> The development of the American educational scheme has been planless, haphazard. We have always suffered because of this planlessness. The price that we are called upon to pay for our lack of forethought and the consequent lack of system becomes heavier year by year. Unified action has always been impossible because there was no unifying agency.... A unifying agency has now at last been established to stimulate discussion, to focus opinion, and in the end to bring about joint action on major matters of educational policy.[43]

From his experience with the Bureau of Education and, more importantly, his experience with the SPEE and the War Department, Capen could well predict what such "major matters of educational policy" might be. He wrote Bureau Commissioner Claxton—that "irresponsible freak"—about a pressing need for more technical training and discipline in the colleges of arts and sciences, and preached "the new gospel of technical education" to a friend at Tufts, calling for "close cooperation with commercial enterprises." "I believe that one of the most important problems before education in this country," he assured his former CEST colleague J. W. Dietz, "is the relation between educational institutions and the problems of industry and commerce."[44]

One of the earliest and most visible manifestations of the ACE's intimacy with industry was its committee for cooperation with the industries, which had been set up to coordinate the educational side of Hollis Godfrey's cooperative plan. The efforts by the council to solidify industry-education cooperation, however, extended far beyond Godfrey's modest scheme. The major goal of the council was the gradual reshaping of American higher education as a whole into an efficient mechanism, one which would make truly effective industrial cooperation possible. This grandiose undertaking was largely the work of two educators: Capen, who remained director until 1922 and a most active member of the executive committee for twenty years thereafter, and Mann, who served as director from 1923 until 1934. Both men shared a profound identification with the corporate reform movement within technical education, and heartily endorsed the "educational results of military training" developed during the war. (Thus, they continued to serve on the Civilian Advisory Board of the United States Army while "steering" the ship of American education, viewing the two roles as but different means to the same end: the creation of an industrially responsive educational system.) In 1934 Mann passed the gavel to George Zook, who, as successor to Kolbe as president of the University of Akron, had carried on the tradition of industrial cooperation early kindled in the municipal universities. By that time the ACE was composed of seventy-nine national and regional educational associations, sixty-four national organizations in fields related to education, and 954 institutional members—universities, colleges, secondary schools, public- and private-school systems, and educational departments of industrial concerns.[45]

The work of the ACE proceeded along three interwoven lines of activity. The first of these was directed toward a perpetuation of the centralized authority achieved during the war, and entailed the extension of both governmental and corporate industrial authority over education; the second involved the standardization of American educational procedures and institutional classification, and constituted a continuation of the prewar survey activities undertaken by Capen at the Bureau of Education; the third aimed toward the extension, within the educational institutions, of the testing, rating, and guidance procedures developed by the science-based industries before the war. Brought up to an unprecedented level of sophistication by the CEST, the CCP in the Army, and the various psychology committees during the war, this work on the "personnel problem" became the basis of a new "science of education," the hallmark of the ACE.[46]

Centralization of educational authority was a top priority for the leaders of the ACE. Although the ACE represented all of the major educational associations in the United States, it did not enjoy nearly the authority over educational matters that these former CEST directors had become accustomed to. From the start, therefore, they sought to gain some legitimate—that is, governmental—power over American education as a whole. "I was convinced before the Council was established," Capen later wrote, "of the need of a representative organ which should view the national enterprise in higher education cosmically." Coupled with this need for broad representation and a "cosmic" perspective was the need for a centralized and powerful agency for educational action. "The thing America needs as a central organization for education," Mann observed, "is an organization for research, for setting standards by defining objectives. . . . This type of organization may be called the General Staff organization of education." It would operate according to "the plan on which the General Staff of the Army operated," involving the allocation of decentralized responsibility after objectives are defined by centralized command. "The same principle," Mann noted, "is recognized as sound in business organization."[47]

One of the ACE's first formal acts along these lines was the establishment of a special commission to survey the public resources available for its corporate work. The commission was charged "to study and discover the extent to which the free educational system of the country can be maintained and developed by the more complete economic utilization of both present and future sources of public revenue." At the same time, the ACE began to push for national legislation which might provide it with leverage. "The interpretation of Federal legislation," Capen wrote in the first director's report, "is a continuing project of the first importance." In 1920 and 1921 Capen worked feverishly to drum up support for the creation of a Department of Education, with a Secretary in the President's Cabinet. In cooperation with business leaders like Edward A. Filene, he was able to marshal the support of the leadership of thirteen national organizations behind the pending Smith-Towner Bill, which was designed to create a Department of Education and to provide considerable governmental appropriations for "educational investigations." In 1921 Capen and his associates submitted a petition to the President urging him "to create a Department of Education with a Secretary in the Cabinet."[48]

Although the Smith-Towner Bill failed to pass, Mann reported optimistically to Henry Suzzalo in 1927 that "a national organization of education is developing. The next session of Congress will, I believe, do something about enlarging the scope of the Federal education office. Suitable cooperative relations between that office and the privately organized associations, like the ACE, must be evolved." Of course no men feared government control of education more than the leaders of the ACE and their fellow corporate reformers. What they wanted, rather, was corporate control of education through governmental means.*

In his announcement of the ACE annual convention for 1929 Mann articulated the "significant industrial tendencies that offer practical approaches for cooperation by education." These were to be the guidelines for American educational progress: "to create practical methods by which each individual may find work he can do best; to find by experiment the critical tests that control individual conduct and to use these to secure voluntary cooperation and stimulate individual responsibility; [and] to stress excellence of service as a determining force in profitable business."[50] Through the medium of the ACE, the corporate

---

*The merely apparent contradiction between fear of government control of education and the use of government for control of education was perhaps best reflected in the activities of Ray Lyman Wilbur, Secretary of Interior under President Hoover. Wilbur, once president of Stanford, was also, like Suzzalo of the University of Washington, a former regional director of the SATC and a close associate of SATC directors Capen and Mann. As nominal supervisor of the Bureau of Education, however, Wilbur declared that "it seems to me that there is a distinct menace in the centralization in the national government of any large educational scheme with extensive financial resources available." But, within months of this pronouncement Wilbur set up an unprecedented National Advisory Committee on Education "to make recommendations concerning Federal organization for education." The chairman of this new presidential advisory board was Charles R. Mann of the ACE; Henry Suzzalo, who had by this time succeeded Henry Pritchett as head of the Carnegie Foundation for the Advancement of Teaching, became advisory committee director. Wilbur, quoted in Charles G. Dobbins, *American Council on Education, Leadership and Chronology, 1918–1968* (Washington, D.C.: American Council on Education, 1968), pp. 7, 9.

creed of service, bolstered by the growing array of personnel-management techniques, had entered the soul and body of American higher education.

In addition to governmental coordination, the ACE worked hard to coordinate education with industry. From the outset the council was committed to industrial service; former CEST directors Dietz and Rees, now personnel directors of Western Electric and AT&T, respectively, regularly attended ACE Executive Council meetings, and the first special committee to be appointed by the ACE was headed by Bishop to cooperate with Hollis Godfrey's Council for Management Education. By the mid-1920s the ACE had gone far beyond this. It participated extensively—through Capen and Mann—in the vocational-guidance work of the NRC and the Army and cooperated with the personnel managers of AT&T, Kodak, and Bausch and Lomb.[51]

In 1926 the ACE amended its constitution to admit industrial corporations into its membership. "It is for the purpose of discovering how industries and education may profitably cooperate in solving the personnel problem," Mann reported, "that this Council is inviting industrial firms to become institutional members of the Council on the same footing with universities and colleges."*

The ACE did not limit itself to cooperation with the giant corporations. In 1927, at Capen's request, the American Management Association (formerly the NACS) appointed a committee to work with the ACE on educational cooperation. In addition, at Mann's request, the United States Chamber of Commerce established a committee on education for similar purposes; this committee facilitated cooperation between the ACE and the Chamber of Commerce at the national level and thereby avoided the duplication of effort which had previously resulted from ACE work with local chambers and, through them, with local industries. (This cooperative undertaking was generously funded by the Twentieth Century Fund, a foundation set up in 1919 by Edward A. Filene and Henry S. Dennison.) "The United States Chamber of Commerce is now cooperating with us on a project of working out suitable cooperative relations between education and business," Mann reported enthusiastically to Suzzalo; "this is a very live and far-reaching undertaking."[52]

*With the passage of the amendment Mann invited fifteen new organizations to join the ACE; among these were the leaders of the corporation-school movement: AT&T, Western Electric, Westinghouse, GE, International Harvester, Standard Oil of New York, Goodyear, Dennison Manufacturing Company, Cheney Brothers Silk Manufacturers, and the four Rochester companies. To encourage industrial corporations to join, the Executive Council reduced the required dues from $500 to $200 per year.

As a catalyst and a focal point for governmental and industrial coordination of educational relations, the ACE readily became the spokesman of American education in international affairs as well. In 1924 the American University Union in Europe turned over its responsibility for international cooperation in higher education to a new ACE standing committee established for that purpose. "I think there is much good in American education, much that might profitably be imitated in foreign systems," the ACE chairman, Dean Virginia Gildersleeve of Barnard, observed; her choice of words was particularly suited to an "age of isolationism" in which the plans for the reconstruction of Europe were formulated by the chairman of the board of the General Electric Company.* Along similar lines, Mann and Capen cooperated with T. Coleman Du Pont, Senator from Delaware, in developing the Junior Year Abroad program; this program was created and introduced into the schools to ensure that there would be "a body of young Americans trained to meet the expanding international responsibilities of the United States."[53]

The leaders of the ACE did not want merely to centralize American higher education; they wanted to centralize it for a purpose. The nation's colleges could provide maximum service to corporate industry only if they were geared for integration into the system; the processes of education had to mesh with the processes of industry. Already a most valuable service was provided to industry by the ACE's Division of College and University Personnel; the brainchild of Parke Kolbe, the division was patterned after the wartime Intercollegiate Intelligence Bureau to provide access to the talents of college faculties. By 1924 some ten thousand teachers from two hundred institutions were registered with the division, which compiled a detailed registry "for use of presidents and executives of institutions who want to find suitable men." "The Council is carrying on a substantial and important business activity in the line of higher education," C. J. Tilden of the ACE explained to members of the SPEE. The Division of College and University Personnel operated "in the nature of an engineering specification of what an institution of higher education should be."[54]

ACE efforts to standardize educational processes were promoted most forcefully by Capen. "The Council proposes to bring about a greater uniformity of procedure among the principal agencies now engaged in defining standards," he announced in the first *Educational Record,* and two years later he deplored "the chaotic state of collegiate

*Owen D. Young: the "Young Plan."

standardization" in which "a dozen or more influential agencies are measuring and marking colleges," arguing that the ACE was in "a better position . . . to coordinate efforts . . . [and] to secure the necessary unification of procedures." Acting as the newly elected chairman of the National Conference Committee on Standards of Colleges and Secondary Schools, Capen called a conference to coordinate standardizing activities. "The time appears to be ripe for a review of the whole standardizing procedure of the country, as it applies to higher education," he announced. "By joint agreement it may be possible to secure the adoption of a more nearly uniform series of requirements for accrediting and classifying colleges.[55] As a result of this conference, and Capen's other promotional activities, the National Conference Committee on Standards formally dissolved itself and transferred its responsibilities to the new Committee of Standards of the ACE; the ACE thus became the central accrediting agency for American higher-educational institutions.

As one of its first activities, the Committee on Standards conducted a survey of the entrance certificates used by 143 colleges and universities which revealed that "the institutions are . . . making increasing use of questionnaires and [are] asking for personal data that were almost ignored a dozen years ago." The committee promptly set up a special subcommittee "to study the present and possible use of such personal data," data first used by Potter at Kansas State and Dooley at Westinghouse. Throughout the 1920s the committee promoted the use of standard terminology in the schools, as well as standard degree and testing procedures, and facilitated the establishment of modern personnel offices in the colleges. In 1927 the ACE published its first *Handbook of American Universities and Colleges,* "an accurate and impressive picture of present conditions in American higher education," and three years later funds were provided by the Julius Rosenwald Fund to set up a Committee on Problems and Plans, charged with "the formulation of a comprehensive program of educational investigations." The chairman of the new committee was Samuel P. Capen, now chancellor of the University of Buffalo.[56]

Interwoven with its centralizing and standardizing activities was the ACE's pioneering work in extending the new "science of education" to the nation's colleges. The war had happily dissolved "ancient prejudices, traditions and habits," Charles Mann observed, "thereby liberating us for a new orientation and a recrystallization of thinking in a new and different pattern." The new "science of education" which emerged after the war was the product of both personnel management and the most advanced teachings of applied behaviorist psychology. Born in the personnel departments of large corporations and the education and psychology departments of universities, it had been put to its first successful test by the War Department during the war. In practice, the "new science of education" was an outgrowth of the industrial "personnel problem," which Mann considered "one of the most significant and characteristic features of the present." The leaders of the ACE were in the vanguard of the new movement—the very men who had conducted the wartime adventure in personnel management, they were most eager to carry it on. "The educational results of military training is a matter of intense interest at present," wrote Capen in a classic understatement. The work of CEST, the CCP in the Army, and Yerkes' Psychology Committee thus provided the foundation for the postwar investigations by the ACE. "There are two ways to effect educational reform," Capen later wrote. "One is by decree. The other is by investigation and experiment. One is the authoritarian way. The other is the way of science."[57]

The disciples of the new "science of education" viewed it as an extension of natural science, and approached it with the rapture of the physicist and the practicality of the engineer. Just as Faraday had established the correlation between invisible electricity and visible copper, the physicist Mann explained, "the same procedure is now beginning to be used in liberating and guiding the energies of man."[58] Henry Suzzalo, new head of the Carnegie Foundation and one of the foremost "scientists of education" in the country, waxed eloquent about the scientific means of creating "the effective American University system":

> The American system of schools has a sanction in public efficiency as well as in equality of personal opportunity. It is a special system of getting brains for the public purpose. [Educators] have an immediate responsibility to make the prospect more effective. . . . Soon we must become as wise in pedagogical method as we have long been in scientific method. The processing of human beings through intellectual experiences is far more important socially than the processing of material things. Yet physical technology holds a place of respectability among us which human technology has not yet won.[59]

The conceptual lens through which this proud new breed of "scientific" educators viewed the "process of education" was identical to the one through which the personnel directors of the science-based industrial corporations viewed it earlier in the century: education was one side of the corporate "personnel problem"; it was the means through which industrial employment specifications were to be met. "Job speci-fications establish mutually intelligible communication between indus-trialists and schoolmen," ACE director Mann asserted. "Job specifications ... yield the information schools need to build better citizens." For the ACE educators, the program for reform was rela-tively straightforward: they had to "decide how education can be orga-nized to meet industrial specifications." In Mann's view, "this is one of the most important movements in the whole country in industry and in education."[60]

Thus, under Mann's leadership, the ACE cooperated extensively with the Army, NRC, the Personnel Research Federation, the National Board for Personnel Classification, and various industrial corporations to develop effective personnel procedures. Such work was not restricted to the traditional "vocational" fields of education, moreover; profes-sional education was viewed as but another form of vocational training. There has been "a sudden reversal of the balance between professional and higher education of the non-vocational type," Capen reported in 1924; professional education has now become "the principal business of American universities," and vocational-guidance programs, testing and rating systems, "job analysis," curriculum reform, and the other trappings of "scientific" education were as applicable in this "business" as in any other.[61]

While the establishment of effective personnel-management proce-dures in educational institutions was a primary interest of the council, work in this area was hampered by lack of funds. In 1926 Mann complained that

> appalling wastes are now occurring on every hand for lack of adequate national machinery of cooperation in this new science of education. The validity of control by facts is now fully recognized. Competent research educators are available. Teachers are eager to be led by facts. But fact-finding and fact-disseminating machinery has not been con-structed. We are trying to drive a 1926 Lincoln with a "one-lunger" of the first vintage.[62]

The ACE was bailed out of its dilemma by John D. Rockefeller, Jr. In 1926 he contributed a total of $20,000 for "study of personnel meth-ods" and "study to define the specific occupational objectives of educa-tion." To direct the research funded by the Rockefeller grant, the ACE established a Committee on Personnel Methods and appointed Walter Dill Scott, president of Northwestern University, to head it; soon to become chairman of the ACE himself, Scott was the man who had set up both the CCP in the Army and the Bureau of Personnel Research at Carnegie Tech.

Perhaps one of the most important contributions of the ACE to the educational solution of the corporate personnel problem was the extension of psychological testing within the educational institutions. In 1924 L. L. Thurstone, the engineer-psychologist who directed the testing program of the SPEE, developed his first ACE Psychological Examination for high-school students and college freshmen. The Thurstone tests were used extensively throughout the 1920s and became a staple of American college education; by 1930, 347 colleges and universities were using them as a regular and vital part of their educational "processing." The testing work of the ACE received a tremendous boost in 1930 when the General Education Board of the Rockefeller Foundation contributed $500,000 for ten years of research "in the field of objective testing." The council immediately established the Cooperative Testing Service to administer the research under this grant and in 1939 extended its testing program to teachers as well as students by developing the National Teacher Examinations—a nationwide program for testing teachers. These testing programs of the ACE were carried on separately until 1948, at which time they were merged to form the Educational Testing Service, the central agency for all educational testing in the country.[63]

In 1926 the Executive Council of the ACE received a request from the Manufacturers and Merchants Federal Tax League to investigate the activities of the Ely Institute for Research in Land Economics and Public Utilities of Northwestern University. The letter charged that the institute, while ostensibly conducting disinterested inquiries into the problem of public utilities, was in fact being funded by NELA, the trade association of the country's privately owned utilities companies. ACE director Mann replied that "though the Council is deeply interested in freedom for research, the proposed investigation does not lie in its province." The investigation did lie, however, within the province of the United States Congress.[64]

In 1928, pursuant to a resolution of the United States Senate, the Federal Trade Commission undertook an investigation of the utilities industry, one which was to last until 1933. One of the focuses of this investigation, and of the hearings which were held during the course of it, was the close relationship between the utilities companies and educational institutions; the hearings revealed that the power companies, particularly through the NELA's Committee on Cooperation with

AMERICA BY DESIGN / 256

Educational Institutions, wielded a strong influence over the colleges. The FTC discovered that the utilities were hiring college professors; subsidizing utilities courses in colleges; subsidizing research work, including that of the important Ely Institute; reviewing and editing textbooks; seconding company personnel to college faculties; controlling university extension work; saturating schools with utilities-company propaganda against public ownership; and even conducting summer schools for faculty members. One participant in the hearings concluded that "the utility corporations, through their various committees, bureaus, individuals and organizations have carried on a far-reaching campaign to influence and control the educational institutions of the country."[65]

The FTC investigation sent a minor shock wave through the educational institutions and the personnel-management agencies of industry. It did little more, however, than slow down momentarily the corporate reform of American education that had already been thirty years in the making. Some firms were for a time reluctant to cooperate with the schools at all, despite persistent encouragement from the ACE to do so. Others—including the utilities companies—went right ahead with the more subtle aspects of their educational programs. Most understood that while they might arouse the indignation of some educators and lend credence to the paranoia of others, government investigations alone could in no way undo the corporate knot of the twentieth century: too much had happened, and the investigators had barely scratched the surface.

A Technology of
Social Production

Modern Management
and the Expansion of
Engineering

Problems in human engineering will receive during the coming years the same genius and attention which the nineteenth century gave to the more material forms of engineering. We have laid good foundations for industrial prosperity. Now we want to assure the happiness and growth of the workers through vocational education and guidance and wisely managed employment departments. A great field for industrial experimentation and statesmanship is opening.[1]

—Thomas A. Edison

We have all become somewhat accustomed to seeing the engineers called upon to perform new and strange duties.... Every day sees the duties and responsibilities of the engineer widened and it is difficult to see where the end will be. Engineering is an integral part of everyday life and necessarily assumes fresh aspects as the complexity of modern life increases. It will be increasingly difficult to set its limits and boundaries.

No one can doubt that the scientist and the engineer are to be the most important industrial figures of the near future.... To accomplish this we must broaden our vision and get about our business, which is the industrial organization of our country.[2]

—Dexter S. Kimball

Modern technology meant different things to different people, for it contained a range of possibilities as well as necessities. To the engineers of corporate industry, certain of these possibilities became necessities. Whether because of educated habit, ideological blindness to alternatives, social constraints, or conscious choice, they tended to seize upon only those technological potentials which promised to further corporate objectives, deemed them necessary (and, thus, historically inevitable), and denied all others. In all of their work they strove at once to stimulate and to tame modern technology, to fashion the means for unlimited technical progress while at the same time working to ensure that such progress would itself be a means to corporate ends.*

Actually, these corporate engineers played a double role. As engineers in a capitalist system, they were professionally charged with the profit-maximizing advance of scientific technology. And as corporate functionaries, they assumed the responsibility for coordinating the human elements of the technological enterprise. It was because of this dual role, and without any great imaginative leap on their part, that they began to view the second of their tasks in the same way they viewed the first, as essentially an engineering project. In their minds, the recognition that modern technology was a process of social production compelled them to try to formulate a scientific way of managing that process, a technology of social production.

Reinforced by (and reinforcing) a growing tendency within positivist social science toward engineering-like application, this new form of engineering gave rise to a dominant, two-sided strain in twentieth-century American social thought. On the one hand, social organization and human behavior became new foci of engineering theory and practice; on the other hand, and less obviously, engineering itself began to expand as, for the first time, stubborn and imprecise social and psycho-

---

*Just as social development is not simply technologically determined, so technological development itself is not automatic, a succession of givens. At every stage, it involves human choice, choice which reflects requirements which are social and historical. If certain aspects of technical work are defined by what we know about the relations and properties of matter and energy, these are nevertheless bound up tightly with social factors, making it extremely difficult to distinguish the one from the other. Where do "technical" imperatives end and "social" imperatives begin? The distinction is all the more difficult to draw because of the tendency by technicians routinely to extend the technical (or scientific) justifications for their work far beyond the realm to which they actually apply. Most historical accounts of technical development, moreover, simply second this tendency, readily granting to the engineers the sanction of destiny which they claim for themselves. But in all technical work there is always a tension between technical and social determinants. In actuality, technical imperatives define only what is *possible*, not what is *necessary;* what *can* be done, not what *must* be done. The latter decisions are social in nature. Unfortunately, this distinction between possibility and necessity is lost on most contemporary observers, and with it a large measure of imagination and social vision. A notable exception is the work of Lewis Mumford, which focuses upon this tension throughout. Recent studies by Stephen Marglin and Katherine Stone, in the *Review of Radical Political Economy,* Vol. VI (Summer, 1974) do likewise. Stone's work, which correlates technological changes in the U.S. steel industry with felt corporate needs to discipline, fragment, and motivate the labor force, is especially valuable. See also a useful conceptualization of the problem of technological development in David M. Gordon, "Capitalist Efficiency and Socialist Efficiency," *Monthly Review,* XXVIII (July–August, 1976), pp. 19–34.

logical variables were deliberately introduced within engineering analysis and design. One name is given to this two-sided development, this new technology: modern management.

A pivotal development in the history of industrial America in general, the emergence of modern management was thus also an important chapter in the history of engineering. Four historical currents converged by the turn of the century to give rise to this new social force. The first was the development of the capitalist mode of production, with its inherent demand for capital accumulation and competitive efficiency. The second was the creation of the far-flung, integrated industrial corporations and the subsequent necessity of realizing potential economies of scale to offset heavy investment in fixed capital. The third, which emerged as an aspect of and in resistance to the first two, was the intensifying "man problem"—the need to discipline and motivate labor, and neutralize opposition which thwarted efficient production, challenged the inevitability of capitalist industrial development, and threatened corporate stability. The fourth was the steady flow of scientifically trained engineers into management, particularly within the machine- and science-based industries.

Men, not machines, produce profits. The fundamental innovation of capitalism was not the introduction of machinery into the production process but, rather, the transformation of human labor into an abstracted means of commodity production and capital accumulation. Karl Marx was not alone in recognizing this fact. Charles Babbage, an inventive British engineer, mathematician, and early capitalist theoretician, drove home the point when he urged that "in order to succeed in a manufacture, it is necessary not merely to possess good machinery. . . . The domestic economy of the factory should be carefully regulated."[3] Babbage understood that profitable enterprise presupposed that the capitalist would dissociate labor from the productive process and reintegrate the two on his own terms, in such a way as to maximize the surplus value produced by the labor force in the productive process. The key complementary factors in this transformation were the increasing capitalist monopolization of the intelligence of, and control over, production, and the diminishing autonomy and cost of labor. These in turn made possible the most efficient reorganization of the production process by the capitalist, and thus the maximization of profitable output. Babbage realized, however, that while the actual design of machinery contributed to this transformation indirectly, the capitalist had also to undertake to bring it about directly, through the deliberate engineering of the workplace and the work activity of labor.

Babbage, it seems, was ahead of his time. Throughout the nineteenth century, capitalists and engineers alike devoted their attention primarily to the introduction of profit-making, labor-saving machinery rather than to the organization of the workplace. At the end of the century, however, under the pressure of intensifying competition and the heightened corporate demand to cut costs and increase production, they were compelled to recognize that "the domestic economy of the factory" had replaced the machinery as the limiting factor of production. Existing machinery could be used at peak capacity only if the human activity of production was organized in a correspondingly efficient manner.* They thereupon deliberately expanded their engineering focus to include the workers. As managers in industry, engineers now undertook to expropriate and systematize the intelligence of production, to place it in the hands and handbooks of management, and to use it to reorganize the production process for maximum output and profit. In the science-based industries, where the intelligence of production was already the monopoly of the engineer-managers, this task was more readily accomplished.

Coupled with the requirements of the capitalist mode of production, and intensifying them, were demands generated by the large, multifunctional, integrated industrial corporation. The impetus behind the corporate consolidations at the end of the nineteenth century was the attempt by industrialists and financiers to check the ravages of unrestricted competition, to control production, stabilize prices, and secure markets. Once formed, the corporations concentrated their newly acquired resources into giant plants in order to exercise some centralized control over manufacturing operations. Such giantism, however, involved a heavy capital investment in plant and equipment which could be profitable only if the facilities were utilized at full capacity. And such maximum use of resources demanded, on the one hand, vastly ex-

*It has often been said that modern management was a necessary product of technological development, that it was called into existence by the demands of large-scale production. This is true, but only if it is understood that such technological development was itself an aspect of the development of the capitalist mode of production. Technical and capitalist imperatives were blended in the person of the engineer and converged in his work, engineering. The engineer designed his machines with profit and reduced labor cost as well as the quality and quantity of product in mind, and with the aim of transmitting management authority into the work process (usually described as merely the "transfer of skill" from craftsman to machine). As the father of modern management, the engineer simply extended his efforts beyond the machinery for the same purposes.

panded and guaranteed markets to absorb the high volume of goods produced, and, on the other, efficient control over the entire process of production. Thus, the corporate empires, once they existed, required for their continued existence the rationalization of total operations, from the extraction and supply of raw materials to the marketing of finished products. Only through such efficient management and administration could they achieve the much-heralded "economies of scale" which subsequently became the rationale for their existence.

Business historian Alfred D. Chandler, who has studied these corporate imperatives and the efforts of corporate leaders to meet them, has found that

> The initial motives for expansion or combination and vertical integration had not been specifically to lower unit costs or to assure a larger output per worker by efficient administration of the enlarged resources of the enterprise. The strategy of expansion had come ... from the desire to assure more satisfactory marketing facilities or to have a more certain supply of stocks, raw materials, and other supplies in order to more fully employ the existing manufacturing plant and personnel. Even after a combination had consolidated, its managers continued to think of control of competition as its primary purpose. Finally, many of the later mergers were inspired and carried out by Wall Street financiers and speculators, anxious to profit from promoters' fees, stock watering, and other financial manipulations.[4]

Thus, it was only after the corporations had been established that the modern management techniques which made them economically viable were adopted. "From the 1890's on," Chandler writes, "one of the basic challenges facing the industrialists was how to fashion the structures essential for the efficient administration of newly won business empires."[5]

The men who were most prepared to meet this challenge were the analytically oriented engineer-managers in the large science-based corporations. They pioneered in formulating rationalized procedures in engineering, manufacturing, finance, and marketing; they quantified and systematized corporate operations, developing methods of cost accounting, statistical controls, forecasting techniques, and the procedures for gathering and processing huge amounts of detailed, accurate data to be used in appraising, planning, and coordinating the operations of extended plant and personnel. Equally important, they created the formal administrative structures for the giant corporations, with carefully defined lines of authority and channels of communication through which to control the process of production.

The efforts to organize and manage the capitalist labor process under corporate command, to increase productivity, cheapen labor, and secure complete management control over production, led inevitably to what management called the "man problem." This was nothing more nor less than the resistance of the worker to management's expropriation of his skill and the fruits of his labor, and to the gradual usurpation of his traditional authority over the work process. It took many forms —stubbornness, "soldiering," sabotage, physical violence, trade-union struggles, and radical politics—and was intensified by the sustained efforts of managers to streamline production and by the constant flow of "pre-industrial" rural Americans and Europeans into the factory workforce.

In all of their work, modern managers were ultimately forced to focus upon this "man problem," to overcome worker resistance to their designs. Moreover, after systematically stripping away all the important incentives for diligent and creative work—the rewards of craftsmanship, collective control over the production process, direct economic return for services rendered—these managers had somehow to motivate workers, to get them to "put their hearts into their jobs," as Wickenden phrased it. This concern with worker motivation was central to management's quest for increased productivity, since the degree of worker commitment to a task was perhaps the most significant factor in determining output. And the concern was made all the more urgent by a chronic labor shortage, most severe during the years before World War I, and a very high, costly rate of labor turnover and absenteeism. "Employers were being forced," Reinhard Bendix has observed, "to concern themselves with labor as a problem rather than solve it by simply dismissing the worker who would not do."[6] Modern managers thus sought ways to elicit the cooperation of the worker as they fitted him into the capitalist production process. They appealed to what they perceived to be his interests and, if that failed, used force to achieve the discipline required. At the same time, they stretched their analytical techniques to try to comprehend the workings of this "human element of production" and to develop engineering techniques for controlling "its" behavior. A writer in the *Review* of 1910 best expressed the spirit that underlay this bold new engineering enterprise.

When we engage the services of the human machine, we always have certain duties laid out which the newcomer is expected to undertake, and we try to get the best man for the place. . . . Now when we purchase a machine tool and find it slightly unfitted for requirements, we can

> usually make a change in construction which will correct the difficulty in it. . . . If the human machine could be controlled by the set rules that govern machine tool operation, the world would be a much different place.[7]

The last and least-recognized factor in the emergence of modern management, the one which dramatically defined the nature of the first three, was the steady influx of engineers into executive positions in industry. Already by 1900 there existed a strata of seasoned engineer-entrepreneurs, men who, in transforming their shops into the engineering and machine industry of the twentieth century, had become attuned to the exigencies of management. In addition, the younger graduates of the engineering schools were working their way up the new corporate ladders into important managerial and executive positions, particularly within the science-based electrical and chemical industries. Modern management was thus not simply the creation of engineers; it was the product of engineers functioning as managers. As the director of the Taylor Society later recalled,

> the first coherent public expression of concern with management problems came from the engineers. They were the initiators of the "management movement" and it was they who continued to give it vigor. The first highly trained minds to go into industry were products of the engineering schools. They were trained as technicians and went into industry as technicians, but many of them speedily became operating executives and focused their trained minds on managerial problems . . . ; the management movement arose out of the impact of a group of highly trained engineering minds on an industrial situation of engineering complexity.[8]

The engineers were the first people in industry to attempt to apply systematically the intellectual methods of science to questions of business management. It is for this reason that the literature of the management movement between 1880 and 1910 is found exclusively in engineering journals, and that the same journals remained the primary forum for management matters well into the 1920s (at which time management became an independent discipline with separate publications). In short, it ought not be surprising that as scientifically trained people became managers, management became more scientific.

Modern management issued from the requirements of machine production in a capitalist mode and provided the social basis for technical developments designed to reinforce that mode. In essence, it reflected a shift of focus on the part of engineers from the engineering of things to the engineering of people. Although engineers had long influenced indirectly the way people worked, through their design of the means of production, they now undertook to do so directly, through the design of social activity and, ultimately, of people themselves.

This shift involved two overlapping, complementary phases. The first, social engineering, was the conscious attempt to exercise managerial prerogatives through the medium of the workplace, through organization of the work activity of labor. The second, human engineering, was the movement to control the human element of production at the individual and group level through the study and manipulation of human behavior. These two phases were but different approaches to the same end and were inextricably linked; advances in one required advances in the other.

The social-engineering phase began in the machine shops in the last years of the nineteenth century in the form of "scientific management" or "Taylorism." The most significant contribution of the scientific-management movement, the one which had the most pervasive and lasting impact, was to secure managerial control over the production process and lay the foundation for the systematic reorganization of work. From the machine shops the principles of scientific management were carried over into electrical manufacture (by men such as Dexter Kimball and Hugo Diemer) and chemical manufacture (by Arthur Little) and ultimately became the basis of works management in all industries. In addition, scientific management methods were extended into such areas as cost accounting, office work, and marketing; as the detailed, systematic approach to work design and efficient administration, scientific management became and remained the foundation of modern capitalist enterprise.

The original scientific-management approach to human engineering, however, was less than adequate. Taylor and his associates essentially viewed the worker as a simple, and simple-minded, "economic man." Although they paid close attention to the details of his work, the Taylorites relied upon elaborate, though psychologically crude incentive-pay schemes to motivate him, or upon the pressures of the labor market and the "shop disciplinarian" to coerce him into "voluntary" cooperation. Moreover, the Taylorites steadfastly refused to deal with workers in association, as in the unions, preferring instead to concentrate on the worker as an individual. Eventually, however, in the wake of a trail of failures which pointed up these shortcomings, the scientific-management movement spawned a revisionism which surfaced after Taylor's death. The revisionists paid much closer attention to the psychological and social dimensions of the worker's being, and began to recognize the important role trade unions could play in achieving industrial stability.

The emergence of scientific-management revisionism complemented the growth of "welfare work" activities within industry. Concerned primarily with easing the plight of the worker in industry, the so-called welfare secretaries strove to "humanize" industry, emphasizing the word "human" in their new phrase "human engineering." It was not very long, however, before such humanitarian efforts were coopted by what was perhaps the most potent force in the higher management circles: corporate liberal management reform. The corporate reformers, with engineers prominent among them, combined the ideals of the welfare secretaries with the more "realistic" contributions of scientific management and the nascent social sciences to forge a comprehensive, sophisticated approach to the problems confronting corporate capitalism. They determined to comprehend and meet the exigencies of corporate growth and stability through sophisticated elaborations of the principles of scientific management, using them to control and anticipate complex affairs on a corporate, national, and international scale. They remade the innovations of welfare workers into elaborate industrial-relations programs, to try to foster a spirit of voluntary cooperation among workers, to transform the energy of potential conflict into a constructive, profitable force within a larger corporate framework. Finally, they shifted the emphasis in human engineering to the word "engineering," integrating the methods of the physical and social sciences to create the new science of "personnel management." Starting within the plant, they eventually extended their activities into the homes and schools of workers, all the while refining their techniques to create a disciplined, loyal workforce for the corporate order.

By World War I the revisionist current within the scientific-management movement had become dominant and had converged with the more sophisticated mainstream of corporate liberal management reform. The convergence signaled the demise of welfare work as an end in itself and the recognition by managers that social engineering and human engineering were but reverse sides of the same process. At the same time, it reflected an extension of the engineer's province and an expanded perception of engineers; as engineers increasingly entered the ranks of management, engineering education was broadened to include more social-science and management training. For engineers in the twentieth century, the machinery of production had become a society of people.

Engineers, of course, had always been professionally concerned with the social and economic problems relating to their technical work; they were regularly called upon to determine the viability of a proposed project with regard to its estimated cost and social consequences, in addition to the purely technical questions of whether, and how, it could be done. At the same time, as project, shop, and production engineers, they were occupied with the day-to-day supervision of "men" and sought effective means of promoting the discipline, morale, and maximum output of their workforce. Such considerations of the engineer, however, had always been external to the actual technical task and, as either means to or consequences of that task, were treated in a rather "unscientific," intuitive way. What was new with the emergence of scientific management was that these considerations would now become, in the mind of the engineer, integral parts of the engineering task, and subject to the same technical attention given the more material aspects. The proper use of machinists and laborers, for example, had always been a problem for the engineer who supervised the shop, but his primary professional concern was the improvement of the machines. Now the activity of the machinists and the laborers would receive the same kind of attention formerly reserved for the machine. This new development meant two things: the human activity of the shop would now be "engineered," and the technical province of the provisional engineer would expand to include the human along with the material world.

The expansion of the engineer's province began in the late 1880s when a few prominent engineers—Fred Halsey, Henry Towne, and others—delivered papers before the ASME on the technical problem of wage incentives; interested primarily in increasing the output per worker, these men developed elaborate pay schemes to motivate the worker toward greater efficiency. While they maintained that management might be able to stimulate the productivity of workers in such ways, however, they nevertheless continued to place ultimate responsibility for any increase in efficiency with the worker himself. Other papers by Oberlin Smith, Henry Metcalfe, and Towne, presented around the same time, departed from this tradition: focusing upon the deliberate expropriation of the skills of workmen by management, they shifted the ultimate responsibility for efficiency, and the means of securing it, to management—as H. S. Person maintained, "where it belongs." Writing about such things as inventory evaluation of machinery and shop-order system of accounts, these men presaged a significant progression in the history of capitalism. Capitalism had already divorced workers from ownership of the means of productive work and then reunited them through the profit-rendering mediation of the capitalist; now it would similarly divorce the workers from their own skill, their own traditional store of knowledge, and likewise reunite them "more efficiently" through the mediation of management.* The article presented by Henry Towne in 1886, entitled "The Engineer as an Economist," is generally considered to have been the first significant articulation of the scientific-management movement. In his paper, Towne, president of Yale and Towne Manufacturing Company and a prominent leader within the ASME, argued that "the management of works is unorganized, is almost without literature, has no organization or medium for the interchange of experience, and is without associations or organizations of any kind. . . . The remedy . . . should originate from engineers."[9]

The "remedy," as it subsequently emerged, involved five successive and overlapping phases: (1) the accumulation by management of all information pertaining to both the machines and the human activity of production, through the use of records and research; (2) the systematization of this information into comprehensible and applicable laws and formulae; (3) the "scientific" determination of optimum standards of performance for both machines and workers; (4) the "transference" of this optimum "skill" through the reorganization of the human and mechanical processes of production; and (5) eliciting the cooperation of the workforce through the "development of contented workers."[10] All contributions to the development of scientific management have pertained to one or more of these phases. Between 1890 and 1915 the

---

*The rationalization of industry involved centralized planning, the basis of which was "the deliberate gathering in on the part of those on management's side of all the great mass of traditional knowledge, which in the past has been in the heads of the workmen, and in the physical skill and knack of the workman, which he has acquired through years of experience." Frederick W. Taylor, *Scientific Management* (New York: Harper and Brothers, 1947), p. 49. In the light of clearly stated objectives like these, it is not difficult to comprehend why the craft unions, which were grounded upon the monopolization of traditional skills, initially opposed Taylorism.

Scientific management, however, was never completely successful in this regard, nor could it have been. Indeed, in encouraging later managers to rely exclusively upon management's own resources, it prevented them from taking advantage of the full potential of the "human element" of production. For workers in all industries have retained a considerable store of skills and experience which they now keep to themselves or use to counter management directives. In forthcoming work, as yet unpublished, Katherine Stone argues that the introduction of "suggestion boxes" in plants constituted a tacit acknowledgment by management of this fact, a means of tapping this potent resource without at the same time forfeiting control over production.

development of the movement consisted primarily in the elaboration of the first four phases; after 1915 and the death of Frederick Taylor, emphasis shifted to the fifth and, once again, to the worker himself. This time, however, the perspective had changed: the worker was not to be rewarded for plying his own trade more efficiently; rather he was to be encouraged, on the basis of "the most thorough comprehension of the human being," to play the game as management called it, to become a member of the team.

Frederick W. Taylor, a mechanical engineer from a well-to-do Philadelphia family, came to the fore of the management movement in 1895 with his paper describing his "differential piece-rate" wage system, and most forcefully in 1903 with his comprehensive system of "shop management." Educated abroad and at Exeter Academy, Taylor had eschewed the study of law at Harvard to become a laborer at Midvale Steel; within six years he had become chief engineer at Midvale, a miraculous accomplishment were it not for the fact that one of the company's owners was a close friend of the Taylor family. As a gang foreman at Midvale, Taylor embarked upon his management career, seeking ways of eliminating "soldiering"—the restriction of output by the workers. He began to gather information about the most efficient way of doing certain jobs, and to compare this with the manner in which it was actually done by the workmen in the plant; using his supervisory authority, he then tried to reconcile the latter with the former. The task of the manager, Taylor found, was to gather into his possession all available knowledge about the work which he oversaw, to organize it, and to use it as the "scientific" basis for prescribing work activity; the task of the workers, on the other hand, was "to do what they are told to do promptly and without asking questions or making suggestions."[11]

Leaving Midvale in 1889, Taylor began his career as a "management consultant"; his most important work involved the reorganization of the machine shops of the Bethlehem Iron (later Steel) Company. It was at Bethlehem that Taylor made his most significant advances in "scientific" shop management; he gathered extensive information about the optimum speeds, feeds, depths of cut, and cutting angles of machines such as lathes, planers, and millers; conducted the first studies of proper belting and belt maintenance, and the optimum use and kind of tools for certain tasks; and, with metallurgist Maunsel White, developed the alloy "high speed" steel which "revolutionized machine production"[12] by facilitating the use of higher machine speeds and, thus, higher rates of production. In addition to his study of machine processes, Taylor analyzed the actual work activity of the workmen in the shop, the necessary "motions" and "time" involved in specific tasks, and the fatigue of the workmen who did them. Beyond merely recording all this information, Taylor sought to systematize it into a useful form for practical management. In this he was helped by his first associates in the scientific-management movement: Henry Gantt drew up charts and tables of information, and the mathematical wizard Carl Barth devised the famous "Barth slide-rule " which allowed for the ready determination of machine settings as a function of the quality of metal used.

Taylor's work at Bethlehem provided him with the groundwork for an entire system of management, one which involved all five phases of the management problem. Through his study of metals and machine processes, he was able to determine the optimum use of the machines in the shop; through his analysis of the work activity and fatigue of workmen he could determine the standards for their most efficient labor; and by the systematic formulation of all this information, he could allow for its ready application in any shop. Taylor then devised the means of, and medium for, the efficient reorganization of the shop: "instruction cards," narrow division of labor, the systematic routing and scheduling of work, and "functional foremanship" would be the mechanisms for the "transfer" of the "skill" collected and coordinated by the central "planning department." To ensure the cooperation of the workmen—including the foremen, who had been stripped of much of their authority—and to motivate them to achieve the standards set by the planning department, Taylor relied upon his improved incentive-pay system; if this failed, he depended upon his new "shop disciplinarian." This system of "shop management," presented to the world by Taylor in two papers delivered before the ASME in 1903 and 1905, constituted the core of the scientific-management movement. Nearly all work by the Taylorites between 1900 and 1915 consisted in elaborations of one or more of these phases. Among the best known were Sanford E. Thompson's time study and "decimal dial" stopwatch; Frank Gilbreth's motion and "micromotion" studies and fatigue studies; Gantt's routing "charts," and his "task and bonus" system of incentive pay; Leon P. Alford's "management handbooks"; and Alexander Hamilton Church's contributions to systematic cost accounting. Taylor claimed that his system eliminated the arbitrary use of authority and constituted the basis for a true "harmony of interest" between the warring classes of capital and labor: the scientific means toward greater production and, thus, both higher wages and higher profits. The Taylor system, he maintained, resting as it did on modern science, was beyond question. All men could do, workers and managers alike, was to accommodate themselves to its scientific imperatives.

Although he was not altogether original in his ideas, Taylor made a lasting impact upon the industrial community simply because of who he was. As Lyndall Urwick has suggested,

> His prime importance lies in the fact that, for the first time, a "practical" man, an engineer of great technical distinction, was applying consciously to the whole process of industrial organization the intellectual methods responsible for the advances in the physical sciences which had made modern machine industry possible. Here was no outsider, no professor, but a man whom—ultimately—neither employers nor statesmen could ignore.[13]

After his departure from Bethlehem Steel in 1901, Taylor devoted all of his energies to spreading the gospel and relied upon his associates —Horace Hathaway, Hollis Godfrey, Dwight Merrick, Gilbreth, and, most importantly, Barth and Gantt—to introduce the "Taylor system" to manufacturers. Among the earliest important industrial concerns to adopt the system, in part or whole, were the American Locomotive Company, Union Typewriter Company, Brighton Mills, Yale and Towne Lock Company, Cheney Brothers Silk Company, Plimpton Press, and the major Philadelphia "demonstration plants" of the Tabor Manufacturing Company and Link-Belt Company. Tabor was owned by a boyhood friend of Taylor's, Wilfred Lewis, and by Taylor himself; Taylor had helped his friend out financially in return for the guarantee that his system would be adopted entirely. The Link-Belt Company was headed by the prominent mechanical engineer James Mapes Dodge, who had been one of the first manufacturers to introduce "high speed" steel into his shops and was able thereby to double the rate of his machines in one weekend. He discovered very shortly thereafter, however, that the higher machine speeds necessitated the complete reorganization of the production process, that the work activity and organization of the plant had become the limiting factors of his operations, and he thus called upon Taylor and his associates to introduce their system of management as well.

Taylor's success in convincing these leading industrial figures of the significance of his system lifted him out of obscurity. Through the efforts of Towne and Dodge, both former ASME presidents, Taylor became ASME president in 1906 and was charged with reorganizing the operations and publications of the association. By 1909 he had secured government contracts to introduce his system in the federal arsenals, and in 1910 the Taylor system, renamed "scientific management," burst into national headlines as a result of the testimony of Louis Brandeis and leading management experts before the Interstate Commerce Commission's hearings on the Eastern Rate Case. In glowing terms, scientific management was offered to the world as the key to prosperity and the solution to industrial strife. As a scientific system, however, it had already begun to fall apart behind the scenes.

Opposition to Taylor and his brand of management formed along three fronts: among owners and managers, among both organized and unorganized labor, and, most importantly for the present story, among engineers. Taylor had conceived his system in the machine shops of large corporations—Midvale and Bethlehem—and saw it as a means of extending the "shop culture" authority of the engineer to the corporate operations as a whole; the "planning department," staffed by professional engineers, was at once the medium of efficient corporate management and the bulwark against complete corporate-managerial usurpation of the engineer's fading supremacy in the shop. Adoption of the entire system by a company meant that corporate management and labor alike would have to adhere to the dictates of the engineers in the planning department.[14] It is significant, then, that only two companies —Tabor and Link-Belt—allowed for the introduction of the whole system; Taylor's attempt to put it into practice at Bethlehem led to his dismissal by the management, and the attempt to bring it to Watertown Arsenal prompted a labor strike. Employers did not readily accept Taylor's approach as an effective answer to trade-unionism, much less as an "answer" to corporate authority. When they did introduce scientific management, they tended to do so not as a "system" but as an arsenal of efficiency techniques, a way of streamlining production.

Before 1911 Taylor and his associates had restricted their activities to shops with a minimum of labor organization, or none at all. Of the early band of Taylorites only Gilbreth, who came to Taylorism from the building-construction industry rather than the shop, had real experience with trade unions. On the whole, Taylor and his associates had nothing but disdain for trade unions and viewed them as antithetical to the essential "harmony of interest" between capital and labor based upon increased productivity. They focused their attention upon the worker as an individual, seeking to improve his efficiency through scientific standardization of his activity and incentive wage schemes, thereby to effect the collective efficiency of all workmen. The workmen responded to Taylor's rather authoritarian methods in various ways:

the unorganized depended upon insubordination and sabotage;[15] the organized used the strike. Between 1905 and 1912 instances of worker refusal to cooperate with "time-and-motion" man Merrick were common, and union workers regularly walked out of shops when scientific management was introduced. The American Federation of Labor, moreover, began an intense campaign against the system, which was viewed by the trade unions as a direct threat to worker control of their trades; in 1911 the bubble burst with the spontaneous strike of the molders at Watertown Arsenal.

The strike at Watertown, a government arsenal employing civilian union labor, led immediately to a full-scale government investigation of "the Taylor and other systems of shop management." After extensive hearings, the investigators concluded that scientific management, while an effective means of "working out details" of production and administration, was not designed to ensure the best welfare of the worker. Even Taylor's dubious claim that his system demanded a "mental revolution" on the part of managers no less than workers was not convincing in the light of much evidence to the contrary.[16] The evidence of abuses uncovered by the inquiry, coupled with articulate and widespread labor opposition to scientific management, led to the banning of Taylor methods in the arsenals, in the Navy yards—on instruction of the new Assistant Navy Secretary, Franklin D. Roosevelt—and in all government-funded operations by 1916. These restrictions remained on the books until 1949, when they were finally removed through the combined efforts of Senators Ralph Flanders of Vermont, a former president of ASME, and Robert Taft of Ohio, co-author of the Taft-Hartley Bill.

The strong labor opposition to scientific management and the government investigation of Taylorism led to more than the banning of Taylor methods from government works; it signaled a major shift within the management movement itself. The government investigation and the investigation of Watertown by Miner Chipman had revealed that scientific management was not nearly so "scientific" as Taylor had claimed. It was discovered, for instance, that the "time and motion" experts frequently knew very little about the proper work activities under their supervision, that often they simply guessed at the optimum rates for given operations. This practice, of course, undercut the entire rationale of scientific management, for it meant that the arbitrary authority of management had simply been reintroduced in a less apparent form. The intense opposition to scientific-management methods, among both unorganized and organized labor, moreover, indicated to the more perceptive proponents of the system—Chipman, Harrington Emerson, Gilbreth, Gantt, Alford, and others—that it failed to elicit the willing cooperation of workers as individuals, even with the use of upgraded incentive-wage schemes, and that it failed to meet the challenge of labor in groups, the unions.

Taylor's fellow engineers had been skeptical about the scientific pretensions of his system from the outset. There was considerable opposition within the ASME among men who preferred to consider management an "art" rather than claim it as a "science," and few engineers outside of the ASME had ever heard of Taylor before 1910. Although Dodge and Towne had helped to get Taylor elected president of the society in 1906, they had never really convinced its membership that his management system was a science worthy of the professional attention of engineers. This became apparent when Taylor tried to elicit ASME endorsement for the statements of Brandeis and the management experts at the ICC hearings in 1910; he failed. He did succeed in having the ASME establish a committee to evaluate his system on its scientific merit, but this effort backfired: the committee also refused to endorse his scientific pretensions. "The term 'scientific management' has been generally and loosely applied to the new system and methods," the committee wrote. "This is commonly taken to mean that there is a science rather than an art of management. A truer interpretation is that it means management using scientific methods. . . ." Even Taylor's friend Dodge, who chaired the committee, ultimately rejected the scientific claims of Taylor's system, on the grounds that Taylor had used arbitrary definitions for the "average" and "first class" worker in standardizing his results, and that the large percentages added to the calculated time for a job to compensate for unavoidable delays in production had been arbitrarily determined. Taylor conceded to Dodge that there were still a few inexact elements in his system, as there were in all sciences, but argued that ultimately all elements of industrial management could be reduced to an "exact science." Dodge and his colleagues would not yield, however, and refused to have the ASME publish Taylor's *Principles of Scientific Management;* that internationally famous tract had to be issued outside of the professional engineering community.[17]

For the engineers in the scientific-management movement the rejection of the scientific pretensions of Taylor's system had two important consequences. First, it meant that the methods already developed required further refinement, to eliminate room for error and arbitrary judgment. Second, it indicated that the mere transfer of traditional engineering techniques to the social problems of management would not suffice; the new human focus of engineering required that the discipline of engineering itself had to expand, to include the new methods of the social sciences. The rejection of the pretensions of scientific management, like scientific management itself, had been based upon too narrow a conception of "science."

Thompson's "decimal dial" stopwatch and Gilbreth's "micromotion" studies based upon moving-picture films were the most important attempts to eliminate error and arbitrariness from the Taylor system; both focused upon one of its weaker points, time-and-motion study. Certainly the most vulnerable point, however, was Taylor's approach to the human problem of worker motivation. Relying exclusively upon incentive-pay schemes which mechanically linked pay to productivity, Taylor had no appreciation of the subtleties of psychology or sociology, and their possible use as management tools. Among the earliest of the Taylorites to develop such an appreciation were Gantt, Gilbreth, and Chipman. The latter, in his investigation of the Watertown strike, noted that the workers had objected less to the time studies and piece-rate system as such than to the "manner" in which these devices had been introduced. Chipman thus argued that effective scientific management had to be grounded upon the "consent" of the worker; the worker had to be made to feel that he was participating in all management decisions that affected him.

Gantt had quite early found Taylor's differential piece-rate system to be too harsh on the worker, and had thus developed his "task and bonus" system as a less severe substitute; rather than punishing a workman for not meeting standard rates, Gantt's system rewarded him if he did, and rewarded him more if he exceeded the rate prescribed. More than most of the other management engineers, Gantt focused upon the worker, the "human factor," from the standpoint of the impact of the industrial organization upon him as a human being. "The one common element in all enterprises is the human element," Gantt maintained, and he accordingly sought, especially in his later years, to "humanize" management by making it more flexible and more responsive to the human needs of the workforce. Toward the end of his life Gantt combined this view of management with the orthodoxy of "planning department" supremacy, which gave science priority over profit, and thus inevitably he became a spokesman for a technocratic radicalism.[18]

The single most important proponent of a broader view of the "human factor" within the scientific-management movement was Frank

Gilbreth. Gilbreth's interest in motion study led him to study the internal as well as the external causes of worker motivation; in his search for the "one best way," he sought to identify the causes of worker fatigue in the worker's mental disposition as well as in his working conditions. The most significant contribution Gilbreth made to the revisionist movement within scientific management, however, was his marriage to Lillian Moller—Lyndall Urwick labeled the union "providential." A psychologist by training, Moller fully appreciated the relevance of the industrial psychology of Hugo Munsterberg and Walter Dill Scott to the problems of scientific management. It was through her, and only indirectly through her husband, that the tools of industrial psychology became the tools of scientific management. Together the Gilbreths attacked the "scientific pretensions" of Taylor's system, and its authoritarianism, and sought to remedy its shortcomings through the use of the new social sciences. With their emphasis upon the welfare and training of the worker, they began to bridge the gap between scientific management and worker psychology, and to bring the Taylor movement into closer contact with corporate liberal personnel management.[19] As one perceptive historian of the movement has observed,

> Among the leaders of scientific management, a different tone was becoming recognizable. For some time prior to the war, the ideology of the . . . movement had been undergoing some subtle, but nevertheless significant changes. The emphasis placed by Lillian Gilbreth on "the human element" and the "psychology of management" reflected the beginnings of a shift of attention from the physical aspects of the job to the worker himself.[20]

When Taylor had argued, before the House Committee investigating his system, that scientific management demanded a "mental revolution" on the part of both workers and managers, he was merely expressing a faith that both would eventually recognize their mutual interest in increased productivity. Through the introduction of industrial psychology, however, Lillian Gilbreth had begun to render the management movement more capable of actually effecting this "revolution." After her husband's death in 1924, she joined A. A. Potter and the Department of Industrial Management at Purdue and continued her work within the realm of corporate liberal management reform.

In addition to the reorientation of scientific management toward securing the "consent" and "contentment" of workers, a parallel drive within the movement aimed at effecting the "mental revolution" among managers: formal management education. Taylor viewed management education as heresy and nonsense; for all his scientific pretensions about objectivity in management, he firmly maintained that managers were born, not made. As a result of Taylor's antipathy, management education emerged full-blown only after his death in 1915. The actual development of management education, however—a uniquely American phenomenon—had begun as early as 1904; at that time, G.E.'s Dexter Kimball began to offer his novel course in works administration at Cornell's Sibley College of Engineering—four years before such a course was offered at the Harvard Graduate School of Business. And in 1909 Hugo Diemer, former production engineer at Westinghouse, established his course in industrial management at Penn State and published the first textbook on scientific management, *Factory Organization and Administration.* Courses in engineering administration, combining instruction in engineering and management, were begun a few years later, first at Carnegie Tech and then at MIT. By 1922 ten schools in the country were offering such instruction in the science of management, and by 1932 the number had jumped to thirty.[21]

The revisionist movement within scientific management was reflected also in a new appreciation of the importance of "industrial relations" to effective management. Recognizing that labor unions were here to stay, farsighted management engineers began to attempt the "mental revolution" through cooperation with labor in groups and programs of industrial welfare. Gilbreth, with his experience in the highly organized building trades, was among the first to make the move, joined by such non-engineers as Robert F. Hoxie, Robert G. Valentine, and Harlow S. Person. The wartime experience of others heightened their appreciation of labor-union cooperation; Gilbreth, Thompson, Gantt, and Hathaway worked with such agencies as the Ordnance Department, the Emergency Fleet Corporation, and the United States Shipping Board, and Morris Cooke directed the labor-relations activities of the Ordnance Department throughout the war. The new positive approach to labor unions became the dominant theme of the scientific-management movement after the war and led, through the efforts of such people as Henry Dennison, Lillian Gilbreth, Sam A. Lewisohn, Dexter Kimball, and the influential editor of *Industrial Management,* Leon P. Alford, to the rapprochement between the management movement and the AFL during the 1920s.

The postwar activities of the scientific-management movement centered in three closely related organizations, which often held joint meetings. The oldest, the Society for the Promotion of Scientific Management, had been started by Gilbreth and Cooke much against the will of Taylor, who preferred to work solely within the ASME. Renamed the Taylor Society after Taylor's death, the society became a revisionist forum under the direction of Harlow Person, with offices in the Engineering Societies Building in New York—also the headquarters of the corporate-liberal Engineering Division of the NRC. By the early 1920s the society had opened its membership to management people outside of the Taylor movement proper, and representatives from GE, Du Pont, Western Electric, and AT&T began to take part in its expanded activities.[22] In 1917, at the prompting of the chairman of the Aircraft Production Board of the CND, Howard E. Coffin, the Society of Industrial Engineers was formed to promote efficient management. Headed by such men as Harrington Emerson, the original "efficiency engineer," Dexter Kimball, and J. W. Roe, management professor at NYU, the new society cooperated very closely with the Taylor Society and eventually merged with it in the 1930s.

The third forum for scientific management was established in 1920 within the association where the movement had first begun, the ASME. Founded and directed by Leon Alford, the new Management Division of ASME also strongly embraced the flexible revisionist tendencies within the scientific-management movement and focused attention upon "industrial relations" and "human engineering," the hallmarks of corporate liberal management. Unlike Taylor's earlier efforts along narrower and more rigid lines, the division proved popular among member engineers and within two years became, and remained, the largest within the ASME. Thus all three agencies of the scientific-management movement reflected the new gospel of revisionism. What was new to the Taylorites, however, was not really new; the scientific-management movement which began in machine shops during the closing hours of the nineteenth century had merely caught up to and merged with the most dynamic current in twentieth-century corporate industry: corporate liberal management reform.

Unlike the scientific-management movement, the corporate liberal reform movement was at no time the exclusive work of engineers. The corporate engineers, however, contributed significantly to its development, joining with other farsighted business leaders, bankers, politicians, trade-union leaders, and academic social-scientists to try to forge a viable corporate order. Corporate liberal management was far more sweeping and flexible than scientific management in its attitude toward the problems confronting modern capitalism. While sharing scientific management's systematic approach to efficient enterprise and fetish for detail and organization, it viewed that organization on a larger scale; whereas the Taylorist management experts confined their attention to the machine shop, the foundry, or the total operations of a single manufacturing plant, the corporate liberal managers embraced the sprawling empires of the giant corporations and, ultimately, the society as a whole. At the core of the corporate liberal management movement at all levels, as Wesley Mitchell characterized it, was the twofold effort "to understand and utilize [the] economic forces which control business activity"[23] and the "psychological forces which control human behavior."

At the level of the individual corporation, the science-based industrial reformers rarely relied upon outside management consultants to streamline their far-flung operations. This did not mean, however, that they failed to appreciate the importance of rational management and administration. Indeed, the corporate engineers who directed GE, Du Pont, General Motors, and a host of other mammoth enterprises were management pioneers in their own right.

At GE, for example, systematic management techniques were developed and practiced by such men as Dexter Kimball and E. W. Rice. Kimball, a mechanical engineer, made important contributions to scientific management as the works manager of the Stanley Electrical Manufacturing Company (which became the Pittsfield plant of GE) and, as professor and later dean of engineering at Cornell, offered the first courses in an American university on the principles of scientific management and wrote some of the classic texts in the new field. Rice began his career as a student of Professor Elihu Thomson at Philadelphia's Central High School and became the manager of the Thomson-Houston company, where he systematized production. When Thomson-Houston became part of GE, Rice became, successively, technical director, vice-president, and president, all the while promoting rational, streamlined methods of operation. The cafeteria system at GE, developed by Rice's staff, indicates the enthusiasm with which this was done. In 1917 the staff applied "engineering principles" to the process of providing food for employees and came up with an automated system in which meals could be served to hungry customers in one minute flat. The procedure was, in effect, no different from any other in the factory; indeed, it was apparently viewed as simply another aspect of the production process: the fueling of the human machinery.

Promptly upon the blowing of the whistle at noon four lines of men form in front of four cash registers to purchase their luncheon tickets. Few of us have ever had the opportunity of watching a cashier make change and sell 29 tickets per minute, yet this is the rate of speed at which each of the four cashiers operates. Anyone who hears the clang of a cash register bell every half second can appreciate how rapid must be the food distribution necessary to keep pace.

After the men file past the cash register, they approach at right angles to the end of one of the four conveyor belts. Adjacent to the nearer end of the belt conveyor the ticket is exchanged for an aluminum tray which is laid on the conveyor belt. The belts travel at the rate of 65 feet per minute and allow 15 seconds for the customer to select his food. . . .

By this time his tray is within five feet of the end of the belt, where the checker o.k.'s the contents of his tray. After removing his tray from the conveyor belt the diner takes it to his seat . . .; many men have finished their luncheon by 12:10 and the process of removing the dishes begins immediately. At 12:10 service again begins in the restaurant.[24]

The man who probably made the most significant contributions to modern management at GE was Gerard Swope. An orthodox Jew and son of a St. Louis manufacturer of watch cases, Swope took a bachelor's degree in electrical engineering at MIT and then joined the Western Electric Company as a design engineer and sales representative. Because of his keen commercial sense, he soon became manager of the St. Louis branch of the company. By 1913 he had become a director of Western Electric and vice-president in charge of domestic commercial activities and all manufacturing, engineering, and commercial activities abroad. After wartime service on the General Staff of the Army helping to plan the Army procurement program, Swope joined GE as the first president of the International GE Company (the consolidation of all of GE's foreign operations). Three years later he succeeded Rice as president of GE.

As head of the GE empire, Swope developed elaborate administrative controls which enabled him to centralize authority and oversee GE's ever expanding operations. Perhaps his most important contribution to GE's growth was his initiation of product diversification. Swope adopted this new policy to make full use of existing resources, particularly the heavy investment in research and development, and to create a greater demand for GE's primary line of products—electrical generating, transmission, and control equipment. Before 1922 the only product the company sold directly to the public was the incandescent electric lamp. By 1930 half of GE's business was derived from products unknown to the public before 1919, including a wide range of household electric appliances.

In addition to his administrative and commercial contributions, Swope initiated some far-reaching labor policies in an effort to secure the cooperation of GE employees. As a young engineer-manager at Western Electric's Hawthorne plant, Swope had taught technical courses at Hull House and gained first-hand experience with industrial welfare work. In addition, he had studied business law with a leading corporate liberal lawyer, Louis D. Brandeis. At GE he translated this experience into elaborate industrial-relations programs—life insurance, unemployment benefits, workmen's compensation, pension plans, profit-sharing, and the like—designed to eliminate the most apparent evils of industrial labor and encourage the loyalty of corporate employees.[25]

What GE was to the electrical manufacturing industry, Du Pont was to the chemical. Here the key figure was Hamilton Barksdale. The son of a physician from a distinguished Virginia family, Barksdale graduated as a civil engineer from the University of Virginia, and spent the early part of his career on river-development and railroad projects. His work on a railroad near Wilmington, Delaware, brought him into contact with the Du Pont family, and he soon thereafter joined the Repauno Chemical Company (which had just been set up by Lammot Du Pont and was independent of the Du Pont Company proper) and married the daughter of Victor Du Pont. Barksdale quickly became an important manager in the new company and began to devise his own "system of management"; in 1893 when Repauno became part of the Du Pont Corporation, Barksdale was made its general manager as a result of his management innovations. When, in 1902, the engineers Pierre S., Alfred I., and Coleman T. Du Pont bought the Du Pont Company and began to chart its new expansionist course, Barksdale was promoted to director, vice-president, and *de facto* general manager of the entire Du Pont operation.

As the cousins Du Pont embarked upon their program of acquisition and diversification, Barksdale simultaneously devised the administrative and managerial procedures needed to supervise, control, and streamline the enlarged, diverse enterprise. In all of his administrative work he emphasized the importance of a rational, systematic approach to business affairs, from production to marketing. He formulated a theory of administration in which he stressed the need for clearly established guidelines and criteria against which to evaluate corporate progress; for the articulation of an overall "philosophy of management" to give direction to short-range policy-making; for the establishment of administrative "permanence" through the ongoing training of "understudies"; and for flexible decision-making processes through the broad delegation of authority, the group approach to management, and the continuous development and refinement of administrative skills. On the practical level, Barksdale put his theory into operation with the establishment of Du Pont's Executive Committee and Development Department, the agencies through which top management could supervise corporate affairs and devise new methods for dealing with immediate and long-range problems.

Barksdale introduced scientific works management as well. He applied systematic methods to the manufacture of explosives, introducing the carefully regulated flow of work, the scheduling of materials, inventory analysis, and the standardization and simplification of processes and products, and he carried these methods into the sales department with the introduction of market analyses, a separate "statistical division," and the integration of marketing and technical services. Like Taylor, he focused upon the work activity of the individual worker, setting up a "labor efficiency division" to establish standards of performance, set rates, and devise incentive-pay schemes to elicit worker cooperation. Unlike Taylor, he recognized the stubborn complexity of human psychology and anticipated the need for what he called "motivational researchers."[26]

Under the innovative leadership of Barksdale and his colleagues, Du Pont became one of the first corporations to adopt a strategy of diversification, primarily through bold research and development activities, in order to make full use of surplus resources. In addition, Du Pont was among the first giant industrial corporations to develop a decentralized, multidivision administrative structure to coordinate the diversified operations. Perhaps Barksdale's most important contribution to the development of management, however, was his training of a group of "understudies," young ambitious engineers who would become major figures in corporate management. Prominent among them were Frank McGregor, Harry Haskell, William Spruance, John Lee Pratt, and F. Donaldson Brown.

McGregor was an MIT engineer who chaired the subcommittee of the Development Committee that made the initial proposals for the adoption of the decentralized structure. In his work he was joined by Haskell, a mining engineer and Du Pont vice-president (whose brother was president of the Repauno Chemical Company), and Spruance, a judge's son with an electrical-engineering degree from Princeton and vice-president in charge of production. Pratt was a University of Virginia civil engineer like Barksdale and perhaps his closest disciple. After a short career at Du Pont, he became assistant to General Motors founder William C. Durant and helped to reorganize that automotive empire. Brown was an electrical engineer from VPI who joined Barksdale's staff (and married his daughter) after a stint as general manager at the Sprague Electric Company. Brown's major innovations were in finance, devising techniques and developing formulae for relating the return on capital invested to the turnover of capital and to the volume of sales as well as to profit. Breaking down company operations into component parts, he provided executives with an accurate standard against which to measure the performance of each unit of operation. As treasurer of Du Pont, he introduced sophisticated statistical controls and economic forecasting methods with which to comprehend, anticipate, and manage the multifarious aspects of corporate enterprise. In 1920, when Du Pont assumed ownership and management of General Motors, Brown joined its new president, Pierre S. Du Pont, as vice-president and participated in the reorganization of that stumbling giant.[27]

In addition to the Du Pont personnel who came to the rescue of General Motors—Pierre S. Du Pont, Brown, and Pratt—there were a few men already at the automotive company who understood the principles of modern management. Prominent among them were two electrical engineers, Charles E. Wilson and Alfred P. Sloan. Wilson, a graduate of Carnegie Tech, started out with Westinghouse and then joined Charles Kettering at Delco, where he helped design the electric starter motor for automobiles. When Delco became part of the GM organization, Wilson began his climb up the corporate ladder. In 1940 he was made president of General Motors.

The key figure at General Motors, of course, was Sloan. Born in New Haven, the son of a prominent coffee and tea importer, Sloan was an early graduate of the electrical-engineering department at MIT (where he was a classmate of Gerard Swope). After graduation, he joined the Hyatt Roller Bearing Company, where as a draftsman he helped to improve the billiard ball (made of the plastic invented by John Wesley Hyatt) and develop the commercial use of ball bearings. Recognizing the great role the Hyatt company could play in the emerging automotive industry, Sloan persuaded his father to buy the company, with the result that by 1897, at the tender age of twenty-two, he became its president and general manager. As an indefatigable promoter of his company's products, he became intimate with the giants of the new industry and acquired a thorough training in all phases of the business: designing, engineering, production, sales, advertising, and executive management. In 1916, when Durant organized the United Motors Company (a consolidation of various parts and accessories manufacturers, including the Hyatt company), Sloan was elected president. Two years later when the United Motors Company became a part of the new General Motors Corporation, Sloan was made director and vice-president in charge of accessories and a member of the executive committee. In that capacity, he began to formulate a plan for reorganizing the corporation to coordinate all its diverse activities. In 1920, when Pierre Du Pont took over the reins, Sloan started to put his plan into operation with the assistance of Brown and Pratt and the encouragement of the new president. Three years later he succeeded Du Pont as president and carried through the adoption of a multidivisional, decentralized structure.[28]

In addition to their work within the science-based electrical and chemical industries proper, the corporate engineers thus played a key role in carrying the gospel of modern management into other industries. Paul Litchfield, an MIT chemical engineer, introduced systematic management at the Goodyear Rubber Company; James Barker, an MIT graduate and onetime professor of civil engineering, joined with Robert E. Wood, a West Point engineer and once assistant director of Du Pont's smokeless-powder plant, and Theodore Hauser, an Iowa State electrical engineer, to devise a decentralized administrative structure at Sears Roebuck; and Frank A. Howard, a George Washington University engineer and patent lawyer, worked with Everett J. Sadler, a Naval Academy engineer, to similarly reorganize Standard Oil of New Jersey.* Not all of the corporate engineers, however, confined their management activities to individual corporations. Some broadened their range even further, to embrace the "social mechanism" as a whole.[29]

*By this point the reader may well be wondering where Henry Ford fits into the picture; it is a difficult question to answer. Although he worked as an engineer for Detroit Edison before entering the automobile manufacturing business, Ford—like Taylor but unlike most of the other men discussed here—was neither a college-trained professional engineer nor one tutored in science. Unlike Taylor, Ford laid the technological foundations for mass production more in the manner of a mechanical-minded inventor than as a proponent of scientific management. Moreover, although he pioneered in industrial relations with his famous "sociology department," he never fully shared the corporate liberal vision, preferring to prevent unionism at all costs, enforcing a strict puritanical code of behavior for his workforce to enhance its productivity, and raising the wage level to ensure its cooperation, loyalty, and ability to buy his cars. Ford, much more than many of his seeming colleagues, resists categorization. Omission of him here is thus by no means simply an oversight, but deliberate.

By the turn of the century, as corporations expanded to encompass a wide range of productive activity, they began to set up "planning departments," special executive committees, and research staffs to grapple with problems of labor turnover, unemployment, market variations, production rates, long-range budgeting, and the like, problems which defied simple analysis and at the same time lay at the foundation of corporate economic viability. They began to pool their resources and establish collective research agencies, through various trade associations, in search of effective ways of understanding and coming to terms with the larger factors which determined their success or failure. It had become clear to many that the determinants of corporate health extended far beyond the isolated plant, or even an entire industry; that the factors ranged as far as the corporations themselves: nationally and globally. Harlow Person, the economist who directed the Taylor Society during the 1920s, proudly proclaimed the expanding field of management; in 1931 he embraced the entire scope of corporate activity, outlining the pyramiding factors upon which its stability depended.

> Stabilization of material forces is not sufficient; human relations must be stabilized; stabilization of production is not sufficient; merchandising must be stabilized. Stabilization of production and merchandising is not sufficient; general administration must be stabilized. Stabilization of an individual enterprise is not sufficient; all enterprises in the industry must be stabilized. Stabilization of one industry is not sufficient; all industries of a nation must be stabilized. . . . Stabilization of national industry alone is not sufficient; international economics must be stabilized. Achievement of any of these ends is a step toward a more balanced and harmonious industrial and social life; each end is but a means to another greater end.[30]

This ever-widening perception of the exigencies of capitalist growth and stability prompted corporate leaders to band together, to pursue governmental regulation, or otherwise to counter the vagaries of the competitive market. It compelled them likewise to seek military protection for their investments overseas, a constant cause for concern, and forced them to create research agencies for the collection and analysis of all data upon which corporate stability depended. While these undertakings involved the participation of experts of various kinds—statisticians, economists, and sociologists, for example—they were also the work of engineers, men with technical training as well as considerable breadth of vision: modern management was no simple affair. Malcolm

Rorty, Magnus Alexander, and Herbert Hoover were representative of this new breed of engineer.

Rorty was an electrical engineer who pursued an impressive career as a production man with the J. G. White Company and the New York Telephone Company, and as a research executive with the Bell Labs. His interests soon extended beyond the limited scope of electrical engineering, however, to statistics and economics. He conducted statistical studies of various social problems for AT&T and in 1916 joined Wesley Mitchell and Magnus Alexander in a major study on the distribution of income in the United States, a study which was interrupted by war service. Shortly after the war, Rorty teamed up with Mitchell and John R. Commons to establish the National Bureau of Economic Research, Inc., an agency charged with conducting sweeping statistical analyses of the problems confronting the corporate order.[31] During the 1920s the staff of the NBER contributed significantly to the seminal studies of "recent economic changes" and "recent social trends" commissioned by Herbert Hoover. Rorty served as president of the NBER in 1922–3, in addition to being vice-president of AT&T and later president of the American Management Association—vanguard of the corporate liberal management movement.

Magnus Alexander was another electrical engineer with broad vision. After working as a design engineer at Westinghouse and GE, he branched out at GE to supervise the technical courses for college-graduate engineers, to organize and direct the apprentice school, and to found the medical department and safety committee. In addition, he began extensive statistical studies of industrial accidents and produced the first comprehensive report on the cost of labor turnover in the United States. In 1916 he collaborated with Rorty and Mitchell on the distribution-of-income study and, in the same year, founded the most important social-research agency at the disposal of industry: the National Industrial Conference Board. From 1916 until his retirement, Alexander directed its activities and, in this capacity, contributed as much as any single individual could to the corporate effort to comprehend and meet the challenges of advanced capitalism.

Joining Alexander in that regard was Herbert Hoover. Despite the Depression-born mythology which casts Hoover as a bungling conservative, the truer-to-life characterization of this former mining engineer as a highly intelligent and sophisticated corporate reformer has begun to emerge. After amassing a large personal fortune as a promoter of mining ventures throughout the world, Hoover entered public life. During the war, in addition to directing the famous food relief programs in Europe, Hoover contributed to the coordination of industrial resources under the auspices of the War Industries Board. Immediately after the war, as president of the progressive Federated American Engineering Societies, he headed the famous Waste in Industry survey which placed much of the blame for industrial inefficiency upon management. As Secretary of Commerce during the 1920s, Hoover completely reorganized the department to meet the growing needs of corporate industry, tirelessly promoting voluntary industrial cooperation and efficiency and the new government creed of service to industry. As President, in addition to anticipating many of the New Deal programs for fighting the Depression, Hoover commissioned the most comprehensive studies of broad social change ever undertaken up to that time, the famous *Recent Economic Changes* and *Recent Social Trends,* to assess what had happened to America in the twentieth century. Finally, Hoover was the engineer's engineer; Morris Cooke described him as "the engineering method personified." In all of his activities he served as a model of corporate engineering; as an inspired Charles Mann observed, Hoover's accomplishments suggested "how engineers do what lawyers talk about."[32]

The extended activities of engineers in the shop, the corporate board rooms, and the various private research agencies and bureaus of government, undertaken to "understand and utilize the economic forces which control business activity," reflected the ever widening horizons of engineering itself. So too did their increasing efforts, at various levels, to understand and utilize the forces underlying those "economic forces": the "psychological forces which control human behavior." In this aspect of their work, the corporate engineers strove to eclipse the industrial strife that challenged corporate stability, to effect a working "harmony of interest" between labor and capital by eliciting the willing participation of workers in their own exploitation.

The corporate liberal management approach to the "man problem" was twofold. "Industrial relations" aimed at improving the lot of workers in order to win their cooperation and loyalty; "personnel management" aimed at the processing of human beings and the engineering of human behavior through the incorporation of the new tools of psychology and sociology within engineering. By the 1920s the former had more or less collapsed into the latter to become "human relations," the catchword for transforming the stubborn "human factor" of production into an efficient, adjusted part of the corporate mechanism. The process of instilling in employees and potential employees a motive of selfless service, and restricting aspirations and human potentials to fit corporate-defined dimensions, involved reform both within and without the industrial plant. Ultimately, it led the corporate reformers into the schools and homes of employees to habituate and train students for industrial employment and stamp out such vices as drink, disorderliness, laxity, and radical politics.

Industrial-relations programs derived historically from the "industrial betterment" or "social welfare" movement which attained its peak influence during the first decade of the century. Motivated by religious calling and *noblesse oblige,* welfare workers achieved a foothold within industry as "welfare secretaries" operating out of "welfare departments" on the periphery of management. The welfare workers focused upon the plight of industrial workers and strove to improve their lot and to "uplift" them, through educational, cultural, medical, and recreational services within the plant. No particular preparation was required for welfare work, and among those who practiced it were people trained in nursing, medicine, architecture, domestic science, and engineering; a considerable proportion, in addition, possessed no special aptitude beyond a devout need to help the "working poor." As industrial employees, the most important contribution of the welfare workers was their intense devotion to the worker as a human being, the first significant expression of such interest within modern industry. While they were themselves responsible for the oppressive-sounding term "human engineering," they placed emphasis upon the word "human" rather than "engineering" and strove to "humanize" the workplace and thereby ease the suffering generated by industrial capitalist development.

As the first real link between the company and the employee, who were being drawn further apart as industrial concerns expanded, the welfare workers injected into management a spirit of human cooperation which would remain. Welfare work as an industrial institution, however, came into disrepute by the second decade of the century, owing to its paternalistic approach to labor problems, its self-righteousness, and its vaguely "humanitarian" rather than practical, businesslike orientation to the problems of industry. One company, for example, sought to "eliminate absolutely all the hysterical elements and 'charity phases' so often found dominating welfare work"; an industrial physician reported hearing the term "welfare" referred to as "hell-fare";[33] and one of the largest coal operators in the country declared: "I hate those words, 'welfare work,' I wish they could be taken from the language."[34] But this rejection of welfare work did not mean that industrial leaders had turned their back on the "human factor" or given up the task of "human engineering." Quite the contrary. The same coal operator taught English to his miners, taught their wives to make American dresses, aided them in starting home gardens, gathered their children into kindergartens, and otherwise provided for their education and encouragement. He did not engage in such activities, however, out of a spirit of charity or guilt-prompted philanthropy; rather, like other industrial leaders, he had realized that "human engineering" and "social uplift" were important new methods of good, profit-making business practice.

"Industrial relations," in Leon Alford's definition, "comprises that body of principle, practice, and law growing out of the interacting human rights, needs, and aspirations of all who are engaged in or dependent upon productive industry." From a different perspective, that of Thorstein Veblen, it constituted "what business men may be expected to do for cultural growth on the motive of profits."* By means of industrial-relations activities, corporate liberal reformers tried to win the friendship, or at least the loyalty, of workers, and to effect industrial stability—the *sine qua non* of future profits. The Ford Motor Company, for example, found by 1916 that "all this investment, profit sharing, factory environment, comfort, educational work, looked at from the cold blooded point of view of business investment is the very best investment it has ever made." Similarly, the General Electric Company noted that "the principles and method" of its Mutual Benefit Association—a group insurance program created for employees at Lynn in 1902—"develop contentment among the members, and relations of mutual loyalty between the employees and the Works' management." "Industrial relations" variously involved recognition of organized labor and collective bargaining, profit-sharing, company

*That Veblen was closer to the mark is made clear in an early lecture by Erwin Schell, professor of engineering administration at MIT. Addressing his students on the trends in management, Schell had this to say about industrial-relations programs: "The accident prevention movement, which endeavors to maintain the employee in a safe working environment, has shown a definite return in the reduction of compensation insurance rates. The value of the company doctor, the industrial dentist, the visiting nurse, and the first aid equipment has been measured definitely in the reduction of absenteeism. The favorable results obtained from good lighting, heating and ventilation and from wholesome foods served in company restaurants and lunch rooms have been reflected in terms of shop production. The solution of transportation and housing problems, the development of shop publications, the introduction of group insurance and mutual benefit plans have proved their worth in many cases by the attending reduction of labor turnover." Lecture One, "The Trend in Management," *Course Record in Business Management* (1920), p. 5, MIT Archives.

magazines, insurance programs, pension plans, safety reform, work-men's compensation, "shop committees" for "joint control," and, espe-cially after the war, restricted work hours and the "living wage." It also carried over such aspects of the industrial-betterment movement as gardens, restaurants, clubs, recreational facilities, bands, and medical departments.[35]

There were important instances of engineer participation in a num-ber of these areas. Magnus Alexander, for example, chaired the Massa-chusetts commission on old-age pensions and sat on the five-man commission on workmen's compensation—one of the first in the coun-try; H. F. J. Porter, a prominent mechanical engineer, established the first "representative shop committee" to allow for "worker participa-tion" at the Nernst Lamp Works of Westinghouse in 1903; and John Henning, a mining engineer in Louisiana, instituted the first voluntary workmen's-compensation plan in 1904. Probably the most significant contribution of engineers to industrial relations, however, came in the area of industrial safety reform. Engineers, as line managers on the production floor and designers of machines, were acutely attuned to the problems of industrial safety. One fifth of the members of the advisory committee of the New York Commission on Industrial Safety were engineers, and the ASME pioneered in the formulation of industrial safety codes. Engineers also played an important part in the creation of a national agency for safety reform, the National Safety Council; the founding convention of the National Safety Council, held in Milwaukee in 1912, was conducted under the auspices of the Association of Iron and Steel Electrical Engineers.[36]

At the first National Safety Council convention, one of the speakers observed that "the subject of accident prevention is one which has, during the past two years, become to nearly all large manufacturers and to a great number of smaller ones, as important as the question of output."[37] The cause for this concern is not hard to guess; as Alford explained, "about 1910 American juries began to award large sums in suits for personal damages, where the plaintiff had been injured by machinery or otherwise in industry."[38] Increasing costs to industry in the form of damage suits constituted another area of waste. No one better appreciated this than Magnus Alexander, who spoke on "The Economic Value of Industrial Safety" at the founding convention of the council. Drawing upon his own experience as chairman of the safety committee at GE, Alexander argued that "the employer ... should have economic considerations, as well as humane considerations as the impelling force for establishing safe and sanitary working conditions.

To the employer, in the long run, it is far cheaper, besides being more humane ... to prevent an accident than to pay for the consequences of such an accident." While he recognized the economic implications of the safety movement, moreover, he fully appreciated that it was a "movement of education," one means among many to effect harmonious industrial relations. His closing remarks at this convention constitute a classic statement of the corporate liberal approach to social reform: if the corporations do it voluntarily, it gets done their way and they enjoy the "applause of the masses" and a degree of social harmony; if the corporations don't do it voluntarily and there is a demand for it, it will eventually be forced upon them by a hostile society not particularly concerned with corporate welfare.

> And one more thought; the regrettable thing to me is, and I want to throw out this thought because it has its bearing on many similar things that are in the making just now, it is a regrettable fact that, barring a few notable exceptions, our employers throughout the United States did not get very busy on this important matter of accident prevention and safeguarding until they were forced by legislation. What they might have done voluntarily, and should have done voluntarily years ago with the applause of the masses ... would ... have prevented a great deal of hostile legislation.
>
> Gentlemen, bear that in mind, because there is a great deal in the relation of employer and employees in improving the welfare of our employees that our employers ought to do and must do, and unless they wake up and go to work and do it soon voluntarily, with all the wonderful effect that it will have on such relationship, legislative action will force them. ... [39]

Industrial-relations measures constituted an indirect attack upon the "man problem"—the root cause, in the eyes of industrial leaders, of social instability. Rather than focusing directly upon the worker as an object for study, they aimed at improving the external conditions of his daily life, on the assumption that better living and working conditions would render him more cooperative, loyal, and content, and, thus, more efficient and "level-headed." Industrial relations, then, was the means by which farsighted industrial leaders strove to adjust—or to give the appearance of adjusting—industrial reality to the needs of workers, to defuse hostile criticism and isolate irreconcilable radicals by making the workers' side of capitalism more livable. The other corporate liberal approach to the "human factor," "personnel management," was considerably more direct in that it focused upon the worker himself, individually and in groups, as an object of scientific investigation and control; it was here that the corporate engineers made their most important contribution to the solution of the "man problem."

Personnel management involved two interdependent aspects of "human engineering": employment management, the increasingly scientific attempt to motivate and utilize efficiently the human material of industry; and industrial education, the systematic habituation and training of the working population for optimum corporate service. It emerged in response to a number of historical developments which came to the fore in the early part of the century. One of these, of course, was the obvious disenchantment of a large segment of the population with the conditions of industrial life, a growing sentiment stemming from the loss of control over its own productive labor. Just as important was the expansion of the corporations. As companies grew in size, the ranks of employees naturally swelled proportionately and became ever more removed from the immediate purview of the central management. The application of scientific-management techniques aggravated this problem because, as Malcolm Rorty noted, it rendered "practically all types of business and industry . . . open to efficiently large-scale corporate control." The expansion of the corporations demanded more effective personnel procedures through which to manage the growing armies of employees. (From another perspective, it meant also, as Rorty observed, that "the young man of capacity and intelligence will have to look forward more than ever before to a career in which he will continue throughout to be a subordinate worker in a large corporate organization."[40])

Personnel management emerged to confront head-on the task of refashioning the living part of industry—from the laborer to the lawyer —into a coordinated industrial army. Its rapid development during the first two decades of the century prompted Wesley Mitchell to observe that by the 1920s what characterized American industry was "not only more production per man, more wages per man, and more horse-power per man," but also "more management per man."[41] Personnel management was "human engineering" with emphasis on the word "engineering"; the reverse of industrial relations, it aimed at the adjustment of the workers to meet the new industrial needs of corporate capitalism.

Engineers, who had had so large a hand in shaping corporate industry, were, as production and corporation-school managers, among the earliest to undertake this new task.* The personnel-management move-ment evolved out of the industrial-welfare department and the corpora-tion school and thus involved both "uplifting" the workers and educating them for more useful industrial service. What distinguished it from its sources was the new emphasis upon, and perception of, the worker as an object of scientific study and control. The employment manager of the Norton Company, home of one of the country's first employment departments, explained what the new "human engineer-ing" entailed:

> It seems to be very largely a question of knowing or judging what given individuals or groups of individuals will do under a given set of condi-tions, and knowing from experience what people have done under such circumstances, provide suitable means so that the circumstances and what follows from them may go along the line which will bring the greatest profit to the company employing the men.[42]

The task of "human engineering," then, appeared to differ little—in the view of early practitioners—from "the more material forms of engineering." It similarly consisted in the prediction of behavior based upon careful observation, and the determination and creation of the conditions necessary to produce a desired end. That end, moreover, remained essentially the same: profit—only in this case the behavior of human beings rather than physical materials would become the object of attention, the means to that end. Underlying this new "branch of engineering," therefore, was the assumption, the faith, that, as one more recent believer phrased it, "Deep as they are, the factors that 'make a man tick' can be described and analyzed with much of the precision that would go into the making of dies for the side of a Sherman tank."[43] It is thus not at all surprising that engineers, people with considerable experience in the use of scientific procedures, "the handling of men," and the employment of personnel, should undertake this new challenge to their discipline. "The human element of labor is a challenge to the engineer," the managing editor of the *Engineering News-Record* wrote in 1918:

---

*Surveying the experience of American employers in dealing with the "human factor in industry," two executives of the Metropolitan Life Insurance Company observed in 1920 that four organizations dominated the field on the national level: the NACS, the National Safety Council, the National Society for the Promotion of Industrial Education, and the National Employment Managers' Association. The first three were organized at the initiative of engineers, and the fourth involved a fair degree of engineering participation as well. Lee K. Frankel and Alexander Fleisher, *The Human Factor in Industry* (New York: The MacMillan Co., 1920), p. 14.

> He has applied the laws of physics to produce efficient machines. . . .
> He must now step in—not as a welfare worker, not as a sociologist,
> but as an engineer—to help labor find its place in the production
> scheme. Cannot scientific analysis resolve the causes of maladjust-
> ment which threaten the life of our institutions? Cannot the engineer-
> ing mind reorganize the human elements of production as it has
> already done with the mechanical and material elements to secure ef-
> ficiency?[44]

The employment-management movement, as it was called by those
who participated, began late in the first decade of the century, largely
in response to newly identified causes of industrial waste—the increasing rates of labor turnover and absenteeism. It developed rapidly during
a period of intense labor shortage before, during, and after the war, and
lost momentum after 1921 in the wake of business stagnation and a
glutted labor market. By that time, however, it had already provided
industry with a hitherto unknown arsenal of human engineering techniques—psychological and trade tests, job analyses, personnel records,
rating scales, systematic training, etc.—and, more important, with
what Henry Dennison termed "a habit in the business mind of considering personnel management, . . . the serious project of human engineering, . . . as a difficult, distinct, and major function of business
management." By 1923, the organizers of the new American Management Association had clearly adopted that habit. In their statement of
purpose for the new organization, they noted with some urgency that
"the day when American management can afford to treat the human
factor as 'taken for granted' has gone by and today emphasis must be
laid on the human factor in commerce and industry and must apply to
it the same careful study that has been given during the last few decades
to materials and machinery."[45]

"No other line of development in industrial relations," Leon Alford
noted in 1919, "has had the rapidity of growth of employment management." The "impelling motive" behind such growth, however, he
added, "has not been entirely that of fostering good industrial relations. . . . The major reason in the minds of most industrial executives
in establishing employment departments has been to secure employees
during the period of labor scarcity and to find out why men leave."
Such problems were certainly real for employers; by 1917, one social
historian has discovered, "workers improved their incomes as much by
moving from job to job as they did by striking. It was common for
workers to accept six to eight jobs in a single day of searching, then to
report to the most promising one." A contemporary economist ob-

served that an annual factory turnover rate "of 1,600 to 2,000 per cent was by no means phenomenal."[46]

As early as 1907 Westinghouse engineer H. F. J. Porter had pointed out to an attentive ASME audience the "evils of labor turnover" as a cause of industrial inefficiency. It was not until 1913, however, that a GE engineer, Magnus Alexander, produced the first statistical study of the problem, documenting its impact upon industrial efficiency. Another major study, which outlined the most promising means of handling the turnover problem—"hiring and holding," as it was termed— was made a few years later by Boyd Fisher. A vice-president of the Detroit Executives' Club, a group of engineer-minded leaders of the nascent automotive and related industries, and later service manager of the Lockwood Greene Engineering Company, Fisher devised a cost system for recording the waste incurred by unnecessary hiring and firing, and pressed for such remedies as care in hiring, testing of applicants and employees, application files, and better working conditions. "A list of men required for the year's predicted production," he argued, "should be just as much a part of the engineering department's specifications as the blue-prints and the routing."* In addition to these in-plant procedures, Fisher argued for "the extension of factory influence into the whole life of the worker"—an innovative notion which he attributed to Henry Ford. Noting that "eighty per cent of the causes of labor turnover lie outside the plant," Fisher urged that the employment manager must become a "co-partner with the teacher, the minister, the social worker in the business of reforming men." Pointing out that in Detroit the vice-president of Ford Motor Company had become police commissioner and an executive of Detroit Edison was president of the school board, Fisher boasted that "it wasn't Billy Sunday, it was the employers of Michigan that put the state in the prohibition column. They wanted to remove the saloon on the route between home and factory." As early as 1916 Fisher had thus outlined the broad scope of employment-management activities. A colleague in this effort, the em-

---

*In his lectures at MIT, Professor Schell, an engineer himself, referred to employment management as "labor maintenance" similar to other types of industrial maintenance functions. "Prior to the coming of the employment manager, the manifold tasks pertaining to the maintenance of funds, of buildings and equipment, of machinery and tools, of materials and supplies, of standards, of records and statistics, and of product demand were given to trained experts for accomplishment. These duties deal with things. The function of labor maintenance deals with men and is therefore the only activity which, when improperly conducted, weakens and endangers the control structure of the production executives who conduct the fundamental activity of the industry." *Course Record in Business Management*, 1920, MIT Archives.

ployment manager of the General Railway Signal Company of Rochester, suggested additional responsibilities. These included the preparation of adequate job specifications; the introduction of new employees; conducting follow-ups on employees; keeping adequate records of all personnel procedures; rendering final decisions on all disputes, discharges, promotions, and transfers; studying earnings of all employees and labor turnover rates; investigating the causes of absenteeism and terminations of employment; and supervising the instruction of all employees. Above all, he added, the employment manager must "aim to give the plant a good name."[47]

B. F. Goodrich established the first employment department in the country, at Akron in 1900, and during the next decade many of the nation's major companies followed suit. In these cases the employment department was intimately related to the "educational departments" which administered the corporation-school activities of the company. Thus, J. W. Dietz of Western Electric, Channing R. Dooley of Westinghouse, Albert Vinal of AT&T, Magnus Alexander of GE, and Mark M. Jones of Thomas A. Edison, Inc, were all involved in the development of employment management in their respective companies. The corporation schools, which trained graduate engineers as well as apprentices, were among the earliest departments of management to utilize procedures such as testing, record-keeping, job specifications, etc., which ultimately came to be used for all employees. In addition to the large corporations, many smaller concerns, particularly those most influenced by scientific management—Yale and Towne, Dennison Manufacturing, Norton Company, Cheney Brothers, Plimpton Press, and Curtis Publishing Company, for example—were among the earliest to set up employment departments.[48]

Around 1910 the first two local employment-managers' associations were established, independently of each other, in Boston and Detroit. The former was an offshoot of the Vocational Bureau of Boston* and

*The Vocational Bureau had been founded two years earlier by engineer Frank Parsons, the "father of vocational guidance." Parsons viewed human material like any other. In his book *Our Country's Need, or the Development of a Scientific Industrialism,* he wrote: "Life can be moulded into any conceivable form. Draw up your specifications for a dog, or man . . . and if you will give me control of the environment, and time enough, I will clothe your dreams in flesh and blood. . . . A sensible industrial system will seek to put men, as well as timber, stone, and iron, in the places for which their natures fit them, and to polish and prepare them for efficient service with at least as much care as is bestowed upon clocks, electric dynamos, or locomotives." Quoted in Joel Spring, *Education and the Rise of the Corporate State* (Boston: Beacon Press, 1973), pp. 92, 95.

included representatives from such firms as GE, Cheney Brothers, Dennison Manufacturing Company, and Norton Company. The latter, a special committee of the Detroit Executives' Club, included managers from Ford, Studebaker, Packard, Dodge Brothers, Michigan State Telephone Company, and Saxon-Solvay Company. By 1917 there were ten such associations on the local level, representing one thousand business concerns.* In all of these associations, engineering-oriented companies tended to dominate the proceedings. For the first few years the associations cooperated through common membership in trade associations, and after 1913 through the NACS, which served as a clearinghouse for all employment as well as educational activities of member concerns. In 1917 employment managers assembled at the University of Pennsylvania to establish a national organization for employment-management activities exclusively; dominating the conference were the leaders of the employment-managers movement in the United States, men like Alexander (GE), Fisher (Detroit Executives' Club), John Bower (Westinghouse), L. B. Ermeling (General Railway Signal Company of Rochester), Henry Dennison (Dennison Manufacturing Company), and scientific management revisionists Morris Cooke and Ordway Tead. Clarence H. Howard, president of the Commonwealth Steel Company, voiced the keynote of the conference. "The most important engineering course today," he declared, "is human engineering." Another conferee, and a familiar member of the NACS, was Professor Walter Dill Scott, whose presence reflected the growing interest in applied psychology as a management tool, an interest most pronounced at AT&T.

Professional psychology was first applied to industry in the realm of advertising by Scott and his associates as early as the late 1890s. By 1915, a more comprehensive formulation of its utility in personnel management had been offered by Professor Hugo Munsterberg of Harvard. According to Munsterberg, a keen observer of the scientific-management movement in industry, psychology could be used to detect "those personalities which, by their mental qualities, are especially fit for a particular kind of economic work"; ultimately, he argued, its use could lead to "overflowing joy and perfect inner harmony" throughout

*Among these were the associations of Chicago (Sears, Commonwealth Edison, Armour, Marshall Field, etc.); Newark (Thomas A. Edison, Inc., Western Electrical Instrument Company, Hyatt Roller Bearing); New York (New York Edison, AT&T, Macy's); Philadelphia (Curtis Publishing Company, Strawbridge and Clothier, American Pulley Company); and Rochester, New York (Kodak, Bausch and Lomb, Taylor Instruments, and General Railway Signal Company).

the realm of industry. Munsterberg had drawn upon the testing techniques of Alfred Binet and had established testing procedures at the Boston Elevated Railway Company, AT&T, and the American Tobacco Company in an effort to solve their labor-turnover problems. At the request of a consultant engineer, Scott joined the work at American Tobacco and, in addition, developed testing programs at Western Electric, the National Lead Company, and other firms. E. L. Thorndike, another pioneer in applied psychology, did similar work for the Metropolitan Life Insurance Company. In 1915 the Carnegie Institute of Technology established the country's first center for such industrial-service activities, headed by Scott, Thurstone, and Walter Bingham, and it carried on testing work for such companies as Westinghouse, Burroughs Adding Machine, Carnegie Steel, and Packard Motor Car. During the war the Carnegie group played a major role in the creation and operation of the CCP in the Army, the Psychology Division of the NRC, and the CEST, and after the war it contributed to the establishment of the Personnel Research Federation.[49]

From the outset, industrial psychology, as formulated by Munsterberg and practiced by Scott and his associates, had been influenced by the management work of engineers. As one of the few historians of the subject has observed, "scientific management not only conditioned the industrial climate for the psychologists, it determined to a large degree the direction, scope, and nature of psychological research. The engineers raised most of the problems with which the later psychologists grappled, and more important, scientific management gave to industrial psychology its purpose, its ethic." Both the revisionists within the Taylor movement and the leaders of the corporate liberal management movement viewed industrial psychology as the solution to their "human factor" problems.* At the same time, in their enthusiastic adoption of its techniques, they informed it with their own particular management orientation; industrial psychology and management merged very early in the history of both. Addressing the representatives of the most management-minded and science-minded companies in the

---

*It is not hard to guess why the engineers so readily adopted the behaviorist approach proffered by the industrial psychologists and later sociologists to select, classify, and motivate workers. It minimized the significance of the major characteristic distinguishing society from nature, people from things—human consciousness, rational purpose; it enabled the engineers to study the human being as if "it" were animated by hidden laws of behavior, laws not altogether different from those which regulated physical bodies. The engineers appropriated the findings and methods of the psychologists with such enthusiasm simply because these enabled them to manipulate people without appeal to reason. (*Continued*, p. 298)

country, therefore, Scott had little difficulty promoting his wares. "When the employment department becomes, as it should, the pivotal department in our commercial and industrial organizations," Scott declared, "every position will be looked upon, not merely as a productive unit, but also as a place for training and for testing for more responsible positions.... If this group of employment managers is really to become professional, I think it must be because you do what other professions do, you utilize all that science has available for you at the time."[50]

The Philadelphia conference appointed an organizing committee charged with the creation of a national group, and the declaration of war caused them to redouble their efforts. In the midst of a serious labor shortage, there was now an unprecedented demand for labor in munitions plants, war-supply industries, and shipbuilding. The war at once dramatized the need for efficient personnel management and precipitated the "marvelous expansion," in Alford's words, of the personnel-management movement. Morris Cooke, chairman of the Storage Committee of the CND and later head of labor relations for the Ordnance Department, immediately recognized the needs and opportunities which the war had generated. Early in 1917 he suggested that the government arrange for the establishment of employment-management training courses in the country's colleges, to prepare men for wartime personnel work. The Army, Navy, Ordnance Department, Quartermaster Corps, and Emergency Fleet Corporation all welcomed the

---

Whereas Taylor, for all his shortcomings, relied upon the reasoning ability of his subordinates to see the value to them of scientific management—and disciplined or dismissed them if they failed to see it—his more sophisticated successors, while claiming to deal more directly with the human factor, in reality sidestepped it altogether. Ignoring the consciousness of the worker—which they came to view as merely epiphenomenal, symptomatic of underlying drives—they focused upon the unconscious, irrational underpinnings of human behavior. "It is clear," Professor Schell lectured his students, "that if we are to usefully interpret human behavior we must appreciate that it is actuated by many forces other than those of reason." "Men do not live by logic alone," Wickenden wrote in 1923. "The student engineer must get some understanding of the complexities that lie back of human psychology," Joseph Willets of Wharton School argued in 1926. "The student should learn not to expect rationality in human responses." The purpose underlying this fascination with worker psychology was made clear by Professor Schell: "If the executive acknowledges the presence of these instincts he may find explanation for many occurrences which are otherwise mystifying. He may also learn to use them in inciting desired behavior, in organizing group will." Joseph Willets, discussion in *The Teaching of Labor Relations in Engineering Schools, An Informal Conference of Engineering Educators at the Home of Sam A. Lewisohn* (privately printed, 1931), MIT Dewey Library. William Wickenden, "The Engineer as a Leader in Business," SPEE *Proceedings,* XXXI (1923), 113. Erwin Schell, *Course Record in Business Management, 1920,* MIT Archives.

suggestion and appointed Boyd Fisher of the Detroit Executives' Club —now a captain in the Ordnance Department—to take charge of the courses. Emergency Training Courses for Employment Managers were subsequently established at the University of Rochester, Harvard, Columbia, University of Pittsburgh, University of Washington, and University of California to provide this necessary service.[51]

In May 1918 the first class of "students" graduated from the University of Rochester. This first course, Secretary of War Baker noted, had been conducted "without cost, either to the government or to the students, and had been taught primarily by representatives of Kodak, Dodge Brothers, and Bausch and Lomb; among the first graduates were managers from Du Pont, GE, Packard, and the Malleable Castings Company. "We have trained employment or service men," the director of the Rochester course explained, "men who will spend their lives handling the labor problem. We have taught our service men to regard the laborer as a human being," the director continued, "a human being with instincts that are essentially human, and with aspirations that are sometimes superhuman."[52] He neglected to add that the "labor problem" was not a problem for labor.

This commencement for the Rochester course was also the occasion for the second national conference of employment managers and the founding convention of the new National Employment Managers' Association. The conference and the new organization once again brought together the revisionists within the Taylor movement and the personnel managers of the large corporations. Typically, the conference was dominated by representatives from such science-based companies as GE, Du Pont, Westinghouse, Kodak, and International Harvester. The National Employment Managers' Association immediately became the dominant national forum for all employment-management matters. Its members cooperated extensively, during the war, with the CCP in the Army, the CEST, the NACS, and the Psychology Committee of the NRC. After the war they continued to work intimately with the NRC, the NACS, and the new Personnel Research Federation. In 1921 the NEMA changed its name to the National Industrial Relations Association and then merged with the NACS (renamed the NACT) to form the National Personnel Association, adopting the word which the war had put on every manager's lips. In 1923 this organization changed its name to the American Management Association.[53]

Personnel managers tried not merely to process industrial labor; they were compelled to produce it as well. Just as it had created the modern engineer himself, industrial capitalism had generated the need for the skilled workers upon whom the engineer could depend to effect his designs. The development of the modern factory system, however, destroyed the traditional form of apprenticeship which had provided such workers. The new industrial system, with its extreme division of labor, broke up the crafts, replacing versatile craftsmen with the cheaper detail-workers who attended the machinery. And those craftsmen who remained—in the machine shop or foundry, for example—were increasingly required to devote their full energies to efficient machine-like production and thus had no time in which to train apprentices. By the turn of the century, therefore, there was already a shortage of skilled workers for industry and the supply was dwindling. At the same time, there was an unprecedented demand for disciplined "unskilled" workers to man the factory production lines and for a new breed of technically trained "semi-skilled" worker whose specialized talents adapted him to the requirements of large-scale capitalist production. The new apprenticeship system of industrial education, promoted by corporate educational reformers early in the century, was designed to meet these demands.

The corporate drive to create a new apprenticeship system meshed nicely with a growing popular demand for the extension of educational opportunity, on the one hand, and a growing concern among educators over the educational requirements of a modern industrial and highly technological society, on the other. The former was reflected in such developments as the university extension movement, the vocational-guidance movement, and the plethora of correspondence schools, continuation schools, and part-time and evening courses for working people. The latter appeared in the form of cooperative education on the public-school level as in Cincinnati, and Fitchburg, Massachusetts; modern trade schools such as those in Worcester, Rochester, New York, Minneapolis, and Philadelphia; and the innovative Gary Plan, which combined shopwork and trade education with the more traditional "cultural offerings" in the public schools. But whereas many of these educational-reform measures were directed toward greater educational opportunity and greater correspondence between education and industrial reality—so-called "education for life"—as ends in themselves (notably in the work of John Dewey), the corporate educational reformers of the day viewed them as means to corporate ends. Greater educational opportunity for working people, for example, was viewed not as opportunity for upward mobility, but rather as the precondition for more efficient and valuable corporate service; inclusion of shopwork in the curricula meant not a broader-based education for the individual, but rather one which better prepared him for industrial employment.[54]

Engineers played a prominent role in the development of corporate educational reform, and for an important reason. "In all industries," President Pritchett of MIT observed in 1902, "the demand is becoming urgent for men and women who have had sufficient training in applied science to grasp the plans of the engineer above them, and who have the practical knowledge to carry them into execution." The growing number of engineers were dependent upon skilled labor to practice their profession. If an engineer had to draw all the plans, produce the machine parts to specifications, and assemble his creations, then he would no longer be an engineer; he would once again be a draftsman, a mechanic, a technician. As Milton P. Higgins, a shop-culture engineer who founded the Norton Company, put it, the apprenticeship system was essential for the supply of "the workmen upon whom the engineer must depend to realize his ideas."[55]

As a leading educator of engineers and a great admirer of corporate accomplishments, Henry Pritchett decried the sorry state of industrial education in the United States. Speaking before the Twentieth Century Club in Boston in 1902, he pointed out the alarming fact that less than one third of one percent of workers between the ages of fifteen and twenty-four were receiving "any formal instruction from the state or from private institutions concerning the arts and sciences which bear directly upon their occupation." While the apprenticeship system of old was breaking down day by day, Pritchett argued, there was no available "technical equipment" to replace it. He pointed to the growing number of private trade schools, the correspondence schools, and the YMCA educational programs as but temporary solutions to the problem and, alluding to Germany's efficient system of trade education, posed the question "whether our plan of popular education can be made to minister to this vast host." America needed a distinct system of trade education, Pritchett asserted, one which could meet the pressing needs of American industry and at the same time extend the advantages of "popular" education to all working people. "We have grown too much accustomed in our schools and in our colleges to hold out the extraordinary rewards of college education [the careers of lawyers, doctors, engineers, etc.] as a reason for education," Pritchett argued; it was high time that popular education was offered to the "vast host" of untrained workers on the grounds that it would make them better workers, more efficient and contented subordinates in industry, rather than because they might be able thereby to "uplift" themselves into the class of professionals and managers. Like a number of his colleagues in the corporate and educational communities, Pritchett endorsed a "dual system" of public education for the United States, one which offered two different kinds of educational "opportunity."* Similar approaches to the problem had been made by Dugald Jackson and Magnus Alexander the year before, and by men like James Mapes Dodge, Charles R. Richards, Charles A. Prosser, David Snedden, and the ubiquitous Frank Vanderlip. Frederick P. Fish, AT&T president and chairman of the Massachusetts State Board of Education during Snedden's tenure as commissioner, called for a vocational-education system paralleling the traditional schools which would meet "the practical needs of life" for the "rank and file" of industry. All of these men argued for the early determination of the destiny of children so that they might be placed on the proper educational track (an approach to education which Dewey caustically labeled "social predestination"). "The 'room at the top' motto," Pritchett told his fellows at the elite Twentieth Century Club, "has been overworked."[56]

Corporate reform efforts in the area of industrial or vocational education fell into three categories: the creation of private trade schools, the establishment of corporation schools for apprentices, and the drive for publicly supported industrial education. Engineers participated in all three areas of activity, and most significantly in the last two. Herman Schneider, for example, was directly responsible for the establishment of the work-study programs in the public schools of Cincinnati and Fitchburg, Massachusetts; Milton Higgins was the prime mover behind the establishment of the famous Worcester Trade School; and men such as MIT-trained Charles R. Richards pioneered in vocational education at places like the Cooper Union in New York. The systematic and coordinated efforts of engineer-managers to set up apprentice-training schools within the industrial plants, however, easily outweighed these isolated developments.

Magnus Alexander established one of the nation's first such schools at Lynn in 1902, designed to train youths "for a life of industrial efficiency." Alert to the fact that "the supply of skilled workmen in this country" was "utterly inadequate," Alexander set up special "training rooms" at GE devoted exclusively to apprenticeship training. Instruc-

---

*The difference between the new industrial education and that which engineers received was neatly summarized by Professor Schell of MIT in one of his weekly quizzes. Whereas the goal of the former was "discipline and reflex action," the aim of the latter was "knowledge." *Lectures in Business Management, 1920–21,* MIT Archives.

tors were selected from the company staff on the basis of their actual teaching ability rather than simply because of their knowledge or experience, and formal lectures on scientific subjects supplemented the trade-shop training. The GE apprentice school, moreover, provided more than mere training in a skilled trade. Like the Test Course for graduate engineers, also directed by Alexander, it was designed to habituate apprentices to the requirements of subordinate corporate employment and "teamwork." Alexander stressed the "great psychological importance" of having apprentices do actual commercial work, work which gave them a feeling for their "place in real industrial life" and an appreciation of "the value of time and money." In addition, the course was designed to promote a "feeling of loyalty" among the boys toward GE, "their alma mater," a feeling which was reinforced through the company-sponsored Apprentice Alumni Association. Noting that nearly eighty percent of apprentice-course graduates either stayed with or returned to GE, Alexander remarked that "this feeling of loyalty is a gratifying assurance of the future personnel . . . of the company."[57]

During the next decade a large number of companies set up apprentice-training schools to meet their need for skilled labor. Among the earliest were International Harvester, Western Electric, Yale and Towne, Westinghouse, and Brown and Sharpe Manufacturing Company. In 1913 these companies joined with others similarly engaged in educational work to form the NACS, which became the clearinghouse for the cooperative solution of employment and training problems.[58]

The corporation-school educators went a long way toward meeting the industrial demand for skilled labor. In 1914 President William Redfield of the NSPIE, the former Secretary of Commerce under Woodrow Wilson, wrote to the NACS that "the training of the worker [is] the largest and most important factor in the whole problem of production." "No institutions have met better the needs of industry," he added, "than have these corporation schools." The approach of the corporation educators was summarized by Charles Steinmetz of GE at the organizing meeting of the NACS (Steinmetz was at the time president of the school board in Schenectady):

> The cogs in the wheels of modern industry are human beings; as an essential factor, then, in the work and in the efficiency of the corporation enters the human element. . . . Our work is more than merely the organization of corporation schools into an association . . . ; it means developing and organizing the essential part of human corporate development, the educational feature of the human element.

Channing R. Dooley, educational director for Westinghouse (which operated the famous Casino Technical Night School), put the same point another way: "Education is not an end, it is a means. It is not an ornament, but a facility to use. It does not make life easier, but on the contrary fits a man for harder work."[59]

For the corporation educators, education was essentially a corporate management problem and, as such, an engineering problem as well. W. W. Kincaid of the Spirella Company, a leader in the personnel-management and educational movements, explained that education was a "management process," whether it was directed toward apprentices, managers, engineers, or salesmen. It did not merely provide for development of the individual but, more importantly, it developed the corporate "team." "One of the most important elements in an educational program today," Kincaid wrote in an early issue of the *American Management Review,* "is the molding of a united work force"; more important than the mere training of individuals, "it is a job of developing teamwork within the organization. Every worker must be shown how to make his efforts fit into and support the work . . . of the whole organization."[60]

Like the other areas of management, education was a problem for engineering analysis and solution. In this case, it was a branch of "human engineering," involving the usual trappings: psychological tests, job specifications, rating systems, and the like, all of which were designed to afford "maximum thoroughness in minimum time."[61] Addressing the corporation educators on the "principles of effective training of employees," Charles Mann explained what most of his audience already well understood. "Job specifications and objective tests are at present the best available measuring device for fitting together men and jobs," he asserted, drawing upon the well-known experience of the Army's CEST.

> They are all that is needed when there are plenty of skilled men and the problem is to place the men most advantageously. . . . When the supply of skilled men is exhausted, green men have to be trained to do the work. The job specifications then define the objective of training. They set the goal. The tests then measure progress and tell when the man has arrived.[62]

The engineering approach to education dominated discussions of educational techniques among the corporation educators. Henry H. Tukey, educational director of the Submarine Boat Corporation, crystallized this common perspective in his call for "educational engineers." "The problems of the educational engineer are somewhat comparable to those of the production engineer," he explained to the members of NACT in 1921.

> He must recognize plant conditions, analyze them to determine the kinds of training needed, establish aims which conform to the needs, select methods suited to aims and training conditions, analyze trade content and arrange it in suitable learning order, establish methods of ascertaining the learners' progress, continually measure effectiveness of training results, estimate and provide records of training costs, promote and maintain the plant interest in training, etc. . . . These and many others are among the hard "nuts" he must crack. . . . The educational engineer must know the actual effect of the efforts of his department upon production and upon labor turnover. He must continually check up his results in terms of dollars and cents and specific educational progress. If he deals in generalities he has no license to the term "engineer."[63]

The corporation-school educators sought to reduce education to a scientific procedure, much as they had sought to reduce the problems of management in general to engineering design. Education thus became an integral mechanism of the corporate production process, geared, like all others, toward efficiency and stability, and hardly reflecting the humanity of the material that was routinely processed. "Endless statistics are available as to machine-hours, costs, etc.," Dooley observed, "but relatively none as to the qualities of men for the various jobs and how to determine them impersonally. . . . It is good business to consider personnel problems impersonally," he noted. "Sounds strange, doesn't it, but it is true."[64]

The scope of corporate educational activities reflected the corporation educators' broad perspective of their function in industry. They hardly restricted themselves to pedagogy, but rather spent the bulk of their time developing techniques of personnel management and vocational guidance: fitting the man to the job through training. In addition, they devoted considerable attention to such "educational problems" as industrial safety and the inefficiency and "menace" of immigrant workers. Focusing upon the worker as the primary cause of "wasteful" industrial accidents, for example, they vigorously supported the "safety first" movement and developed methods of safety instruction. With regard to the "immigrant labor problem"—the need to discipline and train those who were displacing the skilled craftsmen in industry—they strongly supported the movement to "Americanize" the immigrant, "to make loyal American citizens and productive workers out of the great mass of our foreign-born population." Perceiving Americanization as a "problem in human engineering," the editor of the *Engineering News-Record* called upon fellow engineers to take a leading role in the movement; besides being in the "strategic position to take the first steps," he argued, engineers were most capable of developing "a scientific way to make [such] workers an integral part of the [industrial] enterprise." Since America and corporate industry were closely identified in the minds of corporate educators, Americanization meant, in their view, the socialization of aliens to the "American way of life" and the habituation of immigrants to the discipline of industrial employment. And precisely because they viewed the immigrants as "aliens," they were able to manipulate them with considerable "scientific" detachment.*[65]

As was the case with scientific research and engineering education, the corporate reformers never viewed privately financed agencies as the final solution to their labor problems. While they depended upon such activities in the short run, they sought at the same time to gear existing public institutions to provide the necessary service at public expense. Charles Prosser, a leading advocate of separate industrial education in the public school system, pointed out that the corporation schools, however effective, reached only "a small part of the mass of workers

*"Immigrants must be absorbed," Westinghouse's Channing Dooley declared. "The welfare of industry, as well as of community life, demands the Americanization of the foreign born citizen. Industrial managers must make plans to absorb these men into the full spirit of American industry and American life." The engineers, like many others who supported the movement, saw it as a means of adjusting the potentially radical "foreign-born" to American life, a life dominated more every day by the requirements of corporate industry. It is thus not surprising to find that a meeting of the Immigration Committee of the Chamber of Commerce of the United States, called by Vincent Astor for the discussion of Americanization, was heavily attended by corporate engineers, many of them prominent figures in the various educational-reform movements. Among them were Frederick Bishop, Ira Hollis, Dexter Kimball, Dugald Jackson, Arthur Greene, Charles Mann, Frank Jewett, J. J. Carty, Calvin Rice, and Magnus Alexander; the chairman of the meeting was Gano Dunn of the Engineering Foundation. "Memorandum of the Meeting of the Immigration Committee of the U.S. Chamber of Commerce," January 19, 1917, Records of the Division of Industrial Relations, National Research Council Archives; C. R. Dooley, "Education and Americanization," *Industrial Management,* October 1917, pp. 49–50. For a discussion of the broader implications of Americanization, see Herbert Gutman, *Work, Culture, and Society in Industrializing America* (New York: Alfred A. Knopf, 1976), pp. 3–78; for a contemporary critique of Americanization, see Robert Williams Dunn, *The Americanization of Labor: The Employers' Offensive Against the Trade Unions* (New York: International Publishers, 1927).

in the country." "One of the most important factors in this question of scientific management," Prosser argued, "is the human element, and one of the very large factors in the whole problem is to be the school . . .; the question is how the school is to play its part in the selection and training of workers of every kind."[66]

Manufacturers throughout the country had long been interested in the public schools as a supplier of workers, and it was certainly not unusual for men like Steinmetz of GE, Higgins of Norton Company, and Fish of AT&T to sit on city and state boards of education. Indeed, Scott Nearing found in 1917 that in 104 of the country's largest industrial cities, over half of the school-board members were businessmen—merchants, manufacturers, and bankers.[67] While businessmen exercised considerable influence on public education in the United States, however, it was not until 1906 that any concerted attempt was made to earmark public funds for trade-school education. The breakthrough came in 1906 with the Douglass Commission report in the state of Massachusetts. The commission, set up by Governor Douglass (a shoe manufacturer himself) in response to pressure from manufacturers in the state, surveyed the needs for trade education and the means available for meeting them. Finding that some 25,000 children between the ages of fourteen and sixteen (the so-called wasted years) were either at work or idle rather than actively learning a trade, the commission recommended that elementary and high schools be modified so as to provide instruction which would promote the "industrial intelligence" required by industry. The commission set up a permanent Commission on Industrial Education charged with establishing industrial schools independent of and parallel to the public school system under the Board of Education. In 1909 this commission was merged with the State Board of Education, headed by Frederick P. Fish. David Snedden was named commissioner and Charles Prosser was made director of vocational education.[68]

The Douglass Commission report and the subsequent work of the Commission on Industrial Education launched a nationwide "vocational education movement." While the National Association of Manufacturers and local chambers of commerce championed the cause, many teachers' organizations and the craft-based labor unions vehemently opposed it, the latter fearing that the spread of industrial education would provide an unprecedented source of "cheap labor" for industrialists and thus threaten the unions' bargaining position. While debate raged within local school boards over the merits of vocational education, the labor unions and teachers' organizations blocked early attempts to establish "dual systems" in Chicago, Atlanta, New York, and elsewhere, and unions began to set up their own union-run "labor colleges" to circumvent the business-controlled public educational system.[69]

The violent debates within the National Education Association graphically reflected the dimensions of the struggle. Frederick Roman, an economics professor from Syracuse University, strongly denounced the dual system of education, declaring that "our capitalists have already robbed our forests and our mines and the natural resources of the country generally—and we are now asked to accept a system of education which looks to the exploiting of our children."[70] And Ella Flagg Young, superintendent of schools in Chicago and former colleague of John Dewey at the University of Chicago Laboratory School, critically observed that

> We are constantly being confronted by complaints from the outside that our manual training courses and technical work do not fit the children to take the lower types of work and remain satisfied with their jobs. They complain that our training fits them to be foremen; they want us to turn out the kind of labor that they have been importing from Europe. . . .[71]

Young perceived the demands "from the outside" as being antithetical to education: "Not one man in 500 will come out at the top, but every man should want to be one of the 499 others that have tried and failed. . . . It is ridiculous to bind any boy or girl to a life vocation at from ten to fourteen." George Counts, in his study of the school system of Chicago, drove home the same point another way; he noted that of the many businessmen supporters of vocational education, not one "expressed the hope that [his] own children would enter the industrial occupations by way of the proposed system of vocational education."[72]

The most important agency to emerge in support of vocational education reform was the National Society for the Promotion of Industrial Education (NSPIE). While it included spokesmen from teachers' associations and labor organizations and prominent "uplifters" such as Jane Addams, the NSPIE was from the outset the creation of, and vehicle for, corporate liberal educational reformers. The NSPIE grew out of a meeting at the Engineers Club of New York, called immediately after the appearance of the Douglass Commission report by Charles R. Richards, then professor of manual training at Columbia's Teachers College, and James P. Haney, director of manual training in the New York City school system. At this first meeting, an organization committee was appointed to set up a national association to promote the ideas presented in the Massachusetts report. Milton Higgins, president of the Norton Company and pioneer in trade-school education, was named chairman. In January 1907, a half-year after the first meeting, the NSPIE was formally established. Henry Pritchett was its president; Magnus Alexander was vice-president. Among the members of the first board of managers were Higgins, Frank Vanderlip, Frederick W. Taylor, and Frederick P. Fish; included in the first membership roster were J. J. Carty, Charles A. Coffin, president of GE, C. R. Dooley of Westinghouse, W. W. Kincaid of the Spirella Company, Charles Mann, and Dugald Jackson.[73]

In the view of NSPIE leaders, the drive for public industrial education was but another phase in the extension of corporate personnel management. "The underlying purpose which gave birth to this association," Pritchett told its members at the organizing convention, "is the thought that we are no longer fitting our youths for their opportunities in the way in which they must be fitted. In this day, every nation must make of each citizen an effective, economic unit, and then must bring these units into efficient organization." The objective of the society, according to the NSPIE constitution, was "to bring to public attention the importance of industrial education as a factor in the industrial development of the United States." The NSPIE was thus created to extend the work of Massachusetts to the rest of the country, a purpose reflected in the fact that the chairman of the Douglass Commission, Carroll D. Wright (father-in-law of Samuel Capen), succeeded Pritchett as president of the NSPIE.[74]

While one branch of the national organization devoted its energies to studies and surveys, another, under the direction of Magnus Alexander, undertook a propaganda campaign to push the industrial-education idea throughout the country. By the close of 1907 thirty-eight state committees had been established to arouse and crystallize public opinion in favor of the movement, and to pressure state legislators for public appropriations. The committees in Ohio, New Jersey, and Wisconsin, where the efforts were directed by Charles R. Van Hise and Louis E. Reber of the University of Wisconsin, were among the earliest to secure public funds.

While Magnus Alexander, acting as "national commissioner," directed such local activities, others within the NSPIE lobbied for federal support. In 1912 the NSPIE hired Charles Prosser of the Massachusetts Board of Education to direct its lobbying efforts in Washington and, after five years of difficult politicking with labor groups, farmers, and teachers, succeeded in steering the Smith-Hughes Bill through Congress. Authored largely by Prosser, the act stressed the importance of "fitting for useful employment" all persons between the ages of fourteen and eighteen; in addition, it created a Federal Board for Vocational Education, with Prosser as director, charged with establishing and administering a separate vocational-education system for the United States. While its immediate effects were overshadowed by the wartime training activities of the CEST, the act set the pattern for nearly fifty years of federal aid for vocational education.[75]

By the 1920s, modern management, with its focus upon the "human factor" in industry, was firmly in the corporate saddle. Its emergence did not mean a change in industry alone; it signaled as well an expansion of the professional province of the engineer. William Wickenden was fond of quoting an Englishman's retort to the boasts of a French engineer, by way of illustrating the changes within both industry and engineering. "There would be about as much sense in making an engineer the executive head of an industry," the Englishman quipped, "as in making a veterinarian the commander of a regiment of cavalry."[76] Wickenden had then but to survey the realities of modern industry to show how out of touch with the times the Englishman was.

One year before the twentieth century began, Cheeseman Herrick had prophesied that "business now means more than a rule of thumb, it is complex, intricate, scientific and those who are to engage in it need a different equipment than has hitherto been thought sufficient for the businessman."[77] From that time on, engineers—men with the "different equipment" of sound scientific training—had flowed into managerial positions in industry; successive studies conducted between 1904 and 1929 documented that between two thirds and three fourths of engineering-school graduates were becoming managers in industry fifteen years out of school. The impact upon both industry and the engineering profession is most dramatically suggested by the fact that in the 1920s an engineer became President and the chief executives of five of the largest and most dynamic corporations in the country— General Motors, Singer Sewing Machine Company, General Electric, Du Pont, and Goodyear—had been classmates in engineering at MIT. "Business depends more and more on the efficiency, the smooth operation of our huge production machinery," the banker O. H. Cheney observed in 1926; "for this reason more and more of the executives are men with engineering training. More and more are the business owners, the investor, and the banker turning the management of industry over to the engineer."*[78]

"The engineer is a team worker," Wickenden often boasted; he was quick to add, however, that "fortunately for him, his basic training gives him command of the forms of knowledge hardest to improvise— the science and the technology of production—and this is a weighty advantage in competition for the captaincy of the team."[79] There can be no doubt that the engineer's rise to "captaincy" owed more to his technical expertise in the matters of production than to anything else. Once in a commanding position, however, the engineer quickly found his training in science less than adequate preparation for management responsibilities, the "handling of men." During the first three decades of the century, therefore, engineers in industry cooperated with college and technical-school educators to expand the content of engineering training, to focus more upon the "human element" and upon the myriad social factors involved in all engineering practice.

The preamble of the American Engineering Council constitution defined engineering as "the science of controlling the forces and utilizing the materials of nature" and "the art of organizing and directing human activities in connection therewith." "The twentieth century addition to the definition of one hundred years ago," Leon Alford pointed out, "is the words 'and the art of organizing and directing human activities in connection therewith.' " This expanding notion of

*The Wickenden findings of the 1920s have been confirmed by subsequent studies. See, for example, William K. LeBold, Warren Howland, and Robert Perrucci, "The Engineer in Industry and Government," *JEE,* March 1966, p. 239; "What Engineers are Doing Six to Thirty Years after Graduation," *Power Engineering,* August 1955, pp. 100–1; Carolyn C. Perrucci and William K. LeBold, *The Engineer and Scientist: Student, Professional, Citizen,* Purdue University Engineering Bulletin, Engineering Extension Series No. 125 (January 1967), p. 22, Figure 4.4; John B. Rae, "Engineering Education as Preparation for Management: A Study of M.I.T. Alumni," *Business History Review,* XXIX (1955), 64–79.

This advancement of engineers into management has had a profound effect upon the composition of the nation's top corporate executives. A 1964 study of the social and educational backgrounds of top executives of the nation's six hundred largest industrial corporations found that "the professionalization of the big business executive is increasingly correlated with qualification in science and technology" and that "the most rapidly increasing number [of executives] had begun as engineers." Of the top executives in 1900 only seven percent had degrees in science or engineering. By 1920 that proportion had jumped to twenty percent and by 1964 it was a third. See *The Big Business Executive/1964: A Study of His Social and Educational Background* (New York: Scientific American, 1965). This study updates and confirms the earlier findings of Mabel Newcomer, *The Big Business Executive: The Factors That Made Him, 1900–1950* (New York: Columbia University Press, 1955).

engineering practice, reflecting the growth of modern management from within engineering, reflected also what Charles Mann—author of the first major study of engineering education—called "the expansion of the engineering spirit." "The successful engineer today must direct the powers of men as well as the powers of nature," Mann observed; "the problems of the conservation of human resources are as much engineering problems as are those of the conservation of material resources." Charles F. Scott, an early engineering educator at Westinghouse and Yale, similarly reflected in 1926 that "with expanding industry the work of the engineer has expanded wonderfully. Instead of regarding the activities of the engineer as confined to the drafting room, we now have the demand for personnel training." By the 1920s management had clearly become a recognized branch of professional engineering; the engineer, in directing the machinery of industry, had realized that he had also to ensure "that the bearings of our human industrial structure are properly designed, properly constructed, and properly adjusted."[80]

The changing content of engineering was perhaps nowhere better reflected than in the schools of engineering and the curricula they offered. In 1926 one of a series of informal conferences of engineering educators was called by the president of the American Management Association and the educational director of AT&T, to promote the teaching of management in engineering schools. Harry P. Hammond, assistant director of the Wickenden investigation, reflected then that

> Engineering education has made remarkable progress in the direction that Mr. Lewisohn [president of AMA] had in mind in calling this meeting. When I took my course in engineering, about twenty years ago, labor problems, administration problems, management problems were hardly ever thought of or provided for in the curriculum. In looking over college catalogues of that period one can hardly find a single instance of any such subject in the curriculum. Now they are to be found in almost all college curricula in one place or another, either as distinct courses of study [e.g., "industrial engineering," "engineering management"] or as required subjects [in mechanical-engineering or electrical-engineering courses] or as seminars.[81]

Clearly, by 1926 the innovative courses introduced by Dexter Kimball at Cornell and Hugo Diemer at Penn State had become dim memories. Whereas in 1922 at least ten engineering schools were giving formal instruction in management subjects, by 1932 this number had jumped to thirty-five; the range of subjects, moreover, had extended far beyond the problems of production, the original focus, to include the problems of distribution, merchandising, finance, office administration, patent law, cost accounting, and the various aspects of the "personnel problem."[82]

The history of management training at MIT illustrates the scope of such instruction as well as the influence of engineers in this new field of scientific capitalism. MIT's Course XV, Engineering Administration, was established in 1913; in 1932 it became a separate department in the School of Engineering, and in 1952 the independent School of Industrial Management (now the Sloan School).* One of the most important figures in the early management training program was Erwin Schell, an MIT-trained mechanical engineer who returned to his alma mater after a brief stint in industry to teach Course XV and later head the new department. Twenty years later when the School of Industrial Management was founded—owing to the largess of alumnus Alfred Sloan—fully one-third of the top industrial executives who sat on the steering committee had professional engineering degrees. The first dean of the new school, E. P. Brooks, a vice-president of Sears Roebuck, had been a member of the first graduating class of Course XV.

Instruction in Course XV covered all aspects of works management, economics, finance, accounting, business law, and marketing. Students were brought into contact with working executives at dinners and on

*It would be a mistake to view Course XV as a minor curriculum innovation of a single school; it was rather the brainchild of a new breed of industrial leaders intent on reproducing themselves. The MIT course was patterned after the Carnegie Tech course in "commercial engineering" which had been created a few years earlier in response to the "persistent demands on the part of the larger industries [of Pittsburgh] for graduates who have completed a curriculum in accordance with their outline." At MIT, too, the new form of education was established "in recognition of the changes that have come into our industrial and business life, . . . to train men effectively to meet the new conditions." Among the industrial leaders who had a hand in setting up the course in 1913 were the presidents of GE, Du Pont, U.S. Steel, General Chemical, the Northern Pacific Railroad, and Alcoa. In 1926, members of the Course XV "Advisory Committee" who met at the Bankers' Club in New York City included a present and a past president of GE, the presidents of General Motors, Stone and Webster, American International Corporation, Yale and Towne, and the Guaranty Trust Company, and vice presidents of Eastman Kodak, White Motor Co., Singer, Union Carbide, and the Pennsylvania Railroad. Similarly, in 1952, when the School of Industrial Management was founded, the advisory committee members included the chief executives of Du Pont, GM, Westinghouse, Sears Roebuck, Standard Oil of New Jersey, Standard Oil of Indiana, International Harvester, Sprague Electric, and Chicopee Mills, as well as the chairman of the federal reserve bank of Cleveland, a partner of Lehman Brothers investment bankers, and a vice president of AT&T. See "Report of the Committee on Business Engineering of the Alumni Council," May 19, 1913; "Minutes of the Advisory Committee on Course XV," February 25, 1926; and "Minutes of the Advisory Committee of the School of Industrial Management's Advisory Council," 1952. All in Presidential Files, MIT Archives.

field trips and during summer employment, and heard lectures by prominent proponents of scientific management—Gilbreth, Hathaway, Harrington Emerson—on such topics as time-and-motion study, industrial fatigue, employment departments, and shop committees. A major focus of attention throughout was the human factor. "The horizon of scientific management may be limited to that one phase of shop or industrial life which has to do with ——," read a question in one of Professor Schell's periodic quizzes. The correct answer was straightforward enough: "the control of men." Schell repeatedly emphasized to his students that "labor management [was] their great responsibility," that "their success or failure in accomplishing [their] work . . . will not be measured in terms of their technical knowledge nor in terms of their technical skill, but rather in terms of their proficiency in organizing the will of the employees—in handling men and women."[83]

Schell thus urged his students to consult the growing body of scientific knowledge about "the workings of the human mind" and lectured them on the various factors which govern worker behavior. Like many of his colleagues in industry, he stressed the importance of irrational drives, deep-seated instincts which executives could learn to use "in inciting desired behavior and in organizing group-will."* In a lecture entitled "The Workmen: Their Impulses and Desires," Schell discussed the various instincts that he believed lay behind worker behavior and how the skillful manager might use them: "The executive who, by facilitating promotion . . . makes marriage a possibility for a young man"—thereby dealing with the "Sex Impulse"—"stands to receive large dividends in increased loyalty and length of service. . . . The executive who assigns the new employee a locker, a key, a machine and bench, with name affixed, is bringing instinctive satisfactions of proprietorship ["the Wish to Possess"] which show returns in reduced labor turnover." The fact that shop committees provide an outlet for the

---

*Such concern was not just a passing fancy, not just a reflection of lay fascination with Freudian psychology. In 1913, the MIT committee which originally proposed the establishment of Course XV had emphasized the practical import of psychology, in advertising and salesmanship, for example. Most important, they stressed the fact that "the efficiency of labor is largely dependent on the assignment of individuals to the work for which their mental characteristics fit them." Similarly, when Dean Brooks outlined the seven "guideposts" for the new School of Industrial Management in 1953, he topped his list with "the need in industry for greater emphasis upon an understanding of human behavior." This was followed directly by "the need for increased emphasis on developing a better understanding of the American enterprise system and the significance of profits." See "Report of the Committee on Business Engineering of the Alumni Council," May 19, 1913; and "Minutes of the Advisory Committee of the School of Industrial Management," October 6, 1953. Presidential Files, MIT Archives.

workers' "Desire for Self-expression" accounts "for their success in bringing greater satisfaction to the employees as well as in developing a formidable barrier to the spread of trade unionism. . . . The Desire for Leadership . . . is sometimes called the submissive impulse. I like to think of it, however, as the desire to work under good leadership. . . . The Fighting Spirit . . . is found in every normal person. It rests with the leader to divert it into useful channels."[84]

It is significant that while Schell emphasized the essentially irrational, unconscious roots of worker behavior, he took pains to instruct his students, the future managers, to act in an exaggeratedly rational manner. Managers should express no emotion whatever, he urged. "The executive should make it clear that his actions are impersonal and in a sense automatic." "Self-control" was the most important executive trait; "your attitude must always be impersonal." Schell's aim was to foster a code of managerial conduct that would make the control of people by other people appear to be the control of people by an automatic mechanism. Managers, in manipulating the irrational drives underlying worker behavior, were thus to assume the pose of reason itself.[85]

Paralleling the emergence of management training within the engineering schools was the growth of graduate schools of business administration. Among the more prominent of these were the Amos Tuck School at Dartmouth, the Harvard Graduate School of Business Administration, and the Wharton School of Finance. At such schools, established by businessmen and economists rather than by engineers, the focus was placed more upon the commercial aspects of industry— finance, accounting, law, merchandising, etc.—than upon production. While the two types of management schools often found themselves in competition for students and support, they quite commonly cooperated in their efforts. At Harvard, for example, there was a five-year program which formally combined the resources of the business-administration and engineering schools for the teaching of potential managers. In addition, the business schools actively recruited engineering graduates as students, and more and more engineers followed up their technical training with graduate work in business administration before seeking employment.* At the University of Minnesota the business school so

---

*This trend has continued. In 1974 sixty-five percent of the students at Purdue's Graduate School of Industrial Administration held the B.S. degree in engineering, and another thirty-five percent held the same degree in the hard sciences; while these particular figures are unusually high, they do reflect the general trend. Among the class of 1973 of the Harvard Graduate School of Business Administration, the largest single group were B.S. engineers, twenty-seven percent, and hard scientists accounted for another twelve percent. See *The Harvard Business School Bulletin,* November 1972, p. 20.

appreciated the value of engineering training in successful management that it required its students to spend their first two years in the engineering school "in order to get the fundamental engineering atmosphere and viewpoint."[86]

The expansion of engineering education was reflected also in the increased subject requirements in the humanities and social sciences. Throughout the twentieth century the proportion of the four-year curriculum devoted to such studies grew until by the 1950s it had become twenty percent. In the first study of engineering education, in 1918, Charles Mann emphasized the need for more instruction in social science; a decade later Wickenden repeated the message, stressing the point that "the choice of humanities studies is to be governed . . . by their functional relation to engineering pursuits." The SPEE Hammond Report of 1940 finally adopted the minimum of twenty percent for courses in economics, sociology, psychology, government, and history (above and beyond the offerings in accounting, finance, and management proper).[87]

Of course, teachers of humanities and social sciences lobbied strenuously for an extension of their domain within the engineering schools. But the major impetus came neither from them nor from their engineer colleagues; it came from the practicing engineers in industry, who recognized the need for a broader engineering training, one which better prepared graduates for their ultimate managerial responsibilities. Robert Rees of AT&T, for example, called for greater emphasis upon the "human factor" in engineering education. "Our task," he told assembled engineering educators at American Management Association President Sam Lewisohn's house, "is to make the engineer more conscious of the human factors, to indoctrinate him so that he will realize that, after all, the man himself is the all-important problem."[88] "Such technical questions as the frequency and length of rest periods in work carried on at different degrees of speed and intensity," economist Don Lescohier suggested, "are as important in technical training as an understanding of the peculiarities of iron or copper or of energy."[89] The underlying motive behind the expansion of the curriculum was succinctly expressed by Samuel Stratton, first director of the National Bureau of Standards, when he became president of MIT in 1923:

> Our technical schools are training the future brain workers and managers of industry. We may, therefore, well ask ourselves, at this time, if there is anything we can do beyond what we are now doing to train our students to understand more fundamentally and to meet successfully

the gravest of all their future responsibilities, the organization and management of men.[90]

As the evolution of the engineering curriculum suggests, engineers by the 1920s had begun to turn to the social sciences—economics, psychology, and sociology—for new ways of comprehending and carrying out their managerial responsibilities. While they viewed these younger "soft" sciences as considerably less scientific than the "hard" physical sciences, engineers nevertheless began to realize that the social sciences, however imperfect, might provide solutions to pressing problems which seemed to defy traditional engineering methods.

One ardent proponent of this new creed in engineering was Willard Hotchkiss. An economist by training, and the founder of the schools of business administration at both Stanford and Northwestern, Hotchkiss served as president of the Armour Institute of Technology in Chicago before becoming director of humanistic-social studies at Carnegie Tech. In 1935 he discussed the role of "social sciences in engineering schools," in answer to the question "Is the material [of the social sciences] relevant to an engineering education?"

> Successful engineers today are influencing the social aspects of engineering and industrial enterprises as completely as they are the technical aspects. Even purely designing engineers cannot advance very far without having considerable responsibility for handling men, an obviously social activity. It is just as essential that engineers know what they are doing and why, when they deal with these social and economic questions as when they consider the technical side of their work. The fact that engineers in the past have tended to isolate the physical bases upon which their projects rest and deal with the measurable and finite forces and values embodied in them, does not make these measurable and finite forces and values in any sense the exclusively important factors with which engineering and engineers are concerned. There is ample evidence that shying away from the human factors in engineering places a limit upon the success which engineers are able to secure as a result of their work. . . .
>
> Though cause and effect may be more obscure in the social sciences today than they are in mathematics and in physics, it is one of the major tasks of study in this field to clarify these relations. Progress similar to that which was previously made in physical science is now being made in various branches of social science, in proceeding objectively from the known to the unknown. The fact that data are complex is an added reason why engineering students should expand and adapt methods of reasoning which have earned their respect in physical science and engineering.[91]

In a similar vein, and from an engineering perspective, William Wickenden called for more integration of social science within engineering. "Few would suggest that social science is as yet sufficiently advanced to play an instrumental part, that is to supply either tools of analysis or criteria of decision for use in solving actual engineering problems," Wickenden conceded in 1937, in his retirement address as chairman of the engineering section of the American Association for the Advancement of Science. But, he argued, "if we can get engineers and social scientists to doing actual work together the problem will tend to solve itself. . . . Our part is to create a sphere of action for the social scientist in our domain" which will encourage him in his work to "dovetail into our inherently concrete, inductive and instrumental ways of thinking and doing." Meanwhile, Wickenden suggested, the engineers themselves must try to "blend" the physical and social sciences. As he well realized, he was once again suggesting a fundamental change in the engineer's perception of the world, and in his way of dealing with it, a challenge he had made fourteen years earlier, in an SPEE report on business training for engineers:

> The broader conceptions of the engineer's work in adapting energy and matter to social and economic ends can be brought about, not by major or minor changes in the engineering curricula, but by a comprehensive change in the viewpoint of the entire engineering fraternity.

An engineer as well as an educator and employer of engineers, Wickenden fully recognized that the increasing managerial responsibilities of engineers, and the subsequent development of scientific management by engineer-managers, reflected profound changes within engineering. His call for a "comprehensive change" in the engineering viewpoint, to enable engineers to deal effectively with men as well as matter, was an indication of how real that change had already become. Wickenden closed his 1923 report with an acknowledgment of that fact; "the signs of the times," he remarked, "indicate that this change is fast taking place."[92]

One such sign was the work of some of Wickenden's electrical-engineering colleagues in the Engineering Division of the NRC. Early in January 1924 Dugald Jackson and C. G. Stoll of Western Electric met in the office of Frank Jewett to formulate a study of the effect of illumination quality and intensity upon worker productivity. As originally conceived, the study involved a broad range of activities, including the physics of light, the physiology of sight, worker psychology, and illumination engineering, and was to be carried out in scientific laboratories, psychology laboratories, and the plants of such companies as GE, Western Electric, and the Dennison Manufacturing Company—all important sites in the development of modern management. Thomas A. Edison was named honorary chairman of the study committee.[93]

During the next few years the researchers, under Jackson's direction, focused upon the Hawthorne plant of Western Electric; their studies ran the gamut of management techniques then available—mechanical correlations, fatigue studies, and individual psychology—but failed to produce any consistent results. Jackson's reports between 1925 and 1931, to the NRC and the president of MIT, reflect both a mounting frustration and a heightened determination, as he repeatedly had to postpone making any final conclusions. The "nut" was finally cracked not by the engineers or the psychologists, but by Elton Mayo and his associates from the Harvard Business School, who had arrived at Hawthorne in 1927. After careful study, Mayo and his colleagues had discovered that an intricate network of social relations existed among the workers through which they, and not management, regulated output; they also found that a worker's productivity was a function of the attention given him by management (the so-called Hawthorne Effect). The discoveries of Mayo and his colleagues were startling and constituted a revolution in management thought; needless to say, the engineers who had conceived the original project had hardly suspected anything of the kind. They had unknowingly launched what has been called "the first major social-science experiment," "the single most important social-science research project ever conducted in industry."[94]

The Hawthorne experience called into question many of the basic assumptions of scientific management, gave impetus to the infant applied sciences of industrial psychology and sociology, ushered in the new field of "human relations," and provided management educators with a wealth of case-study material. Less obviously, it reflected the extent to which the horizons of engineering had expanded as engineers strove to control the elusive human factor in every engineering problem. For it had been the corporate engineers, with the support of the large corporations of the science-based electrical industry, who had conceived the project and pushed it ahead to its unanticipated conclusion. ("The present big need," Jackson had written in 1928 after repeated frustration at Hawthorne, "is encouragement and support for psycho-physical research in the field until the principles are discovered and verified.")[95] And it was not long after the experiment was over that the same engineers began to incorporate industrial sociology within the engineering curriculum, declaring Mayo's writings to be required reading for all student engineers.[96]

In 1929, in his final report as director of the most thorough study of engineering education in history, William Wickenden again called attention to the "expanding scope of the engineer's work and training." "The rise of the engineer as an organizer and manager," he wrote, "has been a natural evolution covering the last half century.

> The concern of industry has advanced from isolated tools and processes to an organic conception of production and service as a whole, in which pure mechanics could not be segregated from financial, legal, marketing and personnel problems, so that the engineer in his planning, is dealing quite as often with money and men, as with materials and machines.[97]

"Looking to the future," Wickenden concluded, "the schools of engineering can scarcely limit their concern to the mathematical and physical sciences, to problems of design and construction, and to the specific details of engineering economy. Engineering will include in its tools any and all sciences as they become exact enough to yield economically predictable results." Among the more urgent challenges to the profession, Wickenden argued, was the need "to bring together the mechanical, physiological, and psychological factors in human work within the bounds of a predictable science."[98] Thus, Hawthorne was more than merely a chapter in the history of management; it was the latest phase in the evolution of engineering. Born into the world of production, the corporate engineers of science-based industry had taken for their task the production of a world.

Epilogue

Has modern technology in America been tamed? Has the most potent revolution in social production since the invention of agriculture[1] become merely a means to corporate ends, a vehicle of capitalist domination? The social history of engineering in this country lends itself to such a sober conclusion. For the creators of the new scientific mode of production, the self-proclaimed revolutionaries who unlocked the forces of nature and heralded the coming of a new day for mankind, worked a counterrevolution as well. In standardizing science and industry, reforming the patent system, routinizing research, transforming education, and developing modern management, the corporate engineers of science-based industry strove at once to push forward and to stay their revolution, to reap its immediate benefits and yet forestall the coming of that new day which it seemed to portend.

Those who have followed in the footsteps of the men described here have competently and sometimes cleverly continued their work, only now they take for granted the social order which was, not so very long ago, fashioned by design. A description in *Fortune* of what had become of the General Electric Test Course program by 1953 bears this out. "Human relations is now the key," the journalist reported. The modern engineer-manager

> will be a generalist who will not think in terms of specific work but rather in terms of the art of managing other people's work. It is the technique of managing, not the content of what is managed, therefore, that will be his preoccupation. He encourages, or, rather, motivates others to work; he does not create, but moderates and adjusts.

The emphasis is on "well-roundedness," fitting in, getting along. "Brilliance is almost becoming a dirty word," the reporter discovered.

> "I would sacrifice brilliance to human understanding every time," one trainee says. Trainee after trainee makes the same point. For one thing, they don't think brilliance is particularly necessary anymore. "All the basic creative work in engineering," as one young man explains, "has already been done." Is a man like Steinmetz any longer apropos? "I don't think anybody would put up with a fellow like him now," says a G.E. trainee.[2]

This domesticated breed of engineer personifies a social transformation of major proportions, and attests to the success of corporate efforts in bringing about a stable correspondence between forces of production and social relations, "a better balance between technological progress and social control."[3] On the whole, corporate managers since 1930 have not departed significantly from this objective. The most dramatic changes—the massive expansion of military and government participation in the technological enterprise since World War II and the tremendous growth in the number of engineers since Sputnik—have only advanced their ends by providing for the wholesale public subsidization of private enterprise and the increasing proletarianization of technical workers.[4] Moreover, the engineer-managers are now equipped with new methods of analysis and control, some borrowed from the social sciences, and new tools of analysis, developed largely in electrical engineering.*

This fact has prompted not a few observers to proclaim that we are living in a technocracy, run by and for the technical elite, according to the cold compulsions of technical reason. As the present study suggests, such proclamations of technocracy are premature at best, and are but another expression of the general mystification of technology that marks our age. For people—engineers included—do not live by technical reason alone. Although, in the wake of modern engineering, corporate industry has taken on a scientific aura and capitalism has assumed the appearance of reason itself, the engineers have no more replaced the capitalist than science has replaced capitalism. Whether as managers

---

*The statistical procedures vital to economic forecasting as well as psychological and sociological concepts and methods were borrowed primarily from the social sciences, while electrical engineering itself gave birth both to the computer, the cornerstone of operations research, and, through circuit theory, to modern systems analysis. The interdependent social and intellectual histories of engineering and the social sciences in the twentieth century, and their coalescence in scientific management, are the subjects of a separate study, now underway.

or technical experts, the engineers have merely continued to serve capital, wittingly or not, their habits of thinking about problems and formulating solutions constituting for the most part but a highly refined form of capitalist reason. Thus, in their work they have continued to labor, routinely if not always consciously, to resolve in practice the tension between the potentials of modern technology and the dictates of the corporate order. This is not to say, however, that the conflicts inherent in the capitalist mode of production have thereby been altogether overcome. This study, in fact, suggests a contrary view.

A workable once-and-for-all design for America did not spring full-blown from the drawing boards of corporate engineers. The task of designing America has rather been an ongoing social process, one marked as much by human conflict (and design failure) as by smoothly running machinery. Thus, as the piles of scrapped designs grew, the corporate engineers increasingly called upon their colleagues and instructed their successors to be flexible, to try new methods, to be alert to ever new demands and changing social conditions. Society, they slowly discovered the hard way, was not simply a second Nature, to be understood and controlled in the same way one understood and controlled the first. Borrowing heavily from positivist and behavioral social sciences, the engineers strove to "expand and adapt" their methods of reasoning, to seek new ways of solving old problems.

But to what, one must ask, were they forever trying to adapt their methods? What was it that compelled them to try to enlarge their vision, to broaden the scope of their designs, to blur the once rigid distinction between the "hard" and "soft" sciences? The conventional historical account of the activities of men like these pictures them struggling against all odds to forge a workable corporate order, but against no people. To be sure, they have been confronted by developments in science and technology whose implications have not always been readily apparent. And, of course, they have had to face the periodic crises that regularly plague the capitalist economy, to guard against market failure, to seek ways of outmaneuvering competitors, to oversee the routine complexities of mammoth enterprise. But these have never been the major focus of attention, nor are they now. However impersonally the "problems" might be formulated, the main challenge has always been people, people with a different vision perhaps, with equally rational but nevertheless conflicting aims. Thus, no matter how sophisticated their approach, how flexible their methods, how pure their intent, the corporate engineers have consistently encountered considerable difficulties in trying to implement their designs "in the field," opposition which they disparaged as "labor trouble," "personnel problems," or simply "politics." And their trials serve as a reminder that this book describes only one aspect of American history in the twentieth century: other people shaped that history too. However firmly the protagonists of this story (or their contemporary successors) convinced themselves that they served the interests of society as a whole, they in reality served only the dominant class in society, that class which, in order to survive, must forever struggle to extract labor from, and thus to control the lives of, the class beneath it. No myth of classlessness, no "end of ideology" ideology, however comforting, however innocent, can ever obscure this fact. And it is precisely this fact, manifested in the myriad "problems" which must forever be analyzed, engineered, or administered away, which both underlies the evolving corporate design for America and defies it.

Notes
Index

Notes

Introduction

1. Karl Marx, *Grundrisse,* translated by Martin Nicolaus (New York: Vintage Books, 1973), p. 706.
2. See Thorstein Veblen, *The Engineers and the Price System* (New York: B. W. Huebsch, 1921); for a valuable commentary, see Edwin Layton, "Veblen and the Engineers," *American Quarterly,* Spring 1962, pp. 64–72.

   See also Charles P. Steinmetz, *America and the New Epoch* (New York: Harper and Brothers, 1916).
3. See Hans Gerth and C. Wright Mills, eds., *From Max Weber: Essays in Sociology* (New York: Oxford University Press, 1946).

   See Jacques Ellul, *The Technological Society* (New York: Alfred A. Knopf, 1964).

   See Lewis Mumford, *Technics and Civilization* (New York: Harcourt, Brace, Jovanovich, 1934), and *The Myth of the Machine,* 2 vols. (New York: Harcourt, Brace, Jovanovich, 1967, 1970).
4. See Georg Lukács, "Class Consciousness" and "Reification and the Consciousness of the Proletariat" in his *History and Class Consciousness* (Cambridge: MIT Press, 1968).

   See Max Horkheimer and Theodor Adorno, *Dialectic of Enlightenment* (New York: Herder and Herder, 1972). For a more accessible discussion, see William Leiss, *The Domination of Nature* (New York: George Braziller, 1972), or Horkheimer's classic *Eclipse of Reason* (New York: Oxford University Press, 1947).

   Herbert Marcuse, "Industrialization and Capitalism in the Work of Max Weber," in *Negations* (Boston: Beacon Press, 1968).

   Marcuse, *One-Dimensional Man* (Boston: Beacon Press, 1964), p. 29.

   Marcuse, *One-Dimensional Man,* p. 32. See also Jürgen Habermas, "Technology as Ideology," in his *Toward a Rational Society* (Boston: Beacon Press, 1970).

5. Herbert Marcuse, "Some Social Implications of Modern Technology," *Studies in Philosophy and Social Science,* IX (1941), 414.
6. Marcuse, *One-Dimensional Man,* p. 32.
7. Karl Marx, *Capital* (Chicago: Charles H. Kerr and Co., 1926), I, 397. Marcuse, "Some Social Implications of Modern Technology," p. 424.

Chapter 1

1. Benjamin Franklin, quoted in Hugo Meier, "The Technological Concept in American Social History," unpublished Ph.D. dissertation, University of Wisconsin, 1950, p. 347.
   Jacob Bigelow, *Elements of Technology* (Boston: Boston Press, 1829), pp. iii–v.
2. For some further discussion of the origins of scientific technology, see Peter F. Drucker, "The Technological Revolution," *Technology and Culture,* II, 342–9; Lewis Mumford, "Technics and the Nature of Man," *Technology and Culture,* VII, 303; Lynn White, "The Historical Roots of Our Ecological Crisis," *Science,* CLV (March 1967), 1203; Daniel Horowitz, "Insight into Industrialization: American Conceptions of Economic Development and Mechanization, 1865–1910," unpublished Ph.D. dissertation, Harvard University, 1966, p. 52; and George H. Daniels, *Science in America* (New York: Alfred A. Knopf, 1971), p. 271.
3. Karl Marx, *The Grundrisse,* translated by David McLellan (New York: Harper and Row, 1971), p. 140.
4. Harry Braverman, *Labor and Monopoly Capital* (New York: Monthly Review Press, 1974), p. 166.
5. Annual Report quoted in Harold C. Passer, *The Electrical Manufacturers* (Cambridge: Harvard University Press, 1953), p. 54.
6. Edison quoted in Philip Alger, *The Human Side of Engineering* (Schenectady: Mohawk Development Service, 1972), p. 18.
   Thomas P. Hughes, "The Electrification of America, 1870–1930," address delivered at the Bicentennial Convention of the Society for the History of Technology, Washington, D.C., October 18, 1975.
   Passer, *Electrical Manufacturers,* p. 177.
7. Passer, *Electrical Manufacturers,* p. 85.
8. *Ibid.,* pp. 325, 104. "Frederick P. Fish," in *National Cyclopaedia of American Biography,* XXVI, 202; XXXIX, 278. *Proceedings of the Bar Association of the City of Boston and of the District Court of the United States for the District of Massachusetts, "In Memory of Frederick P. Fish, Boston, Massachusetts, December 26, 1931,"* p. 4, Massachusetts Historical Society, Boston. See also Edward C. Kirkland, *Industry Comes of Age* (New York: Holt, Rinehart and Winston, 1961), pp. 191–2.
9. Passer, *Electrical Manufacturers,* pp. 129–76.
10. Malcolm MacLaren, *The Rise of the Electrical Industry During the Nineteenth Century* (Princeton: Princeton University Press, 1943), pp. 105–6.
11. N. R. Danielian, *AT&T: The Story of Industrial Conquest* (New York: Vanguard Press, 1939), p. 94 and Chapter 5. See also Horace Coon,

American Tel & Tel: The Story of a Great Monopoly (New York: Longmans, Green & Co., 1939), pp. 31–93.
12. Theodore N. Vail, quoted in Danielian, *AT&T,* p. 95.
    Frederick P. Fish, "The Patent System," *Transactions of the American Institute of Electrical Engineers,* 1909, p. 335.
13. Gabriel Kolko, *The Triumph of Conservatism* (New York: Free Press, 1963), p. 47.
14. The following discussion of the rise of chemical manufactures in the United States is based primarily upon Williams Haynes' six-volume *American Chemical Industry—A History* (New York: D. Van Nostrand Co., 1954); Williams Haynes and Edward L. Gordy, eds., *Chemical Industry's Contribution to the Nation, 1635–1935,* Supplement to *Chemical Industries,* 1935; *The Chemical Industry Facts Book* (New York: Manufacturing Chemists Association, 1953); Don Whitehead, *The Dow Story* (New York: McGraw-Hill, 1968); and William S. Dutton, *Du Pont* (New York: Charles Scribner's Sons, 1949).
15. Herstein's study, cited by Haynes, *American Chemical Industry,* III, 411.
16. Haynes, *American Chemical Industry,* III, 409. See also Haynes and Gordy, eds., *Chemical Industry's Contribution to the Nation.*
17. Haynes, *American Chemical Industry,* IV, 33.
18. Haynes, *American Chemical Industry,* IV, 10–11.
19. See, for example, *Competition and Monopoly in American Industry,* Temporary National Economic Committee Investigation of the Concentration of Economic Power, Monograph No. 21 (Washington: Government Printing Office, 1938).
20. For the impact of the chemical industry upon other industries, see W. D. Horne *et al.,* "Sugar Chemistry," *Industrial and Engineering Chemistry,* XLIII, 805–9; W. O. Kenyon, "Cellulose Chemistry," *Industrial and Engineering Chemistry,* XLIII, 820–30; R. M. Burns, "Electro-chemical Industry," *Industrial and Engineering Chemistry,* XLIII, 301; H. G. Turley, "Leather Making," *Industrial and Engineering Chemistry,* XLIII, 305; E. W. Tillotson, "Glass and Ceramics," *Industrial and Engineering Chemistry,* XLIII, 311; G. Egloff and M. L. Alexander, "Petroleum Chemistry," *Industrial and Engineering Chemistry,* XLIII, 809–19; R. P. Dinsmore, "Rubber Chemistry," *Industrial and Engineering Chemistry,* XLIII, 795–803; Harold F. Williams *et al., The American Petroleum Industry,* 2 vols. (Evanston: Northwestern University Press, 1959), p. 1963; John B. Rae, *The American Automobile* (Chicago: University of Chicago Press, 1965).

Chapter 2

1. Charles R. Mann, "The American Spirit in Education," *U.S. Bureau of Education Bulletin No. 30* (1919), p. 14.
2. Ian Braley, "The Evolution of Humanistic-Social Courses for Undergraduate Engineers" (unpublished Ph.D. dissertation, Stanford University School of Education, 1961), p. 32.

3. *The Inventor,* quoted in Hugo Meier, "The Technological Concept in American Social History" (unpublished Ph.D. dissertation, University of Wisconsin, 1950), p. 352.

Turner, quoted in Charles R. Mann, *A Study of Engineering Education* (Boston: Merrymount Press, 1918), p. 10. For further discussion of the early history of technical education in the U.S., see William E. Wickenden, "A Comparative Study of the Engineering Education in the U.S. and Europe," *Report of the Investigation of Engineering Education* (Pittsburgh: Society for the Promotion of Engineering Education, 1930), I, 807–24; James Kip Finch, *Trends in Engineering Education* (New York: Columbia University Press, 1948); Braley's dissertation; Daniel Calhoun, *The American Civil Engineer* (Cambridge: MIT Press, 1960); and Monte Calvert, *The Mechanical Engineer in America, 1830–1910* (Baltimore: Johns Hopkins University Press, 1967).

4. Amos Eaton, quoted in Braley, "Evolution," p. 50.
5. Abbott Lawrence, letter to Samuel A. Eliot, June 7, 1847, quoted in Braley, "Evolution," p. 65. See also Samuel Eliot Morison, *Three Centuries of Harvard College, 1636–1936* (Cambridge: Harvard University Press, 1936), pp. 279–80.
6. Mann, *Study of Engineering Education,* pp. 6, 9.
7. For discussion of early university training in chemistry, see Williams Haynes, *The American Chemical Industry—A History* (New York: D. Van Nostrand Co., 1954), I, 393; Alfred H. White, "Chemical Engineering Education," in Sidney D. Kirkpatrick, ed., *Twenty-five Years of Chemical Engineering Progress* (American Institute of Chemical Engineers, 1933).
8. Mann, *Study of Engineering Education,* p. 6.
9. Mansfield Merriman, "Past and Present Tendencies in Engineering Education," *Society for the Promotion of Engineering Education Proceedings,* IV (1896), 17.
10. Quoted in Haynes, *American Chemical Industry,* I, 393.
11. Palmer Ricketts, "Discussion," *Society for the Promotion of Engineering Education Proceedings,* I (1893), 64.

DeVolson Wood, "Presidential Address," *Society for the Promotion of Engineering Education Proceedings,* II (1894), 21.

On the status of scientists within the academy, see George H. Daniels, *Science in America* (New York: Alfred A. Knopf, 1971), p. 280.
12. Samuel Warren quoted in Braley, "Evolution," p. 24.

Francis A. Walker, "The Place of the Schools of Technology in American Education," *Educational Review,* II (October 1891), 209.
13. E. A. Fuertes, "Discussion," *Society for the Promotion of Engineering Education Proceedings,* IV (1896), 22.
14. For discussion of the evolving content of engineering education in this period, see Esther Lucille Brown, *The Professional Engineer* (New York: Russell Sage Foundation, 1936), p. 13; Robert Fletcher, "A Quarter Century of Progress in Engineering Education," *Society for the Promotion of Engineering Education Proceedings,* VII (1899); and William Wickenden, "A Study of Evolutionary Trends in Engineering Cur-

ricula," in *Report of the Investigation of Engineering Education,* I, 522–55.

15. On early electrical engineering training, see Dugald C. Jackson, "The Technical Education of the Electrical Engineer," *Transactions of the American Institute of Electrical Engineers,* IX (1892), 472–86, and James E. Brittain, "B. A. Behrend and the Beginnings of Electrical Engineering, 1870–1920" (unpublished Ph.D. dissertation, Case Western Reserve University, 1969). On chemical-engineering education, see Alfred H. White, "Chemical Engineering Education in the U.S.," *Transactions of the American Institute of Chemical Engineers,* XXI (1928), 55–85, and W. A. Pardee and T. H. Childon, "Industrial and Engineering Chemistry," *Industrial and Engineering Chemistry,* Vol. XLIII.

    Wickenden, "Comparative Study of Engineering Education," p. 821. For further discussion of the emerging scientific bases of engineering, see Edwin Layton, "Mirror-Image Twins: The Communities of Science and Technology," in George H. Daniels, ed., *Nineteenth Century American Science* (Evanston: Northwestern University Press, 1972).

16. Calvert, *Mechanical Engineer in America,* pp. 3–40. See also Joel Gerstl and Robert Perucci, *Profession Without Community: Engineers in American Society* (New York: Random House, 1969), p. 61.

    Ashbel Welch, quoted in Braley, "Evolution," p. 70.

    William H. Burr, "The Ideal Engineering Education," *Society for the Promotion of Engineering Education Proceedings,* I (1893), 2.

17. Wickenden, "Comparative Study of Engineering Education," pp. 819–20. See also *Society for the Promotion of Engineering Education Proceedings,* from 1893 to 1910.

18. Burr, "Ideal Engineering Education," pp. 30, 36, 40.

19. For a brief discussion of the early corporation-school movement, see Berenice Fisher, *Industrial Education: American Ideals and Institutions* (Madison: University of Wisconsin Press, 1967), p. 110.

    For a description of the GE Test Course, see Charles M. Ripley, *Life in a Large Manufacturing Plant* (Schenectady: General Electric Company Publication Bureau, 1919), p. 145.

20. Amos Eaton, quoted in Braley, "Evolution," p. 52.

21. Morrill Act, quoted in Mann, "American Spirit in Education," p. 41. For detailed discussion of the evolution of humanities and social-science courses in engineering schools, see Braley.

22. Burr, "Ideal Engineering Education," pp. 19–20.

    Robert Thurston, "Graduate and Post-Graduate Engineering Degrees," *Society for the Promotion of Engineering Education Proceedings,* II (1894), 78.

23. Henry T. Eddy, "On Engineering Education," *Society for the Promotion of Engineering Education Proceedings,* V (1898), 14.

    Walker, "Place of the Schools of Technology," p. 219.

24. John B. Johnson, "Discussion," *Society for the Promotion of Engineering Education Proceedings,* IX (1901), 76.

25. Merriman, "Past and Present Tendencies," pp. 26–27.

    Burr, "Ideal Engineering Education," p. 20.

Wickenden, *Report of the Investigation of Engineering Education,* I, 46, 47, 232. See also Mann, *Study of Engineering Education,* pp. 107–111.

R. W. Raymond, quoted in Braley, "Evolution," p. 72.

Chapter 3

1. For a brief discussion of early contact between science and manufactures, see Tom Burns, "The Social Character of Technology," *Impact,* VII (September 1956), 155–64.
2. Ralph E. Flanders, "The New Age and the New Man," in Charles A. Beard, ed., *Toward Civilization* (London: Longmans, Green & Co., 1930), p. 22.
3. Dugald C. Jackson, *Present Status and Trends in Engineering Education* (New York: Engineers' Council for Professional Development, 1939), p. 97.
4. Daniel Horowitz, "Insight into Industrialization: American Conceptions of Economic Development and Mechanization, 1865–1910," unpublished Ph.D. dissertation, Harvard University, 1966, p. 238.
5. Henry Towne, Foreword to "Shop Management" in Frederick W. Taylor, *Scientific Management* (New York: Harper and Brothers, 1947), pp. 5–6; quoted in Harry Braverman, *Labor and Monopoly Capital* (New York: Monthly Review Press, 1974), p. 200.

    Towne, "Industrial Engineering," *American Machinist,* July 20, 1905, p. 100.

    A. A. Potter, interview with author, Purdue University, 1974.

    President of Stevens Institute Alumni Association, quoted in Monte Calvert, *The Mechanical Engineer in America, 1830–1910* (Baltimore: Johns Hopkins University Press, 1967), p. 213.
6. Benjamin Wright, *American Railroad Journal,* I (1832), 563, quoted in Daniel Calhoun, *The American Civil Engineer* (Cambridge: MIT Press, 1960), p. 87.

    For a fuller discussion of civil engineering in the nineteenth century, see Calhoun. For an interesting description of early canal and railroad engineering, see Elting E. Morison, *From Know-how to Nowhere* (New York: Basic Books, 1974), Part One.
7. Calhoun, *American Civil Engineer,* pp. viii, 77, 194, 199.
8. U.S. Bureau of Education, *Biennial Survey of Education 1926–28,* p. 698.
9. Edwin Layton, *The Revolt of the Engineers* (Cleveland: Case Western Reserve University Press, 1971), pp. 41–50; Calvert, *Mechanical Engineer,* p. 213.
10. The following discussion of the mechanical engineering profession is based primarily upon Calvert, *Mechanical Engineer,* and Layton, *Revolt of the Engineers.*
11. Alexander Holley, quoted in Lyndall Urwick, "Management's Debt to Engineers," *Advanced Management,* December 1952, p. 9.

12. Charles Steinmetz, "Discussion," *Transactions of the American Institute of Electrical Engineers,* XXV (1906), 266.

    See also Wickenden, *Report of the Investigation of Engineering Education* (Pittsburgh: Society for the Promotion of Engineering Education, 1930), I, 821; Calvert, *Mechanical Engineer,* pp. 215–20; Layton, *Revolt of the Engineers,* pp. 41–50.

13. Samuel P. Sadtler, "Presidential Address," *Transactions of the American Institute of Chemical Engineers,* I (1908), 36.

    See also Alfred H. White, "Chemical Engineering Education," in Sidney D. Kirkpatrick, ed., *Twenty-five Years of Chemical Engineering Progress* (American Institute of Chemical Engineers, 1933). W. A. Pardee, "The American Chemical Society Division of Industrial Chemistry and Chemical Engineering," *Industrial and Engineering Chemistry,* XLIII, 309.

14. See Edward Gross, "Change in Technological and Scientific Developments and Its Impact upon the Occupational Structure," in Robert Perrucci and Joel Gerstl, eds., *The Engineers and the Social System* (New York: John Wiley and Sons, 1969), p. 17.

    Historical Statistics of the United States, U.S. Department of Commerce, Bureau of the Census, 1961, p. 75; and *Biennial Survey of Education 1926–28,* p. 698.

15. Jay Gould, *The Technical Elite* (New York: Augustus M. Kelley, 1966), p. 172; Carolyn Cummings Perrucci, "Engineering and the Class Structure," in Perrucci and Gerstl, eds., *Engineers and the Social System,* p. 284; Wickenden, *Report of the Investigation,* I, 162; Layton, *Revolt of the Engineers,* p. 9.

16. Calvert, *Mechanical Engineer,* p. 242.

17. Wickenden, *Report of the Investigation,* I, 232. See especially Fig. 2.

18. Calvert, *Mechanical Engineer,* p. 231.

19. Charles Steinmetz, "Individual and Corporate Development of Industry," *General Electric Review,* XVIII (1915), 816.

20. Ronald Tobey, *The American Ideology of National Science 1919–1930* (Pittsburgh: University of Pittsburgh Press, 1971), p. 6.

21. Layton, *Revolt of the Engineers,* p. 19.

22. Wickenden, *Report of the Investigation,* I, 229.

23. Onward Bates, "Discussion," *Transactions of the American Society of Civil Engineers,* LXIV (September 1909), 573.

    "Commercialized Engineering," *American Machinist,* January 24, 1907, p. 573.

24. DeVolson Wood, "Presidential Address," *Society for the Promotion of Engineering Education Proceedings,* I (1893), 21.

    H. F. J. Porter, "Discussion," *Transactions of the American Society of Mechanical Engineers,* XIV (1893), 1004.

25. Matthew Elias Zaret, "An Historical Study of the Development of the American Society for Engineering Education" (unpublished Ph.D. dissertation, New York University, 1967), pp. 68, 87. See further discussion and citations in Chapter 8.

26. Steinmetz, "Individual and Corporate Development," p. 813.

27. Charles R. Mann, "The American Spirit in Education," *U.S. Bureau of Education Bulletin,* No. 30 (1919), p. 50.
28. Alfred H. White, "Chemical Engineering Education in the U.S.," *Transactions of the American Institute of Chemical Engineers,* XXI (1928), 85.
29. Paul Baran and Paul Sweezy, *Monopoly Capital* (New York: Monthly Review Press, 1966), p. 16.
30. "William E. Wickenden," *National Cyclopaedia of American Biography,* XXXVI, 391. See also Charles F. Scott, letter to Elmer Lindset, published in *The Case Alumnus,* 1929. Wickenden Papers, Case Western Reserve University Archives, Cleveland, Ohio.
31. William E. Wickenden, *A Professional Guide for Junior Engineers* (New York: Engineers' Council for Professional Development, 1949), pp. 1, 20, 21, 29.
32. *Ibid.,* p. 31.
33. *Ibid.,* p. 47.

Chapter 4

1. Christopher Lasch, "The Moral and Intellectual Rehabilitation of the Ruling Class," in *The World of Nations* (New York: Alfred A. Knopf, 1974), pp. 80–102.
2. Magnus W. Alexander, *The Economic Evolution of the United States: Its Background and Significance* (New York: National Industrial Conference Board, 1929), pp. 28–50.
3. *Ibid.,* pp. 37, 38.
4. Alexander, quoted in George Soule, *Prosperity Decade* (New York: Holt, Rinehart, and Winston, 1947), p. 293.
   Alexander, *Economic Evolution,* p. 46.
5. The following brief description of the economic history of the period is based upon *Recent Economic Changes* (New York: McGraw-Hill Book Co., 1929); Paul Baran and Paul Sweezy, *Monopoly Capital* (New York: Monthly Review Press, 1966); Alfred D. Chandler, "The Large Industrial Corporation and the Making of the Modern American Economy," in Stephen E. Ambrose, ed., *Institutions in Modern America* (Baltimore: Johns Hopkins University Press, 1967), pp. 71–101; Alfred D. Chandler, *Strategy and Structure* (Cambridge: MIT Press, 1962); Willard L. Thorp, *The Integration of Industrial Operation,* Department of Commerce, Bureau of the Census Monograph III (Washington: Government Printing Office, 1924); Edward C. Kirkland, *Industry Comes of Age* (New York: Holt, Rinehart and Winston, 1961); Soule, *Prosperity Decade;* Harold U. Faulkner, *The Decline of Laissez-Faire, 1897–1917* (New York: Holt, Rinehart and Winston, 1951); Martin J. Sklar, "On the Proletarian Revolution and the End of Political-Economic Society," *Radical America,* Vol. III, No. 3; and Gabriel Kolko, *The Triumph of Conservatism* (New York: Free Press, 1963).
6. See, for example, Stuart B. Ewen, "Advertising as Social Production," *Radical America,* III, 42–56, and his *Captains of Consciousness: Advertis-*

ing and the Social Roots of the Consumer Culture (New York: McGraw-Hill, 1976).

7. See, for example, Charles A. Conant, "Can New Openings Be Found for Capital?" *Atlantic Monthly,* LXXXIV (July-December 1899), 601–8 (courtesy of Martin J. Sklar). See also Sklar, "On The Proletarian Revolution," and Baran and Sweezy, *Monopoly Capital.*
8. Faulkner, *Decline of Laissez-Faire,* p. 31.
9. The following brief discussion of labor history draws primarily upon Faulkner, *Decline of Laissez-Faire,* pp. 280–314; Soule, *Prosperity Decade,* pp. 187–200; Kirkland, *Industry Comes of Age,* pp. 356–81; David Montgomery, "The 'New Unionism' and the Transformation of Workers' Consciousness in America, 1909–1922," unpublished paper, 1972; and Graham Adams, *Age of Industrial Violence,1910–1915* (New York: Columbia University Press, 1966).
10. Faulkner, *Decline of Laissez-Faire,* p. 281.
11. Adams, *Age of Industrial Violence,* p. 228.
12. Montgomery, "New Unionism," pp. 7, 8.
13. Faulkner, *Decline of Laissez-Faire,* pp. 101–14.
14. Adams, *Age of Industrial Violence,* p. 228.
15. See, for example, Magnus W. Alexander, "The Development and Major Aspects of the Immigration Problem," *Proceedings of the National Immigration Conference,* Special Report No. 26 (New York: National Industrial Conference Board, 1924), pp. 25–44.
16. James Weinstein, *The Decline of Socialism in America, 1912–25* (New York: Random House, 1967), p. ix.
17. Herbert G. Gutman, "Work, Culture, and Society in Industrializing America, 1815–1919," *American Historical Review,* July 1973, pp. 542–3.
18. Weinstein, *Decline of Socialism,* p. ix.
19. For further discussion of business responses to industrial strife and political challenge, see Robert Wiebe, *Businessmen and Reform* (Cambridge: Harvard University Press, 1962); Kolko, *Triumph of Conservatism;* Samuel P. Hays, *Conservation and the Gospel of Efficiency* (Cambridge: Harvard University Press, 1959) and *The Response to Industrialism, 1884–1914* (Chicago: University of Chicago Press, 1957); Samuel P. Hays, "The Politics of Reform in Municipal Government in the Progressive Era," *Pacific Northwest Quarterly,* LVIV (October 1964), 157–69; and James Weinstein, "Organized Business and the City Commission and Manager Movements," *Journal of Southern History,* XXVIII (May 1962), 166–82; Samuel Haber *Efficiency and Uplift* (Chicago: University of Chicago Press, 1964).
20. Edwin Layton, *Revolt of the Engineers* (Cleveland: Case Western Reserve University Press, 1971), pp. 64–72, and "Frederick Haynes Newell and the Revolt of the Engineers," *Journal of the Midcontinent American Studies Association,* VIII (Fall 1962), 18–26. See also his perceptive "Veblen and the Engineers," *American Quarterly,* XIV (Spring 1962), 64–72.
21. On engineers and progressivism, see Bruce Sinclair, "The Cleveland

Radicals: Urban Engineers in the Progressive Era, 1901–1917" (unpublished paper).

Layton, *Revolt of the Engineers,* pp. 64–72.

22. William Ernest Akin, "History of Technocracy" (unpublished Ph.D. dissertation, University of Rochester, 1971). See also Haber, *Efficiency and Uplift,* pp. 130–60.

23. Roosevelt and Potter letters published in *The New York Times,* October 23, 1936, p. 25, and October 25, 1936, p. 1.

Alexander, *Economic Evolution,* p. 48.

Chapter 5

1. Lyman Gage, quoted by Henry S. Pritchett, "The Story of the Establishment of the National Bureau of Standards," *Science,* XV (February 21, 1902), 282.

2. This brief discussion of the early standardization movement is based upon Bruce Sinclair, "At the Turn of the Screw: William Sellers, the Franklin Institute, and a Standard American Thread," *Technology and Culture,* XI (1969), 20–34; Monte Calvert, *The Mechanical Engineer in America, 1830–1910* (Baltimore: Johns Hopkins University Press, 1967), pp. 170–5; Dugald C. Jackson, "The Relation of Standards and of Means for Accurate Measurement to Effective Development of Industrial Production," advance copy of a lecture to be delivered at Tokyo, October 1935, pp. 7–9, 18, Jackson Papers, Department of Electrical Engineering, MIT Archives, Cambridge, Massachusetts; and Comfort A. Adams, "Industrial Standardization," *Annals of the American Academy of Political and Social Science,* LXXXII (1919), 289–99.

3. Jackson, "Relation of Standards," pp. 14, 15.

John Perry, *The Story of Standards* (New York: Funk and Wagnalls Co., 1955), pp. 127, 131.

Pardee, "The ACS Division of Industrial and Engineering Chemistry," p. 309.

Comfort A. Adams, "The National Standards Movement—Its Evolution and Future," in Dickson Reck, ed., *National Standards in a Modern Economy* (New York: Harper and Brothers, 1952), pp. 22–4.

4. "Henry S. Pritchett," *National Cyclopaedia of American Biography,* XXIX, 124.

5. Calvert, *Mechanical Engineer,* p. 181.

6. Pritchett, "Story of the Establishment of the National Bureau of Standards," p. 282.

7. Charles S. Peirce, quoted by A. Hunter Dupree, *Science in the Federal Government* (Cambridge: Harvard University Press, 1957), p. 271.

NAS quoted in Perry, *Story of Standards,* p. 128.

8. Henry S. Pritchett, "A Tale of Two Presidents," *Technology Review,* February 1923, p. 199.

Pritchett, "Story of the Establishment of the National Bureau of Standards," p. 282. See also John A. Brashear, "The Evolution of Standards of Measurement," *Cassiers Magazine*, XX (May-October 1901), 417.

9. Pritchett, "Tale of Two Presidents," p. 200; Dupree, *Science in the Federal Government*, p. 272; Perry, *Story of Standards*, p. 128; Pritchett, "Story of the Establishment of the National Bureau of Standards," p. 282.

10. Pritchett, "Story of the Establishment of the National Bureau of Standards," p. 282.

11. *Ibid.*, p. 282.

12. Pritchett, "Tale of Two Presidents," p. 200.
   Dupree, *Science in the Federal Government*, p. 274.
   C. L. Warwick, "The Work in the Field of Standardization of the American Society for Testing Materials," *Annals of the American Academy of Political and Social Science*, CXXXVII (May 1928), 49.
   H. G. Boutell, "The National Bureau of Standards," *Tech Engineering News* (MIT), VI (October 1920), 4–7.
   Pritchett, "Tale of Two Presidents," p. 200.

13. George V. Thompson, "Intercompany Technical Standardization in the Early American Automobile Industry," *Journal of Economic History*, XIV (Winter 1954), 2.
   Sinclair, "At the Turn of the Screw," pp. 20–34.
   Adams, "Industrial Standardization," pp. 289–99.

14. Calvert, *Mechanical Engineer*, pp. 169–86.

15. Charles E. Skinner, "The Present Status of Standards in the Electrical Industry," *Annals of the American Academy of Political and Social Science*, CXXXVII (May 1928), 151.

16. Adams, "National Standards Movement," p. 23. See also Adams, "Industrial Standardization," pp. 290–5; Skinner, "Present Status of Standards," pp. 151–7.

17. W. A. Pardee, "The American Chemical Society Division of Industrial and Engineering Chemistry," pp. 309–10.

18. Thompson, "Intercompany Technical Standardization," pp. 1–20.

19. P. G. Agnew, "The Work of the American Engineering Standards Committee," *Annals of the American Academy of Political and Social Science*, CXXXVII (May 1928), 13.

20. Adams, "National Standards Movement," p. 23.

21. "Howard Coffin," *National Cyclopaedia of American Biography*, XXX, 3; Lloyd N. Scott, *The Naval Consulting Board of the U.S.* (Washington: Government Printing Office, 1920), pp. 27, 37–42.

22. Leon P. Alford, "Technical Changes in Manufacturing Industries," in *Recent Economic Changes* (New York: McGraw-Hill Book Co., 1929), I, 116.

23. Perry, *Story of Standards*, pp. 131, 132.
   Magnus W. Alexander, *The Economic Evolution of the United States: Its Background and Significance* (New York: National Industrial Conference Board, 1929), p. 34.
   Dexter S. Kimball, "Changes in New and Old Industries," in *Recent Economic Changes*, I, 89;
   Alford, "Technical Changes," p. 116.

Alexander, *Economic Evolution,* p. 37. See also Willard L. Thorp, "The Changing Structure of Industry," *Recent Economic Changes,* I, 167–218; and E. W. McCullock, "The Relation of the Chamber of Commerce of the United States to the Growth of the Simplification Program in American Industry," *Annals of the American Academy of Political and Social Science,* CXXXVII (May 1928), 9–12.

24. Alexander, *Economic Evolution,* p. 34.
25. For more on Taylor, see, for example, Hugh G. J. Aitken, *Taylorism at Watertown Arsenal* (Cambridge: Harvard University Press, 1960), pp. 22, 30; Harry Braverman, *Labor and Monopoly Capital* (New York: Monthly Review Press, 1974), pp. 85, 86; and Chapter 10 below.
    Jackson, "Relation of Standards," p. 18–19.
    Kimball, "Changes in New and Old Industries," p. 90.

Chapter 6

1. E. F. W. Alexanderson, "Inventors I Have Known," in Philip Alger, *The Human Side of Engineering* (Schenectady: Mohawk Development Service, 1972), p. 137.
2. Abraham Lincoln, quoted by Frederick P. Fish, "The Patent System," *Transactions of the American Institute of Electrical Engineers,* 1909, p. 315.
3. Bernhard J. Stern, "The Corporations as Beneficiaries," *American Scholar,* XVIII (1949), 112.
    Robert A. Brady, "Not Patents But the Patent System," *American Scholar,* XVIII (1949), 106.
4. L. W. Moffett, "A Big Handicap to Industry: How an Antiquated System of Patent Laws Puts a Brake on Progress," *Iron Trade Review,* LVI (March 8, 1915), 558.
5. Fish, "Patent System," p. 315.
6. *Ibid.,* pp. 320–2, 324.
7. Floyd L. Vaughan, *The United States Patent System: Legal and Economic Conflicts in American Patent History* (Norman: University of Oklahoma Press, 1956), pp. 19, 21, 25.
8. Francis B. Crocker, "Discussion of Fish Paper," *Transactions of the American Institute of Electrical Engineers,* May 18, 1909, p. 340.
9. Vaughan, *U.S. Patent System,* pp. 33, 252.
10. *Ibid.,* pp. 252, 33.
11. Fish, "Patent System," p. 322.
    Vaughan, *U.S. Patent System,* pp. 43, 39, 34, 35, 69.
    Alexanderson, "Inventors I Have Known," p. 137.
12. Vaughan, *U.S. Patent System,* pp. 43, 69, 40, 134, 70.
13. Edwin J. Prindle, "Patents as a Factor in a Manufacturing Business," *Engineering Magazine,* XXXI (September 1906), 809–10; XXXII (October 1906), 90.
14. *Ibid.,* XXXII, 168, 407.
15. *Ibid.,* XXXII, 415.
16. *Ibid.,* XXXII, 415.

17. N. R. Danielian, *AT&T: The Story of Industrial Conquest* (New York: Vanguard Press, 1939), p. 94.
18. Vail, quoted by Danielian, *AT&T,* pp. 95–6.
19. *Ibid.,* pp. 99–100.
20. Vaughan, *U.S. Patent System,* pp. 73–5; Danielian, *AT&T,* pp. 99–100.
21. Vaughan, *U.S. Patent System,* p. 73.
22. *Ibid.,* p. 75. See also the report by the Federal Communications Commission, *Investigation of the Telephone Industry in the U.S.,* 76th Congress, 1st Session, House Report No. 340, p. 214.
23. Vaughan, *U.S. Patent System,* p. 268.
24. *Ibid.,* p. 75.
25. *Incandescent Electric Lamps,* Tariff Commission Report No. 133 (1937), p. 36. Cited in Vaughan, *U.S. Patent System,* p. 75.
26. *United States vs. General Electric Company,* 272 U.S. 476 (1926), pp. 480–1. Cited in Vaughan, *U.S. Patent System,* p. 109.
27. *United States vs. General Electric Company,* 82 F. Supp. 753, 798 (D.N.J. 1949), pp. 815, 817. Cited in Vaughan, *U.S. Patent System,* p. 75.
28. Danielian, *AT&T,* pp. 75–6.
29. Ibid., p. 109.
30. Vaughan, *U.S. Patent System,* pp. 75, 76; Danielian, *AT&T,* pp. 108, 110.
31. Vaughan, *U.S. Patent System,* p. 76. See also "Weapons of Monopoly," *New Republic,* February 14, 1944, p. 199, and Alexander Morrow, "The Suppression of Patents," *American Scholar,* XIV (Winter 1944–5), 210–19.
32. Otterson, quoted in Danielian, *AT&T,* p. 115.
33. Danielian, *AT&T,* p. 117. See also Vaughan, *U.S. Patent System,* p. 153.
34. *U.S. vs. G.E.* (1949), p. 905. Cited in Vaughan, *U.S. Patent System,* p. 154.
35. Vaughan, *U.S. Patent System,* p. 33.
36. Danielian, *AT&T,* pp. 101–2, 92.
37. Carty memorandum, quoted in Danielian, *AT&T,* p. 104.
38. Jewett, quoted in Danielian, *AT&T,* p. 196.
39. Otterson memorandum, quoted in Danielian, *AT&T,* pp. 114–15.
40. Vaughan, *U.S. Patent System,* pp. 261–5.
41. *Ibid.,* pp. 261–262.
42. President of Thomas A. Edison, Inc., cited in Vaughan, *U.S. Patent System,* p. 265.
    Alexanderson, "Inventors I Have Known," p. 135.
    Alger, *The Human Side of Engineering,* p. 43.
    Alexanderson, "Inventors I Have Known," p. 135.
43. *Pooling of Patents,* Hearings Before House Committee on Patents on House Resolution 4523, 74th Congress, Part I, p. 860. Quoted in Vaughan, *U.S. Patent System,* p. 267.
44. Louis Brandeis, quoted in Vaughan, *U.S. Patent System,* p. 267.
    Alexanderson, "Inventors I Have Known," pp. 135, 136.
45. Stern, "Corporations as Beneficiaries," p. 112.
46. L. Sprague de Camp, *The Heroic Age of American Invention* (Garden City: Doubleday and Co., 1961), pp. 251–9.

47. Vaughan, *U.S. Patent System,* p. 101.
48. Alvin D. Keene, former GE electrical engineer, interview with author, Rochester, New York, 1973.
49. De Camp, *Heroic Age,* p. 258.
50. Jewett, *Pooling of Patents,* Part I, p. 276. Quoted in Vaughan, *U.S. Patent System,* p. 283.
51. De Camp, *Heroic Age,* pp. 257, 258.
    Alexanderson, "Inventors I Have Known," p. 135.
    Stern, "Corporations as Beneficiaries," p. 112.
    Fish, "Patent System," p. 336. For a discussion of the impact of patents upon industrial development, see *Technology in Our Economy, Temporary National Economic Commission Investigation of the Concentration of Economic Power,* Monograph No. 22 (1941), pp. 212–7.
52. Conway P. Coe, "Factual Data Supplied by the Commission of Patents," reprinted in George E. Folk, *Patents and Industrial Progress: A Summary Analysis and Evaluation of the Record of Patents of the Temporary National Economic Commitee* (New York: Harper and Brothers, 1942), p. 126. See also Jacob Schmookler, *Patents, Inventions, and Economic Change* (Cambridge: Harvard University Press, 1972), Chapter 3.
53. See, for example: "Proposed Patent Legislation," *Scientific American,* April 18, 1896; "Proposed Changes in the Patent Laws," *Iron Age,* February 6, 1896; "Amendments to the Patent Statutes," *Scientific American,* March 20, 1897; Philip Mauro, "Justifying the Past Practice of Liberality in Granting Patents," *Transactions of the American Institute of Electrical Engineers,* February, March and April 1894; John Richards, "Patent Laws: Discussion of the Abolishment or Modification," *Journal of the Associated Engineering Societies,* XIV, p. 474; "The U.S. Patent Office," *Century Magazine,* LXI (1901), 346; Ludwig Guttmann, "Our Antiquated Patent System," *Electrical World,* December 26, 1908; "The Need of a New Building for the U.S. Patent Office," *Railroad Gazette,* April 13, 1900.
    Harry Kursh, *Inside the U.S. Patent Office* (New York: W. W. Norton and Company, 1959), pp. 32, 147.
    Prindle, "Patents as a Factor"; Fish, "Patent System."
    The Story of the U.S. Patent Office (Washington: Government Printing Office, n.d.), p. 20.
54. Fish, "Patent System," pp. 326–7.
55. *Story of the U.S. Patent Office,* pp. 21, 25.
56. L. H. Baekeland, "The U.S. Patent System, Its Uses and Abuses," *Industrial and Engineering Chemistry,* December 1909, p. 204.
57. *Ibid.,* p. 204. See also Baekeland's "Presidential Address" in *Journal of Industrial and Engineering Chemistry,* V (1913), 51.
58. See *Story of the U.S. Patent Office,* pp. 21, 25; Vaughan, *U.S. Patent System,* p. 19; *Revision and Codification of the Patent Statutes,* Hearings Before the House Committee on Patents on House Resolution 23417, Parts I–XXVII, 62nd Congress, 2nd Session (Oldfield Hearings of 1912); Williams Haynes, *The American Chemical Industry—A History* (New York: D. Van Nostrand Co., 1954) III, 410–13.
59. Fish testimony, quoted in Folk, *Patents and Industrial Progress,* p. 258.

Edison testimony, quoted in *ibid.,* p. 258.

Baekeland, "Presidential Address," p. 51.

Moffett, "Big Handicap to Industry," p. 558.

See also *Metallurgical and Chemical Engineering,* X (1912), 276, 326.

60. Ewing, quoted by Moffett, "Big Handicap to Industry," p. 558. See Harry Kursh, *Inside the U.S. Patent Office,* p. 33;

L. H. Baekeland, "Report of the Economy and Efficiency Commission on the Patent Office," *Electrical World,* May 10, 1913; Howard R. Bartlett, "The Development of Industrial Research in the United States," Section II of *Research—A National Resource* (National Resources Planning Board, 1941), p. 36.

61. *Story of the U.S. Patent Office,* p. 21.

62. *Report of the Patent Committee of the NRC,* Reprint and Circular Series No. 1 (1919), p. 1. On patent legislation, see Edwin Prindle, "The Patent Situation," *Chemical and Metallurgical Engineering,* XXV (August 31, 1921), 418. See also "The Patent Situation in the U.S.," *Mechanical Engineering,* XLI (February 1919), 147–9.

63. Prindle, "Patent Situation," p. 418. On the Stanley Bill, see *Journal of Engineering Chemistry,* XIV (1922), 573. See also Glenn B. Harris, "Needed Improvements in Our Patent Office and System," *American Machinist,* LII (April 18, 1920), 798–803.

64. Folk, *Patents and Industrial Progress,* p. 259; *Story of the U.S. Patent Office,* p. 23.

65. "Herbert Hoover and the Patent Office," *Scientific American,* CXXXII (June 1925), 373.

66. *Story of the U.S. Patent Office,* p. 22.

67. Robert S. Lynd, "You Can't Skin a Live Tiger," *American Scholar,* XVII (1949), 109.

68. Conway Coe, "Statement of the Commission of Patents," quoted in Folk, *Patents and Industrial Progress,* p. 143.

69. *Economic Power and Political Pressure,* Temporary National Economic Committee Monograph No. 26, pp. 22–3, quoted by Lynd, "You Can't Skin a Live Tiger," p. 109.

Lynd, "You Can't Skin a Live Tiger," pp. 109–10.

Chapter 7

1. Elihu Root, "The Need for Organization in Scientific Research," *National Research Council Bulletin,* I (October 1919), 8.

2. Howard R. Bartlett, "The Development of Industrial Research in the United States," Section II of *Research—A National Resource* (National Resources Planning Board, 1941), pp. 25–8.

3. Dexter S. Kimball, "Changes in New and Old Industries," in *Recent Economic Changes* (New York: McGraw-Hill Book Co., 1929), I, 107. For further information on the emergence of organized industrial research, see Editors of *Fortune, The Mighty Force of Research* (New York: McGraw-Hill Book Co., 1953), p. 13; Kendall A. Birr, "Science in

American Industry," in David Van Tassel and Michael G. Hall, eds., *Science and Society in the United States* (Homewood, Ill.: Dorsey Press, 1966), pp. 35–80; and Birr's Introduction to his *Pioneering in Industrial Research* (Washington, D.C.: Public Affairs Press, 1957).

4. J. J. Carty, "The Relation of Pure Science to Industrial Research" (Presidential Address, American Institute of Electrical Engineers), *Science,* New Series XLIV (October 13, 1916), 7.

5. Henry Ford, *Edison as I Knew Him,* quoted in Gilman M. Ostrander, *American Civilization in the First Machine Age* (New York: Harper & Row, 1970), p. 217.

   Norbert Wiener, quoted in Ostrander, *American Civilization,* p. 218.

   Elihu Thomson, quoted in W. Rupert McLaurin, *Invention and Innovation in the Radio Industry* (New York: MacMillan Co., 1949), p. 164.

   GE Annual Report, quoted in Bartlett, "Development of Industrial Research," pp. 51–2.

6. Willis Whitney, quoted in Bartlett, "Development of Industrial Research," p. 52.

7. Willis Whitney, quoted in McLaurin, *Invention and Innovation,* pp. 153, 164.

8. Birr, "Science in American Industry," p. 69. Lillian Hartmann Hoddeson, "The Beginnings of Solid State Physics at the Bell Telephone Laboratories, 1900–1947," Technology Studies Colloquium, December 3, 1975, MIT, Cambridge, Mass.; Bartlett, "Development of Industrial Research," pp. 49–50.

9. J. J. Carty, "Science and Business" (address to the Chamber of Commerce of the U.S., Cleveland, May 8, 1924), *NRC Reprint No. 55,* p. 4.

10. Frank B. Jewett, testimony before the Temporary National Economic Committee, quoted in George E. Folk, *Patents and Industrial Progress: A Summary Analysis and Evaluation of the Record of Patents of the Temporary National Economic Committee* (New York: Harper and Brothers, 1942), p. 153.

11. *Ibid.,* p. 153.

12. Bartlett, "Development of Industrial Research," p. 49; McLaurin, *Invention and Innovation,* p. 156. See also John Mills, "The Line and the Laboratory," *Bell Telephone Quarterly,* Vol. XIX (January 1940), and W. S. Gifford, "The Place of the Bell Telephone Laboratories in the Bell System," *Bell Telephone Quarterly,* Vol. IV (April 1925).

    McLaurin, *Invention and Innovation,* pp. 156–8.

13. "Du Pont—How to Win at Research," *Fortune,* XLII (October 1950), 115–34; Bartlett, "Development of Industrial Research," p. 34; Birr, "Science in American Industry," pp. 57–8.

14. Bartlett, "Development of Industrial Research," pp. 35–50. Editors of *Fortune, Mighty Force of Research;* Birr, *Pioneering in Industrial Research,* p. 21.

15. Birr, "Science in American Industry," pp. 58–65, 73; Birr, *Pioneering in Industrial Research,* pp. 22–4; Maurice Holland, *Industrial Explorers* (New York: Harper and Brothers, 1928).

16. Jewett, testimony before TNEC, in Folk, *Patents and Industrial Progress,* p. 153.

Philip Alger, *The Human Side of Engineering* (Schenectady: Mohawk Development Service, 1972), p. 7.

17. Jewett, testimony before TNEC, in Folk, *Patents and Industrial Progress,* p. 153.

18. Joseph Schumpeter, quoted by McLaurin, *Invention and Innovation,* p. 154.

19. E. B. Craft, *Bell Educational Conference, 1925* (New York: Bell System, 1925), pp. 25, 40.

20. *Ibid.,* pp. 43–4.

21. *Ibid.,* p. 47.

22. Jewett, quoted in McLaurin, *Invention and Innovation,* p. 156.

23. Craft, *Bell Conference, 1925,* p. 47.

24. *Technology in Our Economy,* TNEC Investigation of the Concentration of Economic Power, pp. 211–12.

William D. Coolidge, testimony before TNEC, quoted in Folk, *Patents and Industrial Progress,* p. 152.

25. Frank B. Jewett, "Motive and Obligation: Engineering, Industrial Research, Research Without Utilitarian Objective and the Interdependence of the Fields to Which They Pertain," *NRC Reprint No. 68,* pp. 9, 10.

26. "John B. Crouse," *National Cyclopaedia of American Biography,* XXXIII, 564; Holland, *Industrial Explorers;* Birr, *Pioneering in Industrial Research,* p. 24; Bartlett, "Development of Industrial Research."

27. "Robert Kennedy Duncan," *National Cyclopaedia of American Biography,* XXI, 331.

28. Robert Kennedy Duncan, "On Industrial Fellowships," *Journal of Industrial and Engineering Chemistry,* August 1919, p. 600.

29. *Ibid.,* pp. 601–2.

30. Bartlett, "Development of Industrial Research," pp. 71–2; Birr, *Pioneering in Industrial Research,* p. 24; Williams Haynes, *The American Chemical Industry—A History* (New York: D. Van Nostrand Co., 1954), III, 600–3.

31. Haynes, *American Chemical Industry,* III, 393.

32. "Arthur Dehon Little," *National Cyclopaedia of American Biography,* XV, 64.

33. Arthur D. Little, "Chemical Engineering Research," in Sidney D. Kirkpatrick, ed., *Twenty-Five Years of Chemical Engineering Progress* (New York: D. Van Nostrand and Co., 1933).

Holland, *Industrial Explorers,* p. 149.

Arthur D. Little, "The Handwriting on the Wall" (Cambridge: A. D. Little, 1925); see also Arthur D. Little, *The Earning Power of Chemistry, A Public Lecture to Businessmen* (Boston, 1911).

34. A. Hunter Dupree, *Science in the Federal Government* (Cambridge: Harvard University Press, 1957), p. 283; see also Samuel P. Hays, *Conservation and the Gospel of Efficiency* (Cambridge: Harvard University Press, 1959).

35. Dupree, *Science in the Federal Government,* p. 285.

36. *Ibid.,* p. 287.

37. *Ibid.,* p. 325.

38. *Ibid.,* p. 339; *Research—A National Resource* (National Resources Planning Board, 1941), Section I, pp. 8, 91.
39. "Robert S. Woodward" (president of the Carnegie Institution), *National Cyclopaedia of American Biography,* XIII, 108. See also Ronald C. Tobey, *The American Ideology of National Science, 1919–1930* (Pittsburgh: University of Pittsburgh Press, 1971), p. 5.
40. Raymond B. Fosdick, *The Story of the Rockefeller Foundation* (New York: Harper and Brothers, 1952), pp. ix, 146.
41. "Relations Between the Engineering Foundation and the National Research Council," unpublished compilation of excerpts from NRC and Engineering Foundation Files, compiled November 1936. NRC Archives, National Academy of Science, Washington, D.C.
42. Haynes, *American Chemical Industry,* III, 409, 271; Williams Haynes and Edward L. Gordy, eds., *Chemical Industry's Contribution to the Nation, 1635–1935,* Supplement to *Chemical Industries,* 1935. See also Carroll Pursell, "The Farm Chemurgic Council and the U.S. Department of Agriculture, 1935–1939," *Isis,* LX, 308; Bartlett, "Development of Industrial Research," p. 36.
43. William E. Wickenden, "The Place of the Engineer in Modern America," unpublished speech, March 20, 1936, Wickenden Papers, Case Western Reserve University Archives, Cleveland, Ohio.
    Dugald C. Jackson, quoted in *Consolidated Report upon the Activities of the National Research Council, 1919–1932* (mimeographed), April 1932, p. 129, NRC Archives.
44. Frank B. Jewett, "Industrial Research," paper read before the Royal Canadian Institute, February 8, 1919, *NRC Reprint No. 4,* pp. 6–7.
    Dugald C. Jackson, "Types of Practices, Processes and Products That Flow into American Industry Directly from University Research," advance copy of lecture to be delivered in Kyoto, October 1935, Jackson Papers, Electrical Engineering Department, MIT Archives. See also "Cooperation Between the Technical Industries and Technical Education in America," advance copy of lecture to be delivered in Tokyo, Jackson Papers (both papers courtesy of Professor Karl L. Wildes).
45. Carty, "Science and Business," p. 1.
46. *Ibid.,* p. 2.
47. *Ibid.,* p. 4.
48. Jewett, "Motive and Obligation," pp. 9–10.
49. *Ibid.,* pp. 9–10. See also Herbert Hoover, "The Vital Need for Greater Financial Support of Pure Science Research," *NRC Reprint No. 65;* James R. Angell, "The Development of Research in the United States," *NRC Reprint No. 6.*
50. Laurence R. Veysey, *The Emergence of the American University* (Chicago: University of Chicago Press, 1965), pp. 158, 166, 171.
51. Veysey, *Emergence of the American University,* pp. 133–4, 179; Tobey, *American Ideology of National Science,* p. 5.
52. James Creese, *The Extension of University Teaching* (New York: American Association for Adult Education, 1941), p. 5053.
53. Charles Russ Richards, "The University and the Development of the Technological Sciences," in *The University and the Commonwealth: Ad-*

dresses at the Inauguration of Lotus Delta Coffman as Fifth President of the University of Minnesota (Minneapolis: University of Minnesota Press, 1921), p. 129.

54. *Ibid.,* p. 133.

55. Ellery B. Paine, "The Engineering Experiment Station at the University of Illinois," *Proceedings of the American Institute of Electrical Engineers,* 1915, p. 2421.

   A. A. Potter, "Engineering Experiment Stations," *Society for the Promotion of Engineering Education Bulletin,* VI (1916), 619. John R. Morton, *University Extension in the United States* (Birmingham: University of Alabama Press, 1953), p. 23.

56. Creese, *Extension of University Teaching,* p. 98.

   "Charles Van Hise," *National Cyclopaedia of American Biography,* XIX, 19; "Louis E. Reber," *Ibid.,* XXXVII, 402.

57. Potter, "Engineering Experiment Stations," p. 617, and interview with author, October 1973, Purdue University. For a biography of Potter, see Robert Eckles, *The Dean* (West Lafayette: Purdue University Press, 1973).

58. Potter, "Engineering Experiment Stations," p. 619.

59. A. A. Potter, *Experiment Stations in Connection with State Colleges, Memorandum Relative to S. 4874,* Senate Document No. 353, 64th Congress, 1st Session (Washington, 1916) (Courtesy A. A. Potter).

60. Willis R. Whitney, *Industrial Research Stations: Letter of Dr. W. R. Whitney, Chairman of the Committee on Chemistry and Physics of the U.S. Naval Consulting Board, transmitting a copy of a circular letter sent by Dr. Whitney to various scientists; also extracts from replies received thereto, together with certain other matters relative to the bill (S. 4874) proposing to establish Industrial Research Stations in connection with the Land-Grant colleges in several states;* Senate Document No. 446, 64th Congress, 1st Session (Washington, 1916), pp. 3, 4, 5, 7.

   For further discussion of this bill, see Daniel J. Kevles, "Federal Legislation for Engineering Experiment Stations: The Episode of World War One," *Technology and Culture,* XII, 182–90.

61. Raymond M. Hughes, "Research in American Universities and Colleges," in *Research—A National Resource,* Section I, p. 190.

62. Karl L. Wildes, "Electrical Engineering at MIT," unpublished manuscript; "Report of the Visiting and Advisory Committee of the Electrical Engineering Department, 1925," Presidential Papers, MIT Archives.

63. Henry Pritchett, quoted in Wildes, "Electrical Engineering at MIT., pp. 4–6.

64. Edwin Layton, *Revolt of the Engineers* (Cleveland: Case Western Reserve University Press, 1971), p. 161.

   Vannevar Bush, *Pieces of the Action* (New York: William Morrow and Company, 1970), p. 243.

   Dugald Jackson, letter to Francis R. Hart, May 20, 1908, Presidential Papers, MIT Archives.

65. "Report of the Advisory Committee of the Electrical Engineering Department to the President of the Corporation, 1922," Presidential Papers, MIT Archives.

66. Dugald Jackson, letter to Louis A. Ferguson, February 15, 1910, Presidential Papers, MIT Archives.

   Jackson, "Advanced Instruction and Research in the Electrical Engineering Sciences at the MIT," Pamphlet, 1910, Presidential Papers, MIT Archives.

67. "Report of the Advisory Committee," 1912.

68. *Ibid.,* 1914.

69. Dugald Jackson, letter to Theodore N. Vail, October 5, 1915, Presidential Papers, MIT Archives.

70. *Ibid.*

71. Dugald Jackson, "Memorandum on Electrical Engineering Staff at Technology, for Gerard Swope," January 7, 1925, Presidential Papers, MIT Archives.

72. "Report of the Advisory Committee," 1922.

   Frank B. Jewett, letter to Dugald Jackson, April 6, 1925, Presidential Papers, MIT Archives.

   Jackson, "Memorandum on Gift of General Electric," June 1925, Presidential Papers, MIT Archives.

   Jackson, letter to Samuel W. Stratton, December 1, 1924, Presidential Papers, MIT Archives.

73. "William H. Walker," *National Cyclopaedia of American Biography,* Vol. A, 167.

74. William H. Walker, "Chemical Research and Industrial Progress," *Scientific American Supplement 72* (July 1, 1911), p. 14.

75. Jewett, testimony before the TNEC, quoted in Folk, *Patents and Industrial Progress,* p. 153.

76. Wildes, "Electrical Engineering at MIT," pp. 4–14, 4–35.

   Samuel C. Prescott, *When M.I.T. Was "Boston Tech"* (Cambridge: MIT Press, 1954).

   Maclaurin quoted in Samuel P. Capen, "Survey of Higher Education, 1916–18," *U.S. Bureau of Education Bulletin No. 22* (1919), p. 21.

77. Maclaurin, quoted in editorial, *Tech Engineering News,* February 1920, p. 2.

78. Dugald Jackson, letter to F. A. Molitor, American Institute of Consulting Engineers, January 8, 1920, Presidential Papers, MIT Archives.

79. William H. Walker, "Division of Industrial Cooperation and Research of the M.I.T.," *Journal of Industrial and Engineering Chemistry,* XII (April 1920), 394.

80. Dugald Jackson, letter to F. A. Molitor.

   William H. Walker, "The Technology Plan," *Chemical and Metallurgical Engineering,* XXII (March 10, 1920), 464.

81. "The Technology Plan," *National Association of Corporation Schools Bulletin,* VII (1920), p. 2.

82. "The University in Industry," *Scientific American* (March 27, 1920), p. 328.

83. Samuel Stratton, letter to Everett Morss, January 18, 1924, Presidential Papers, MIT Archives.

84. Karl Compton, quoted in "New Administrative Organization," *Technology Review,* April 1932.

85. Martin J. Sklar, "On the Proletarian Revolution and the End of Political-Economic Society," *Radical America,* Vol. III, No. 3, p. 20; *Recent Social Trends in the United States, Report of the President's Research Committee on Social Trends* (New York: McGraw-Hill Book Co., 1933), p. xlvii; Henry G. Badger, "Historical Summary of Higher Education: 1889–90 to 1949–50," *Higher Education,* IX (December 15, 1952), 88.
86. Morton, *University Extension,* pp. 6–7.
87. Nicholas Murray Butler, quoted in *National Association of Corporation Schools Bulletin,* III (December 1916), 5.
88. "Arthur H. Fleming," *National Cyclopaedia of American Biography,* XXX, 400.
89. William E. Wickenden, Memorandum on Case–Western Reserve Coordination, 1924–1932, p. 3, Wickenden Papers, Case Western Reserve University Archives.
90. William E. Wickenden, letter to C. R. Saben, Cleveland Engineering Society, March 29, 1932, Wickenden Papers.
91. Thomas H. Wickenden, letter to William Wickenden, June 27, 1939; Wickenden Papers.
92. Samuel P. Capen, "Inaugural Address," October 28, 1922, Capen Papers, State University of New York at Buffalo Archives.
93. Vernon Kellogg, "The University and Research," *Science,* New Series LIV (July 8, 1921), 19.
    Basic Research, A National Resource (Washington: National Science Foundation, 1957), p. 29.
94. Dupree, *Science in the Federal Government,* p. 305.
95. Thomas A. Edison, interview in *The New York Times Magazine,* May 30, 1915, p. 6.
    Daniels, quoted in Lloyd N. Scott, *Naval Consulting Board of the United States* (Washington: Government Printing Office, 1920), p. 10.
    Scott, *Naval Consulting Board,* p. 220.
96. Dupree, *Science in the Federal Government,* pp. 306–7.
    Robert Cuff, "The Cooperative Impulse and War: The Origins of the Council of National Defense and the Advisory Commission," in Jerry Israel, ed., *Building the Organizational Society,* (New York: Free Press, 1972), pp. 235–6.
97. Hollis Godfrey, Testimony, *Expenditures in War Department Select Committee Hearings Before Subcommittee 2, Military Camps,* S. Congress, H. R. 66th Congress, 1st Session (October 20, 1919), Serial 3, Part 15, p. 880.
    First Annual Report of the Council of National Defense (Washington: Government Printing Office, 1917), p. 97.
98. "Minutes of the Advisory Commission of the Council of National Defense," December 6, 1916, and April 17, 1917; Record Group 62, 1–B1, Box 25, National Archives, Suitland, Maryland.
99. Tobey, *American Ideology of National Science,* pp. 21–30, 35.
100. Helen Wright, *Explorer of the Universe* (New York: E. P. Dutton, 1966), pp. 287–8.
101. Nathan Reingold, "World War I, The Case of the Disappearing Laboratory," paper delivered at the annual convention of the Organization of

American Historians, April 1975, Boston, Massachusetts. See also Daniel J. Kevles, "George Ellery Hale, The First World War and the Advancement of Science in America," *Isis* LIX (Winter 1968), 427–37.

102. Wright, *Explorer of the Universe,* p. 287.

103. *Ibid.,* p. 288.

104. Robert A. Millikan, *Autobiography* (New York: Prentice-Hall, 1950), pp. 132–3.

105. Tobey, *American Ideology of National Science,* pp. 36–8; Dupree, *Science in the Federal Government,* pp. 309, 312; *Proceedings of the National Academy of Sciences,* II (1916), 51.

106. Dupree, *Science in the Federal Government,* p. 312; Millikan, *Autobiography,* pp. 137–46; "Minutes of the Advisory Commission of the Council of National Defense," April 17, 1917.

107. Dupree, *Science in the Federal Government,* pp. 313–15, 323.

108. *Proceedings of the National Academy of Sciences,* II, (1916), 508.

109. George E. Hale, "Origin and Purpose of the National Research Council," unpublished memorandum, quoted by Tobey, *American Ideology of National Science,* p. 52. For further discussion of the extrademocratic and extrapolitical nature of these and similar developments, see David W. Eakins, "The Origins of Corporate Liberal Policy Research, 1916–1922: The Political-Economic Expert and the Decline of Public Debate," in Jerry Israel, ed., *Building the Organizational Society,* (New York: Free Press, 1972), pp. 163–79.

110. Vincent letter, quoted in Millikan, *Autobiography,* p. 180.

111. Tobey, *American Ideology of National Science,* pp. 53–8; Dupree, *Science in the Federal Government,* pp. 326–9; Reingold, "World War I, The Case of the Disappearing Laboratory." See also Myron J. Rand, "The National Research Fellowships," *Scientific Monthly,* LXXIII (August 1951), 71–80.
    Millikan, *Autobiography,* p. 184.

112. Tobey, *American Ideology of National Science,* p. 56; Dupree, *Science in the Federal Government,* pp. 311–12.

113. George E. Hale, mimeographed letter, April 20, 1918, Division of Engineering Files, NRC Archives.

114. George E. Hale, "Memorandum of Dinner Given by Dr. George Hale," May 29, 1918, Division of Industrial Relations Files, NRC Archives.

115. Whitney, Root, Hale, Pritchett, Maclaurin, and Swasey quoted in *ibid.*

116. Root, "Need for Organization in Scientific Research," pp. 8, 10.

117. Jewett, "Industrial Research," p. 3.

118. Henry Pritchett, "The Function of Scientific Research in a Modern State," *NRC Bulletin,* I, 11.

119. Theodore N. Vail, "Relations of Science to Industry," *NRC Bulletin,* I, 13.

120. Pritchett, "Function of Scientific Research," p. 11.

121. Wilson's executive order establishing the permanent NRC, published in *Consolidated Report,* p. 8.

122. Vernon Kellogg, "Isolation or Cooperation in Research," *NRC Reprint No. 67* (1925), pp. 1, 5.

123. Comfort A. Adams, "Report of Division of Engineering of the NRC,

April 26, 1921," cited in "Relations of the Engineering Foundation and the National Research Council," NRC Archives.

124. "Division of Anthropology and Psychology" and "Division of Educational Relations," *Consolidated Report,* pp. 237–46; 66–7.

125. Charles L. Reese, "Informational Needs in Science and Technology," *NRC Reprint No. 33,* p. 7.

126. George E. Hale, National Academy of Sciences, Annual Report for 1918, pp. 41–50, quoted in Dupree, *Science in the Federal Government,* p. 313.

127. "Research Information Services," *Consolidated Report,* pp. 255–63. See also Gordon S. Fulcher, "Scientific Abstracting," *Science,* LIV (1921), 291–5.

128. "Report of the NRC to the Council of National Defense" (1919), quoted in Albert L. Barrows, "The Relationship of the NRC to Industrial Research," in *Research—A National Resource* (National Resources Planning Board, 1941), Section II, p. 367; Gano Dunn, letter to A. L. Barrows, June 12, 1922, Division of the Advisory Committee, NRC Archives.

129. Theodore N. Vail, letter to John Johnston, December 17, 1918, NRC Archives; Minutes of the Division of Research Extension May 1920, NRC Archives; "Division of Research Extension," in *Consolidated Report,* pp. 77–81.

130. See "Relations of the Engineering Foundation and the National Research Council," NRC Archives.

131. Willis Whitney, letter to Howe, April 25, 1920, Division of Engineering Files, NRC Archives.
    Mees, quoted by Howe in letter to Johnston, August 26, 1919, NRC Archives.

132. Comfort A. Adams, quoted in "Relations of the Engineering Foundation and the National Research Council," NRC Archives.

133. Minutes of the Division of Engineering, April 1919–December 1921, NRC Archives. *Consolidated Report,* pp. 109–37.

134. Minutes of the Division of Engineering, February 20, 1923, January 14, 1924, NRC Archives.

135. "Division of Engineering" in *Consolidated Report,* pp. 109–37; Minutes of the Division of Engineering, October 13, 1925; September 17, 1930; February 1931, NRC Archives; Barrows, "Relationship of the NRC to Industrial Research," p. 368.

Chapter 8

1. William E. Wickenden, "Industry and Education Approach Each Other," General Motors Institute Commencement Address, August 23, 1946, Wickenden Papers, Case Western Reserve University Archives.

2. *New York Times* editorial, quoted in the *National Association of Corporation Schools Bulletin,* VII (1920), 437.

3. Frank B. Jewett, "Dinner Address," *Bell System Educational Conference,* 1924, p. 192.
4. *Proceedings of the First Annual Convention,* National Association of Corporation Schools, 1913. See also Berenice Fisher, *Industrial Education: American Ideals and Institutions* (Madison: University of Wisconsin Press, 1967), p. 110.
5. Charles Steinmetz, *National Association of Corporation Schools Proceedings,* I, 156.
6. Steinmetz, "Minutes of the Organizing Convention of the NACS," *National Association of Corporation Schools Proceedings,* I, 406.
   Charles M. Ripley, *Life in a Large Manufacturing Plant* (Schenectady: General Electric Company Publication Bureau, 1919), pp. 141, 143.
7. *Ibid.,* pp. 16, 147, 137.
8. Magnus W. Alexander, "The New Method of Training Engineers," *Transactions of the American Institute of Electrical Engineers,* XXVII (1908), 1462–4.
9. *Ibid.*
10. Ripley, *Life in a Large Manufacturing Plant,* pp. 152, 161.
11. *Ibid.*
12. Charles F. Scott, "The Engineering College and the Electric Manufacturing Company," *Society for the Promotion of Engineering Education Proceedings,* XV (1907), 465, 468.
13. Ibid., p. 468. "Discussion of Scott Paper," *Society for the Promotion of Engineering Education Proceedings,* XV (1907), 485.
14. Scott, "Engineering College," p. 468.
15. Charles F. Scott and Channing R. Dooley, "Adapting Technical Graduates to the Industries," *Society for the Promotion of Engineering Education Bulletin,* II (1911), 140.
16. Charles F. Scott, "Professional Engineering Education for the Industries," *Canadian Engineer,* XLII (1922), 44.
    A. A. Potter, interview with author, September 1973, Purdue University; A. A. Potter, "Personnel Work as Applied to a College of Engineering," *Proceedings of the Thirty-Seventh Annual Convention of the Association of Land-Grant Colleges,* November 1923, pp. 398–412.
    Channing R. Dooley, "Adapting the Technical Graduate to Industry," *National Association of Corporation Schools Proceedings,* I, 163.
17. Albert C. Vinal, *National Association of Corporation Schools Proceedings,* III (1915), 417.
    J. Walter Dietz, *National Association of Corporation Schools Proceedings,* I (1913), 168, 169.
18. J. W. Dietz, "The Graduate Apprentice Course of the Western Electric Company," *Society for the Promotion of Engineering Education Bulletin,* II (1911), 426, 427.
    Frank B. Jewett, "Dinner Address," *Bell System Educational Conference,* 1924, pp. 193–5.
19. F. C. Henderschott, "The National Association of Corporation Schools," *Transactions of the American Institute of Electrical Engineers,* 1913, p. 1413. Henry C. Metcalf, "The National Association of Corporation Schools," *American Economic Review,* June 1913, p. 538; "Minutes of

the Meeting for Organization of the NACS," *National Association of Corporation Schools Proceedings*, I, 409.

20. E. St. Elmo Lewis, *National Association of Corporation Schools Proceedings*, I, 42.

    E. A. Deeds, *ibid.*, p. 38.

    Arthur Williams, *ibid.*, p. 416.

21. F. C. Henderschott, *ibid.*, p. 74.

    C. D. Brackett, *ibid.*, p. 39.

22. Metcalf, "NACS," p. 538.

23. Lee Galloway, "Minutes of the Meeting for Organization of the NACS," *National Association of Corporation Schools Proceedings*, I, 343, 350–1.

24. *National Association of Corporation Schools Proceedings*, Vols. I–IV.

    Galloway, "Minutes of the Organization Meeting," p. 353.

25. *Ibid.*, p. 343.

26. M. S. Sloan, *Corporation Training*, March 1922, p. 15.

27. Henderschott, "NACS," p. 1416.

    E. J. Mehren, "Report of the Proposed Activities of the NACS for Its First Year," *National Association of Corporation Schools Proceedings*, I, 357.

28. Wilfred Lewis, "The Place of the College in Collecting and Consulting the Data of Scientific Management," *Society for the Promotion of Engineering Education Bulletin*, III (1912), 182.

29. Frederick L. Bishop, "The Cooperative System of Engineering Education at the University of Pittsburgh," *Society for the Promotion of Engineering Education Bulletin*, II (1911), 141.

30. Karl L. Wildes, "Cooperative Courses—Their Development and Operating Principles," *Transactions of the American Institute of Electrical Engineers*, 1930, pp. 1086–7.

31. Herman Schneider, "The Cooperative Course in Engineering at the University of Cincinnati," *Society for the Promotion of Engineering Education Proceedings*, XV (1907), 391–8.

32. *Ibid.*, pp. 391–8.

33. Charles S. Gingrich, "The Cooperative Engineering Course at the University of Cincinnati from the Manufacturer's Standpoint," *Society for the Promotion of Engineering Education Proceedings*, XV (1907), 399–411.

34. Wildes, "Cooperative Courses," p. 1090.

35. Herman Schneider, "Selecting Young Men for Particular Jobs," *National Association of Corporation Schools Bulletin*, September 1914, pp. 9, 11, 18.

36. Herman Schneider, "Fitting the Individual to His Life's Work," *National Association of Corporation Schools Bulletin*, IV (1917), 20.

37. Magnus Alexander, letter to Henry Pritchett, June 10, 1907, Presidential Papers, MIT Archives.

    Elihu Thompson, letter to Henry Pritchett, June 10, 1907, Presidential Papers, MIT Archives.

    Willis Whitney, letter to A. A. Noyes, November 10, 1907, Presidential Papers, MIT Archives.

38. Alexander, "New Method of Training Engineers," pp. 1465, 1467, 1468, 1470.
39. *Ibid.*, pp. 1470, 1493, 1495.
    Magnus Alexander to Frederick P. Fish, December 22, 1907, Presidential Papers, MIT Archives.
40. S. M. Basford to Everett Morss, November 21, 1907, Presidential Papers, MIT Archives.
41. Albert W. Smith to Dugald Jackson, December 14, 1907, Presidential Papers, MIT Archives.
42. Charles F. Scott to Dugald Jackson, December 18, 1907, Presidential Papers, MIT Archives.
43. Dugald Jackson, "Cooperative Course—Summary of Views Expressed" (typescript), May 1907, Presidential Papers, MIT Archives.
    Arlo Bates to Jackson, November 11, 1907, Presidential Papers, MIT Archives.
    Jackson to A. A. Noyes, May 15, 1908, Presidential Papers, MIT Archives.
44. Magnus Alexander, "Proposed Cooperative Course" (typescript), May 4, 1917, Presidential Papers, MIT Archives.
45. E. W. Rice to Richard Maclaurin, May 16, 1917; Maclaurin to Rice, June 5, 1917; Maclaurin to W. C. Fish, June 15, 1917; Fish to Maclaurin, June 18, 1917; Dugald Jackson to Maclaurin, June 15, 1917, Presidential Papers, MIT Archives.
46. Dugald Jackson to Administration Committee, April 6, 1920, Presidential Papers, MIT Archives.
47. William H. Timbie, "Cooperative Course in Electrical Engineering at the M.I.T.," *Journal of the American Institute of Electrical Engineers,* 1925, pp. 613–15, 617. See also Magnus Alexander and Dugald Jackson, "Requirements of the Engineering Industries and the Education of Engineers," *Mechanical Engineering,* XLIII (1921), 391–5.
48. Karl L. Wildes, "Electrical Engineering at M.I.T.," unpublished manuscript, pp. 4–123, 4–112, 4–26; Dugald Jackson to Gerard Swope, March 22, 1927; Jackson to Samuel Stratton, May 19, 1927; Jackson to Everett Morss, January 5, 1925; Jackson to Samuel Stratton, July 6, 1929, Presidential Papers, MIT Archives.
49. Alfred H. White, "Chemical Engineering Education in the U.S.," *Transactions of the American Institute of Chemical Engineers,* XXI (1928), 55.
50. Burgess cited in Harrison Hale, "Chemical Education," *Industrial and Engineering Chemistry,* XLIII, 1036.
    J. H. James, "Chemical Education for the Industries," *Society for the Promotion of Engineering Education Bulletin,* II (1911), 79.
    Alfred H. White, "Chemical Engineering Education," in Sidney D. Kirkpatrick, ed., *Twenty-Five Years of Chemical Engineering Progress,* (American Institute of Chemical Engineers, 1933), pp. 350–60.
51. Arthur D. Little, "Report of the AIChE Committee on Chemical Engineering Education (1922)," quoted by White, "Chemical Engineering Education in the U.S." p. 56. See also William H. Walker, "Chemical Engineering Education," *Chemical Engineers,* II (1905), 1; R. T. Has-

lam, "The School of Chemical Engineering Practice at M.I.T.," *Journal of Industrial and Engineering Chemistry*, XIII (1921), 465–8.

52. Haslam, "School of Chemical Engineering Practice at M.I.T."; White, "Chemical Engineering Education in the U.S."

53. Charles M. A. Stine, "Chemical Engineering in Modern Industry," *Transactions of the American Institute of Chemical Engineers*, XXI (1928), 45.

54. Dugald Jackson to Gerard Swope, March 22, 1927; Jackson to Samuel Stratton, July 6, 1929, Presidential Papers, MIT Archives. Wildes, "Electrical Engineering at M.I.T.," pp. 4–26, 4–112, 4–123.

55. E. A. Holbrook, "Frederick Lendall Bishop, An Appreciation," delivered at the Memorial Service, November 13, 1947, University of Pittsburgh Archives, Pittsburgh.

56. Bishop, "The Cooperative System," p. 141. See also "The Cooperative Plan of Engineering Education in the Pittsburgh Industries," School of Engineering Brochure, 1924–5, University of Pittsburgh Archives.

57. William J. Boaton to John Hallock, July 29, 1916, reprinted in "The Cooperative Plan of Engineering Education," School of Engineering Brochure, 1916, University of Pittsburgh Archives. Channing R. Dooley to Frederick L. Bishop, August 17, 1916, reprinted in "The Cooperative Plan," 1916.

58. S. M. Kintner, quoted by Bishop, "The Cooperative System," p. 144.

59. William Wickenden *et al.*, "Cooperation with Collegiate Institutions in Pittsburgh," *National Association of Corporation Training Bulletin*, April 1922, p. 10; William M. Davidson, "Industry and Public School Relations," *National Association of Corporation Training Bulletin*, April 1922, pp. 7–8.

60. "Chancellor McCormick Defines Educational Needs," *National Association of Corporation Schools Bulletin*, III (September 1916), 19.

61. "Utilizing the Municipal Universities to Stimulate Industrial Welfare," *National Association of Corporation Schools Bulletin*, 1919, pp. 486–7; see also C. W. Park, "The Cooperative System of Education," *U.S. Bureau of Education Bulletin*, No. 37, 1916; Nathaniel Peffer, *Education Experiments in Industry* (New York: The MacMillan Co., 1932), pp. 173–6.

62. Dugald Jackson, "Correlations of Industry and Education" (typescript), Jackson Papers, Electrical Engineering Department, MIT Archives (courtesy Karl L. Wildes).

63. Walter S. Ford, "Faculty Committee on Employment," *Society for the Promotion of Engineering Education Bulletin*, II (1911), 478; "Training Students for Their Life Careers," *Personnel Administration*, January 1923, p. 10.

64. A. A. Potter, "Personnel Work as Applied to a College of Engineering," *Proceedings of the Twenty-Seventh Annual Convention of the Association of Land-Grant Colleges*, November 1923, pp. 398–412.

65. *Ibid.*

66. *Bell System Educational Conference*, 1926, p. 250.

67. Frank Vanderlip, *Business and Education* (New York: Duffield and Company, 1907), p. 29.
68. "Education Plan of the National City Bank" (1916), Suzzalo Papers, University of Washington Archives, Seattle, Washington. W. S. Kies to Henry Suzzalo, September 9, 1916, Suzzalo Papers.
69. Arthur E. Morgan, "The Antioch Plan," *Engineering News-Record,* LXXXVI (January 1921), 108–11.
70. *Ibid.*
71. C. R. Young, "Antioch Plan Omits Public Service Training," *Engineering News-Record,* LXXXVI, 480; Arthur E. Morgan, "Letter to the Editor," *Engineering News-Record,* LXXXVI, 480.
72. A. A. Potter to Matthew Zaret, April 20, 1966, cited in Matthew Elias Zaret, "An Historical Study of the Development of the American Society for Engineering Education," unpublished Ph.D. dissertation, New York University, 1967, p. 69.
73. See Zaret, "An Historical Study"; Loyall Osborne, "Proper Qualifications of the Electrical Engineering School Graduates from the Manufacturer's Standpoint," *Society for the Promotion of Engineering Education Proceedings,* 1903, p. 290.
74. *Society for the Promotion of Engineering Education Proceedings,* XV (1907), 17; Zaret, "An Historical Study," p. 85.
75. Charles R. Mann to Henry Pratt Judson, May 28, 1906; Mann to William Rainey Harper, February 21, 1902, Presidential Papers, University of Chicago Archives.
      Charles R. Mann, "A New Movement Among Physics Teachers," *School Review,* XIV (June 1906), 660; Charles R. Mann, "Science in Civilization and Science in Education," *School Review,* XIV (June 1906), 664. See also Mann's *The Teaching of Physics—Purposes of General Education* (New York: The Macmillan Co., 1917).
76. Charles R. Mann, "Address," *National Association of Corporation Schools Proceedings,* 1915, pp. 431–4.
77. "Report of the Joint Committee," *Society for the Promotion of Engineering Education Proceedings,* XXIII (1915), 70; Charles R. Mann, *A Study of Engineering Education,* Carnegie Foundation for the Advancement of Teaching Bulletin No. 11 (Boston: Merrymount Press, 1918); Frederick L. Bishop, "Engineering Education," *U. S. Bureau of Education Bulletin Number 19* (1919), p. 7.
78. "Report of the Committee to Evaluate the Report of the Joint Committee on Investigation of Engineering Education," *Society for the Promotion of Engineering Education Proceedings,* XXVII (1919), 103; Harry P. Hammond, "Promotion of Engineering Education in the Past Forty Years," *Society for the Promotion of Engineering Education Proceedings,* XLI (1933), 60; Frederick L. Bishop, "Society for the Promotion of Engineering Education: Aims and Purposes of the Society," November 1923, University of Pittsburgh Archives; "Editorial," *Society for the Promotion of Engineering Education Bulletin,* III (1912), 354.
79. Frederick L. Bishop, "A Unique Opportunity," *Society for the Promotion of Engineering Education Bulletin,* IX, 2.
80. William Appleman Williams, "A Profile of the Corporate Elite," in

Ronald Radosh and Murray N. Rothbard, eds., *A New History of Leviathan* (New York: E. P. Dutton & Co., 1972), p. 5.

81. Charles R. Mann, "The Effect of the War on Engineering Education," *Society for the Promotion of Engineering Education Bulletin,* VIII (February 1918), 231–2.

82. "Minutes of the Advisory Commission of the Council of National Defense," December 6 and 7, 1916, National Archives. William McClellan to Henry Suzzalo, February 17, 1917, Suzzalo Papers, University of Washington Archives; Robert M. Yerkes, "Report of the Psychology Committee of the National Research Council," *Psychological Review,* XXVI (March 1919), 83–149.

83. *The Personnel System of the Army* (Washington, D.C.: Committee on Classification of Personnel in the Army, 1919), I, 2–25; II, 7, 33–40, Chapter 41; Records of the CCP on the Army, National Archives. See Loren Baritz, *The Servants of Power* (Middletown: Wesleyan University Press, 1960) for a closer look at the early work of Scott and his associates. See also Daniel Kevles, "Testing the Army's Intelligence: Psychologists and the Military in World War One," *Journal of American History,* LV (December 1968), 565–81.

84. Walter Dill Scott, "Memorandum on Plan for Organizing a Committee on Classification of Personnel in the Army," August 14, 1917, War Department Records, Office of the Chief of Staff, National Archives. Walter Dill Scott to Newton D. Baker, August 3, 1917, War Department Records, Office of the Chief of Staff, National Archives. *Personnel System of the Army,* p. 7.

85. "Minutes of the Advisory Commission of the Council of National Defense," April 23, 1917; October 1, 1917; November 5, 1917; December 3, 1917, National Archives. See also *First Annual Report of the Council of National Defense* (Washington, D.C.: Government Printing Office, 1917), pp. 72–4; and *Second Annual Report of the Council of National Defense* (1918), p. 83.

Park Kolbe, *The Colleges in War Time and After* (New York: D. Appleton and Co., 1919), pp. 26, 47.

86. "Minutes of the Advisory Commission," May 21, 1917; Hollis Godfrey to Henry Suzzalo, May 19, 1917, including a copy of "Report of a Conference Held at Washington, May 3, 1917, Under the Auspices of the Committee on Engineering, Education, and Scientific Research of the Advisory Commission of the Council of National Defense," Suzzalo Papers, University of Washington Archives.

87. Samuel P. Capen, "Autobiographical Sketch" (typescript), submitted to the trustees of the University of Buffalo, 1922, Capen Papers.

88. Capen to Mrs. Capen, October 19, 1910; Capen to College Presidents, October 8, 1910, Capen Papers.

89. G. C. Anthony to Samuel P. Capen, June 13, 1912; Samuel Earle to Capen, January 24, 1912, Capen Papers.

90. Samuel P. Capen to Mrs. Capen, February 22, 1914; April 10, 1914; February 13, 1917, Capen Papers.

91. Capen to Mrs. Capen, April 10, 1914; June 26, 1914; September 2, 1917; July 10, 1917, Capen Papers.

92. Capen to Mrs. Capen, July 6, 1917.

93. "Minutes of the Educational Section of the Committee on Science and Research, Including Engineering and Education of the Advisory Commission of the Council of National Defense," May 6, 1917; May 16, 1917; May 26, 1917; July 2, 3, 4, 1917; September 11, 1917. The only existing records of Godfrey's education committee are located in the Suzzalo Papers at the University of Washington Archives. Suzzalo, a member of the committee, was unable to attend the Washington, D.C., meetings and therefore had Godfrey send him the minutes. Samuel P. Capen to Suzzalo, August 6, 1917; Hollis Godfrey to Suzzalo, September 24, 1917; September 13, 1918, Suzzalo Papers. See also Samuel P. Capen, "Survey of Higher Education, 1916–1918," *U.S. Bureau of Education Bulletin No. 22* (1919), pp. 49–52.

94. "Minutes of the Educational Section," May 26, 1917, Suzzalo Papers. Minutes of the Advisory Commission, January 8, 1917, National Archives.

95. Hollis Godfrey to Henry Suzzalo, September 13, 1918, Suzzalo Papers. Capen, "Survey of Higher Education, 1916–1918," pp. 49–52.

96. Charles R. Mann, "The Committee on Education and Special Training," February 1918, Records of the War Department Committee on Education and Special Training (CEST Records), National Archives.

97. Newton D. Baker, Appointment of the Committee on Education and Special Training, February 10, 1918, CEST Records, National Archives.

98. Records of the Vocational Division, Committee on Education and Special Training Records, National Archives. Kolbe, *Colleges in War Time,* pp. 89, 62; Yerkes, "Report of the Psychology Committee"; *Personnel System of the Army,* Chapter 41; A. A. Potter, "Notes on the History of the Vocational Training in Connection with the Student Army Training Corps," CEST Records, National Archives.

99. "Minutes of Committee on Coordination and Needs," CEST Records, National Archives.

J. E. Smith to C. R. Dooley, January 14, 1919, CEST Records, National Archives.

Dooley, quoted in William T. Bawden, "Conference of Specialists in Industrial Education Formerly Connected with the Student Army Training Corps," March 25, 1919, CEST Records, National Archives.

100. Kolbe, *Colleges in War Time,* pp. 70–4. Richard Maclaurin to Henry Suzzalo, August 3, 1918 (draft of SATC regulations), Suzzalo Papers, University of Washington Archives.

101. Samuel P. Capen to Mrs. Capen, July 5, 6, 13, 1918; April 17, 1919, Capen Papers, SUNY Buffalo Archives.

102. *American Council on Education, Leadership and Chronology, 1918–1968* (Washington, D.C.: ACE, 1968), p. 104.

"Minutes of the Organizing Committee of the Emergency Council on Education," ACE Archives, Washington, D.C.

Samuel P. Capen to Mrs. Capen, July 17, 1918, Capen Papers.

103. Carol Gruber, "Mars and Minerva: World War One and the American Academic Man," unpublished Ph.D. dissertation, Columbia University, 1968, p. 381 (courtesy Eugene D. Genovese).

104. Samuel P. Capen to Mrs. Capen, July 13, 1918; July 18, 1918. Capen Papers.
105. Capen to Mrs. Capen, August 14, 1918; July 13, 1918; September 11, 1918, Capen Papers.
106. Capen to Mrs. Capen, July 6, 1918, Capen Papers.
107. Capen to Mrs. Capen, July 20, 1918, Capen Papers.
108. Capen to Mrs. Capen, July 3, 1918; August 24, 1918, Capen Papers.
109. Richard Maclaurin to Henry Suzzalo, August 3, 1918, Suzzalo Papers.
    Kolbe, *Colleges in War Time,* pp. 60–80.
110. *Ibid.,* pp. xix, 194.
    Walter Dill Scott, quoted by Samuel P. Capen, letter to Mrs. Capen, September 1, 1918, Capen Papers.
111. Bishop, "A Unique Opportunity," p. 2.
112. Robert I. Rees to Henry Suzzalo, September 20, 1918 (mimeo), "Observance of October 1, 1918—Official Establishment of the Student Army Training Corps," Suzzalo Papers.
    Kolbe, *Colleges in War Time,* p. 75.
113. *Ibid.,* pp. 70–5; Yerkes, "Report of the Psychology Committee."
114. "Final Report of the Committee on Education and Special Training to the Presidents of Institutions," December 11, 1918, CEST Records, War Department, National Archives. See also *War Department Annual Report,* Vol. I (1919), Part I, pp. 320–1.

Chapter 9

1. Henry Suzzalo, "The Effective American University System," *Educational Record,* IV, 157–61.
2. Charles R. Mann, "The American Spirit in Education," *U.S. Bureau of Education Bulletin No. 30* (1919), p. 57.
    Samuel P. Capen, "Survey of Higher Education, 1916–18," *U.S. Bureau of Education Bulletin No. 22* (1919), p. 623.
3. Charles R. Mann, "Education in the Army, 1919–1925," *Educational Record,* Supplement No. 1 (April 1926), p. 8.
    "Final Report to Presidents of Institutions," CEST Records.
    "Report of the Chief of Staff," *War Department Annual Report,* I (1919), 322.
4. Mann, "Education in the Army," pp. 8, 10, 18–19, 23, 25, 38–9.
5. Charles R. Mann, "National Standard Terminology for Occupations," November 1922 (typescript), Records of the Anthropology and Psychology Division, National Research Council Archives.
6. Charles R. Mann, "The Effect of the War on Education," *U.S. Bureau of Education Bulletin No. 58* (1919), p. 91. See also Charles R. Mann, "The Effect of the War on Engineering Education," *Society for the Promotion of Engineering Education Bulletin,* VIII (February 1918), 230–5.
7. Robert M. Yerkes, "Report of the Psychology Committee of the National Research Council," *Psychological Review,* XXVI (March 1919), 92, 94, 149.

8. Alfred Flinn to John Merriam, May 29, 1919, and June 9, 13, 16, 1919, Records of the Executive Committee, NRC Archives.

9. "Problems of Industrial Personnel; tentative draft of proposed research," August 20, 1919, Executive Committee, NRC Archives.

10. Max Farrand (Commonwealth Fund) to Alfred Flinn, April 9, 1920; Flinn, "Memorandum on Conferences," August 22, 1919; Mark Jones to Charles F. Rand, April 16, 1920, NRC Executive Committee, NRC Archives.

11. Samuel P. Capen, Alfred Flinn, Beardsley Ruml, memorandum, May 1, 1920, minutes of meeting, Committee on Present Status of Personnel Research, November 12, 1920, NRC Executive Committee, NRC Archives.

12. Mann, "National Standard Terminology for Occupations."

13. John W. Weeks to Vernon Kellogg, November 23, 1922. "Report on the Conference on Standardization of Vocational Terminology," January 6, 1923, Anthropology and Psychology Committee, NRC Archives.

14. Mann, "National Standard Terminology for Occupations."

15. Records of the Anthropology and Psychology Division of the NRC; "Report on Conference on Vocational Guidance with Special Reference to Personnel Service for College Students," Executive Committee, May 1924, NRC Archives. See also L. L. Thurstone and Charles R. Mann, "Vocational Guidance for College Students," *Journal of Personnel Research,* III (April 1925), 421–48.

16. "Minutes of Executive Committee of the Division of Educational Relations," May 7, 1921; April 10, 1923; January 14, 1924; November 4, 1925; November 25, 1929; and April 3, 1936, NRC Archives.

17. P. F. Walker, "Tendencies in Engineering Education: New Influences as Indirect Result of War May Bring Changes," *Industrial Management and Engineering Magazine,* LVIII (1919), 447.

18. Mann, "Effect of the War on Engineering Education," pp. 232–3, 235.

19. R. E. A. Anderson, "Broad Training in the Fundamentals Should Be a Part of the Engineer's Education," *American Machinist,* LIV (June 30, 1921), 1108.

20. Hollis Godfrey, "The Teaching of Scientific Management in Engineering Schools," *Society for the Promotion of Engineering Education Bulletin,* III (1912), 22–34; McDonald and Hinton, *Drexel Institute of Technology, 1891–1941* (Philadelphia: Drexel Institute, 1942), pp. 68, 69.

   "First Report of the Special Committee on Cooperation with the Industries," ACE Archives; Samuel P. Capen, "A Plan for Cooperation Between the Colleges and Industry," *Society for the Promotion of Engineering Education Proceedings,* 1920, pp. 52–8.

21. Hollis Godfrey, "The Council for Management Education," *Textile World,* LIX (1921), 864, 884. See also Godfrey's articles in: *American Machinist,* LIII (October 7, 1920), 698; *Gas Age,* XLIII (July 11, 1921), 323; his pamphlet *A Practical Plan for Educating Management Men* (CME Pamphlet, August 14, 1920), Harvard University, Education Library; and *New York Times,* December 27, 1920, p. 22.

   Samuel P. Capen to John A. Cousens, July 23, 1920; John A. Cousens to Capen, July 9, 1920, Capen Papers.

22. Alexander quoted in Clyde L. Rogers, *Let There Be Light—The Conference Board History, 1916–1967* (New York: Conference Board, 1967), p. 7.

23. Rogers, *Let There Be Light,* p. 11. See also "The Intolerable Burden and Cost of Needless and Senseless Labor Strikes," *Industrial Management, The Engineering Magazine,* LII (January 1917), 433; W. J. Arnold, "Famous Firsts: Shangri-la of the Tycoon Era," *Business Week,* October 3, 1964, p. 142, and "A New National Industrial Organization," *Iron Age,* XCVIII (1916), 1118.

24. Rogers, *Let There Be Light,* pp. 15–25.
    New York Times, May 17, 1916, editorial.

25. See "The Need for Executives," *New York Times,* November 30, 1923, p. 2.

26. L. L. Thurstone, "Vocational Guidance Service," Anthropology and Psychology Division, NRC Archives.
    Arthur M. Greene, Jr., "Presidential Address," *Society for the Promotion of Engineering Education Proceedings,* 1920, pp. 28–39; Roy D. Chapin, "Cooperation Between Education and Industry from the Viewpoint of the Manufacturer," *Society for the Promotion of Engineering Education Proceedings,* 1920, pp. 41–51; William E. Wickenden, "The Engineer as a Leader in Business," *Society for the Promotion of Engineering Education Proceedings,* 1920, p. 477.

27. Charles F. Scott, "Report of the Chairman of the Board of Investigation and Coordination," in *Report of the Investigation of Engineering Education* (Pittsburgh: Society for the Promotion of Engineering Education, 1930), I, 1–16.
    See *Society for the Promotion of Engineering Education Proceedings,* XXX (1922), 10.

28. Magnus Alexander, telegram to the Society for the Promotion of Engineering Education, *Society for the Promotion of Engineering Education Proceedings,* 1923, p. 34. See also "Minutes of Meeting of the Division of Deans and Administrative Officers," *Journal of Engineering Education,* XV (1925), 459–79; Magnus Alexander, "The Objective in Engineering Education," *Transactions of the American Society of Civil Engineers,* LXXXVI (1923), 1256–9; and Magnus Alexander, *Eighth Annual Report of the N.I.C.B.* (1924), pp. 30–1.

29. Charles F. Scott, "Professional Engineering Education for the Industries," *Canadian Engineer,* XLII (January 3, 1922), 44.

30. Scott, "Report of the Chairman of the Board of Investigation and Coordination," pp. 1–16; Charles F. Scott, letter to Elmer Lindset, 1929, Wickenden Papers, Case Western Reserve University Archives.

31. *Report of the Investigation.* See also Matthew Elias Zaret, "An Historical Study of the Development of the American Society for Engineering Education," unpublished Ph.D. dissertation, New York University, 1967, pp. 125–40; and Ian Braley, "The Evolution of Humanistic-Social Courses for Undergraduate Engineers," unpublished Ph.D. dissertation, Stanford University School of Education, 1961, pp. 123–32.

32. See, for example, A. C. Jewett, "The Engineering Graduate in Industry," *Society for the Promotion of Engineering Education Proceedings,* XXXII

(1924), 424; and Magnus Alexander, "The Problem of Engineering Education from the Standpoint of American Industry," *Society for the Promotion of Engineering Education Proceedings,* XXXIV (1926), 586.

33. On NELA, see Carl D. Thompson, *Confessions of the Power Trust* (New York: E. P. Dutton and Co., 1932), pp. 330, 392, and *Survey of the Instruction in Public Utilities in Colleges and Universities, of the Industry's Interest in College Graduates, and the Willingness and Ability of Utilities to Cooperate with Higher Educational Institutions,* Report of the Educational Members of the Cooperation with Educational Institutions Committee, NELA, 1929, pamphlet, Massachusetts Historical Society. On Bell, see the Series of Bell System Educational Conferences: *Electrical Engineering and Communications* (1924), *Mechanical Engineering* (1925), and *Economics and Business* (1926). On instruction in industrial relations, see *The Teaching of Labor Relations in Engineering Schools: An Informal Conference of Engineering Educators at the Home of Sam A. Lewisohn* (privately printed, 1928), Dewey Library (Sloan School), MIT.

34. Robert I. Rees, "The Professional Development of the Engineer," *Electrical Engineering,* LII (February and December 1933), 129, 932; Robert I. Rees, "The Cooperation of the Society for the Promotion of Engineering Education with Other Engineering Societies," *Journal of Engineering Education,* XXIII (September 1932), 36–42; "The Engineers Council for Professional Development," *Transactions of the American Society of Mechanical Engineers,* LIV (1932), R-1-6.

35. "Cooperation Between Industry and the Schools," *National Association of Corporation Training Bulletin,* 1921, pp. 447–9.

    Charles R. Mann, "American Council on Education Director's Report, 1927" (typescript), ACE Archives.

36. Thorstein Veblen, *The Higher Learning in America* (New York: B. W. Huebsch, 1918), pp. 220, 222, 224.

37. Mann, "American Spirit in Education," pp. 59.

38. *Ibid.,* pp. 59–63.

39. *Ibid.,* p. 63.

40. "Report of the Organization of the Emergency Council on Education" (typescript), ACE Archives.

    Samuel P. Capen to Mrs. Capen, July 22, 1918.

41. "Minutes of the Meeting of the Emergency Council on Education, March 26–27, 1918," ACE Archives.

    Charles G. Dobbins, *American Council on Education, Leadership and Chronology, 1918–1968* (Washington, D.C.: American Council on Education, 1968), pp. 7–9.

42. Samuel P. Capen to Mrs. Capen, July 21, 1919.

43. Capen, quoted in Dobbins, *American Council on Education,* pp. 7–8.

44. Capen to P. P. Claxton, October 8, 1919; Capen to Cousens, July 23, 1920; Capen to J. W. Dietz, July 21, 1919, Capen Papers.

45. Dobbins, *American Council on Education;* see also *A Brief Statement of the History and Activities of the American Council on Education, 1918–1953,* ACE pamphlet, 1953, ACE Archives.

46. Samuel P. Capen, Director's Report, 1922, quoted in Dobbins, *American Council on Education,* p. 9.

47. Charles R. Mann, "The American Council on Education," *Educational Record,* I (1920), 134.
48. Dobbins, *American Council on Education,* p. 8.
    Samuel P. Capen, quoted in *ibid.,* p. 7.
    Capen to J. H. MacCracken, April 7, 1921, Capen Papers; Dobbins, *American Council on Education,* p. 7.
49 Charles R. Mann to Henry Suzzalo, April 7, 1927, Suzzalo Papers, University of Washington Archives.
50. Charles R. Mann, Director's Report, American Council on Education, 1929, quoted in Dobbins, *American Council on Education,* p. 17.
51. Minutes of the Executive Council, American Council on Education, March 13, 1926; January 8, 1927; September 4, 1927; March 16, 1928, ACE Archives.
52. Mann, Director's Report, 1926, ACE Archives.
    Charles R. Mann to Henry Suzzalo, April 7, 1927, Suzzalo Papers.
53. Gildersleeve, quoted in Dobbins, *American Council on Education,* p. 14.
54. *Ibid.,* p. 11.
54. C. J. Tilden, "The American Council on Education," *Society for the Promotion of Engineering Education Proceedings,* XXXII (1924), 539–40.
55. Samuel P. Capen, "Director's Report," *Educational Record,* I (1920), 33; Samuel P. Capen, "Educational Standardization," *School and Society,* XIII (February 19, 1921), 223.
56. Dobbins, *American Council on Education,* pp. 7, 10, 12, 15, 18.
57. Charles R. Mann, "Director's Report, 1927," ACE Archives.
    Samuel P. Capen to George Zook, March 16, 1920, ACE Archives; Capen, quoted in *Colleague,* University of Buffalo, V, 7; Capen Papers.
58. Mann, "Director's Report, 1927," ACE.
59. Suzzalo, "Effective American University System," pp. 157–61.
60. Charles R. Mann, "Job Specifications," *Educational Record,* Supplement No. 5 (October 1927).
61. Capen, "Educational Standardization," p. 224.
62. Mann, quoted in Dobbins, *American Council on Education,* p. 14.
63. *Ibid.,* pp. 9–18; Minutes of the Executive Council of the American Council of Education, January 15, 1926.
64. Minutes of the Executive Council, American Council of Education, September 25, 1926.
65. Thompson, *Confessions of the Power Trust,* p. 392.

Chapter 10

1. Thomas Edison, quoted in an editorial, *Industrial Management and Engineering Magazine,* October 1920, p. 4.
2. Dexter Kimball, "The Relation of Engineering to Industrial Management," *Journal of the American Society of Mechanical Engineers,* February 1918, pp. 559, 563.
3. Charles Babbage, quoted by Leon P. Alford, "The Present State of the

Art of Industrial Management," *American Society of Mechanical Engineers Transactions,* XXXIV (1912), 1135.

4. Alfred D. Chandler, *Strategy and Structure: Chapters in the History of Industrial Enterprise* (Cambridge: MIT Press, 1962), p. 37.

5. *Ibid.,* p. 23.

6. Reinhard Bendix, *Work and Authority in Industry: Ideologies of Management in the Course of Industrialization* (New York: Wiley, 1956), p. 266.

7. *The Review,* October 1910, p. 35, cited in *ibid.,* p. 271.

8. Harlow S. Person, "Basic Principles of Administration and Management: The Management Movement," in Henry Metcalf, *Scientific Foundations of Business Administration* (New York: Williams & Wilkins Co., 1926), p. 201.

9. Henry Towne, "The Engineer as an Economist," *American Society of Mechanical Engineers Transactions,* VII, 428.

10. The five-phased scheme used here to describe scientific management is a combination and modification of two schemes, one offered by Leon Alford and Alexander H. Church in "The Principles of Management," *American Machinist,* XXXVI, 857, and another by Harlow S. Person in *Graphical Analysis of Scientific Management* (New York, 1944) and in his "Basic Principles of Administration and Management," p. 202. The following history of the scientific management movement is based upon: Hugh G. J. Aitken, *Taylorism at Watertown Arsenal* (Cambridge: Harvard University Press, 1960); Samuel Haber, *Efficiency and Uplift* (Chicago: University of Chicago Press, 1964); Milton Nadworny, *Scientific Management and the Unions, 1900–1932* (Cambridge: Harvard University Press, 1955); Lyndall Urwick, *The Making of Scientific Management* (London: Management Publication Trust, 1945).

11. Frederick W. Taylor, "Why Manufacturers Dislike College Students," *Society for the Promotion of Engineering Education Proceedings,* XVIII (1909), 87. Quoted in Nadworny, *Scientific Management and the Unions,* p. 9.

12. Nadworny, *Scientific Management and the Unions,* p. 10.

13. Urwick, *Making of Scientific Management,* I, 170.

14. Monte Calvert, *The Mechanical Engineer in America, 1830–1910* (Baltimore: Johns Hopkins University Press, 1967), pp. 235–43.

15. See, for example, the provocative unpublished studies by David Montgomery, "The 'New Unionism' and the Transformation of Workers' Consciousness in America, 1909–1922," and "Immigrant Workers and Scientific Management." Montgomery is at the University of Pittsburgh.

16. See Taylor's testimony before the Special House Committee in Frederick W. Taylor, *Scientific Management* (New York: Harper and Brothers, 1947), pp. 79–85.

17. American Society of Mechanical Engineers Committee, quoted in Layton, *Revolt of the Engineers* (Cleveland: Case Western Reserve University, 1971), p. 141. Layton has a good discussion of the engineers' view of Taylorism.

18. Gantt, quoted in Lyndall Urwick, "Management's Debt to Engineers," *Advanced Management,* December 1952, p. 8.
    For more biographical information on scientific management leaders,

see Lyndall Urwick, *The Golden Book of Management* (London: New-man Neame, 1956).

19. Urwick, "Management's Debt to Engineers," p. 11.

   See, for example, Lillian M. Gilbreth, *The Psychology of Management* (New York, 1914); Lillian M. and Frank B. Gilbreth, "The Three Position Plan of Promotion," *Annals of the American Academy of Political and Social Science,* 1916; and Lillian M. and Frank B. Gilbreth, "Discussion," *American Society of Mechanical Engineers Transactions,* XLIV (1922), 1286.

20. Nadworny, *Scientific Management and the Unions,* p. 111. See also Bendix, *Work and Authority in Industry,* p. 285.

21. See Leon P. Alford, "Ten Years' Progress in Management," *American Society of Mechanical Engineers Transactions,* 1922.

22. Haber, *Efficiency and Uplift,* p. 163.

23. Wesley Mitchell, "A Review," *Recent Economic Changes* (New York: McGraw-Hill Book Co., 1929), II, 864. For further discussion of the corporate liberal world-view, see James Weinstein, *The Corporate Ideal in the Liberal State* (Boston: Beacon Press, 1968), and William Appleman Williams, *The Contours of American History* (Cleveland: World Publishing Company, 1961), Part III, "The Age of Corporation Capitalism."

24. Charles M. Ripley, *Life in a Large Manufacturing Plant* (Schenectady: General Electric Company Publication Bureau, 1919), pp. 97–100.

25. See David Loth, *Swope of GE* (New York: Simon & Schuster, 1958), and Chandler, *Strategy and Structure,* pp. 46, 365, 368.

26. Ernest Dale and Charles Meloy, "Hamilton M. Barksdale and the Du Pont Contribution to Systematic Management," *Business History Review,* XXXVI (1962), 127–52.

27. Chandler, *Strategy and Structure,* Chapter 2.

28. Chandler, *Strategy and Structure,* Chapter 3. See also Alfred P. Sloan, *Adventures of a White-Collar Man* (New York: Doubleday, Doran and Co., 1941) and *My Years with General Motors* (New York: Doubleday & Co., 1964).

29. Chandler, *Strategy and Structure,* pp. 314–20.

30. Harlow Person, "Principles and Practice of Scientific Management," quoted in "Ten Years' Progress in Management," 1932, *American Society of Mechanical Engineers Transactions,* p. MAN-1-9.

31. David W. Eakins, "The Origins of Corporate Liberal Policy Research, 1916–1922: The Political-Economic Expert and the Decline of Public Debate," in Jerry Israel, ed., *Building the Organizational Society* (New York: Free Press, 1972), pp. 166–7, 176, 179.

32. Morris L. Cooke, "Public Engineering and Human Progress," *Journal of the Cleveland Engineering Society,* IX (January 1917), 252. Quoted in Layton, *Revolt of the Engineers,* p. 179.

   Charles R. Mann, "Director's Report, 1927," ACE Archives. For more on Hoover, see Murray M. Rothbard, "Herbert Hoover and the Myth of Laissez-Faire," in Ronald Radosh and Murray N. Rothbard, eds., *A New History of Leviathan* (New York: E. P. Dutton & Co., 1972), pp. 111–45; William A. Williams, "Some Presidents," *New York Review of Books,* November 5, 1970; Ellis W. Hawley, "Herbert Hoover: The Commerce Secretariat, and the Vision of an 'Associative State,' 1921–1928," *Journal of American History,* LXI (June 1974), 116–40.

33. Quoted in Henry Eilbert, "The Development of Personnel Management in the United States," *Business History Review,* XXXIII (1959), 351.

34. Quoted in Paul H. Douglas, "Plant Administration of Labor," *Journal of Political Economy,* XXVII (1919), 545. See also Louise C. Odencrantz, "Personnel Work in America," *Personnel Administration,* 1922, pp. 5–15; Haber, *Efficiency and Uplift,* p. 64; and "Welfare Work for Employees in Industrial Establishments in the U.S.," *U.S. Bureau of Labor Statistics Bulletin No. 250* (1919).

35. Leon P. Alford, "The Status of Industrial Relations," *American Society of Mechanical Engineers Transactions,* 1912, p. 164.

    Thorstein Veblen, *The Theory of Business Enterprise,* quoted by Loren Baritz, *The Servants of Power* (Middletown: Wesleyan University Press, 1960), p. 1

    Ford Times, July 1916, p. 549. Quoted in Samuel M. Levin, "Ford Profit-Sharing, 1914–1920," *Personnel Journal,* VI (1927), 82.

    Ripley, *Life in a Large Manufacturing Plant,* p. 45.

36. See *Proceedings of the Third National Conference on Industrial Accidents, and Workmen's Compensation,* Chicago, 1910, p. 5; "Workmen's Compensation Laws in the United States and Foreign Countries," *U.S. Bureau of Labor Statistics Bulletin, No. 126* (1914), p. 12.

    Alford, "Status of Industrial Relations," pp. 170–1.

    Proceedings of the Second Industrial Safety Congress of New York State, 1917, p. 3; *Proceedings of the First Cooperative Safety Congress,* Milwaukee, 1912, p. 205–8.

37. *Ibid.,* p. 227.

38. Alford, "Status of Industrial Relations," p. 174.

39. *Proceedings of the First Cooperative Safety Congress,* p. 227.

40. Malcolm Rorty, quoted by Mitchell, "A Review," p. 864.

41. Mitchell, "A Review," p. 865.

42. E. H. Fish, "Human Engineering," *Journal of Applied Psychology,* I (1917), 174.

43. Ted Cox, "Fitting the Right Man to the Right Job," *American Business,* XX (February 1950), 44.

44. Henry D. Hammond, "Americanization as a Problem in Human Engineering," *Engineering News-Record,* 1918, p. 1116.

45. Henry S. Dennison, "Management," *Recent Economic Changes* (McGraw-Hill Book Co., 1929), I, 518.

    "The Personnel Content of Management," *American Management Review,* April 1923, p. 5.

46. Alford, "Status of Industrial Relations," p. 172.

    David Montgomery, " 'New Unionism,' " p. 8; Leon C. Marshall, "The War Labor Program and Its Administration," *Journal of Political Economy,* XXVI (May 1918), 429, cited in Montgomery, "New Unionism," p. 8.

47. H. F. J. Porter, cited in Alford, "Status of Industrial Relations," p. 172.

    Magnus W. Alexander, "Hiring and Firing," *American Industries,*

August 1915. (Also found in *Annals of the American Academy of Political and Social Science,* May 1916.)

Boyd Fisher, "How to Reduce the Labor Turnover," *Annals of the American Academy of Political and Social Science,* LXXI (May 1917), 10, 14–16, 19. See also Douglas, "Plant Administration of Labor," p. 545.

N. D. Hubbell, "The Organization and Scope of the Employment Department," *U.S. Department of Labor Statistics Bulletin No. 227* (1917), pp. 97–111; see also "Proceedings of the Employment Managers' Conference, Philadelphia," *U.S. Bureau of Labor Statistics Bulletin No. 227* (October 1917), p. 30.

48. Alford, "Status of Industrial Relations," p. 515; Eilbert, "Development of Personnel Management," pp. 351–4; Douglas, "Plant Administration of Labor," pp. 545, 547–8.

49. Joseph H. Willits, "Development of Employment Managers' Associations," *Monthly Review of the Bureau of Labor Statistics,* V (September 1917), 497–9.

"Proceedings of the Employment Managers' Conference, Philadelphia."

Hugo Munsterberg, quoted in Baritz, *Servants of Power,* pp. 26–7.

Ibid., pp. 35–41.

50. Baritz, *Servants of Power,* pp. 35–41.

Walter Dill Scott, "Vocational Selection at the Carnegie Institute of Technology," *U.S. Bureau of Labor Statistics Bulletin No. 227* (1917), pp. 115–19.

51. Alford, "Status of Industrial Relations," p. 172.

Clarkson, *Industrial America in the World War* (Boston: Houghton, 1924), p. 130.

52. "Proceedings of the Employment Managers' Conference, Rochester, New York, May, 1918," *U.S. Bureau of Labor Statistics Bulletin No. 247* (1919), pp. 6–8.

Ibid. See also *Report of the President, 1917–1918,* University of Rochester Archives, Rochester, New York; *The Campus,* April 18, 1918, University of Rochester Archives; *The Democrat and Chronicle,* Rochester, New York, May 9, 1918.

53. See *Personnel Administration,* May 1922, pp. 1–21; "Mental Science the Basis for Individual Efficiency," *National Association of Corporation Training Bulletin,* VIII (January 1921), 1–5; "The Personnel Content of Management," *American Management Review,* April 1923, pp. 1–5.

54. For more on the history of industrial education, see Charles A. Prosser, "The New Apprenticeship as a Factor in Reducing Labor Turnover," *U.S. Bureau of Labor Statistics Bulletin No. 196* (1916), pp. 45–52; Berenice Fisher, *Industrial Education: American Ideals and Institutions* (Madison: University of Wisconsin Press, 1967); Arthur G. Wirth, *Education in the Technological Society: The Vocational-Liberal Studies Controversy in the Early Twentieth Century* (Scranton: International Textbook Co., 1972); Melvin F. Barlow, *History of Industrial Education in the United States* (Peoria: Charles A. Bennett Co., 1967); Charles A. Bennett, *History of Manual and Industrial Education, 1870–1917* (Peo-

ria: Manual Arts Press, 1917); Abraham Flexner *et al., The Gary Public Schools* (New York: General Education Board, 1918), 2 vols. See also Joel Spring, *Education and the Rise of the Corporate State* (Boston: Beacon Press, 1973).

55. Henry S. Pritchett, "The Place of Industrial and Technical Training in Popular Education," *Educational Review,* XXIII (March 1902), 282.

Milton P. Higgins, "Address," *Bulletin No. 1* of the National Society for the Promotion of Industrial Education (January 1906), p. 15.

56. Pritchett, "Place of Industrial and Technical Training," pp. 292, 294.

Dugald C. Jackson, "The Availability of Correspondence Schools as Trade Schools," *Society for the Promotion of Engineering Education Proceedings* (1901), pp. 97–111; James Mapes Dodge, "The Money Value of Technical Training," *Transactions of the American Society of Mechanical Engineers,* XXV, 40–8; Charles R. Richards, "The Problem of Industrial Education," *Manual Training Magazine,* VIII (April 1907), 125–32; Frank Vanderlip, "The Urgent Need of Trade Schools," *World's Work,* XII (June 1906), 761–824; Frank Vanderlip, "The Economic Importance of Trade Schools," *National Education Association Proceedings,* 1905, pp. 141–5; Magnus W. Alexander, "Plans to Provide Skilled Workmen," *Transactions of the American Society of Mechanical Engineers,* XXVI (November 1906), 487–502; Magnus W. Alexander, "The Needs of Industrial Education from the Standpoint of the Manufacturer," *Social Education Quarterly,* I (June 1907), 196–201.

Frederick P. Fish, "The Vocational and Industrial School," *National Education Association Proceedings,* 1910, pp. 367–78.

John Dewey, "Industrial Education—A Wrong Kind," *New Republic,* II (February 20, 1915), 72.

57. Magnus W. Alexander, "The Apprenticeship System of the General Electric Company," *Annals of the American Academy of Political and Social Science,* XXXIII (January 1909), 141–50.

58. See Bennett, *History of Manual and Industrial Education,* Chapter 13.

59. William C. Redfield, letter to National Association of Corporation Schools, June 8, 1914, *National Association of Corporation Schools Bulletin,* II (1914), 686.

Charles Steinmetz, "Address," *National Association of Corporation Schools Bulletin,* I, 425.

Channing R. Dooley, "Education and Americanization," *Industrial Management,* October 1917, p. 50.

60. W. W. Kincaid, "Education and Training as a Management Process," *American Management Review,* May 1923, pp. 3–4. See also John Van Liew Morris, *Employee Training: A Study of Educational Training Departments in Various Corporations* (New York: McGraw, 1921), and Nathaniel Peffer, *Educational Experiments in Industry* (New York: The MacMillan Co., 1932).

61. Charles R. Mann, "Principles Underlying Effective Training of Employees," *Corporation Training,* March 1922, pp. 3–5.

62. Charles R. Mann, "Personnel Methods and Training," *Personnel Administration,* September 1922, p. 3.

63. Henry H. Tukey, "Educational Engineers," *National Association of Corporation Training Bulletin,* 1921, p. 386.

64. Channing R. Dooley, "Employment on Merit," *American Management Review,* December 1923, p. 3.

65. Hammond, "Americanization as a Problem," p. 1116. See also "Making Americans out of Aliens," *National Association for Corporation Schools Bulletin,* V (1918), 348; "A Survey of Our Association's Activities," *National Association for Corporation Schools Bulletin,* IV (March 1917), 8; Kimball, "Relation of Engineering to Industrial Management," p. 562; Frederick L. Bishop, "Editorial," *Society for the Promotion of Engineering Education Bulletin,* VII (January 1917), 511–12.

66. Prosser, "New Apprenticeship," pp. 45, 51.

67. Scott Nearing, "Who's Who on Our Boards of Education," *School and Society,* V (January 1917), 89–90. See also George S. Counts, *The Social Composition of Boards of Education: A Study in the Social Control of Public Education,* Supplementary Monograph No. 33, July 1927 (Chicago: University of Chicago Press, 1927).

68. Bennett, *History of Manual and Industrial Education,* pp. 539–41; "Report of the Massachusetts Committee on Industrial and Technical Education," *School Review,* XIV (June 1906), 438–48.

69. See, for example, George S. Counts, *School and Society in Chicago* (New York: Harcourt, Brace, and Co., 1928); Wayne J. Urban, "Teachers, Politics, and Progressivism: The Early Years of the Atlanta Public School Teachers' Association, 1905–1919," unpublished paper presented to the Mid-West History of Education Society, Chicago, 1973 (courtesy of Wayne Urban); Ernest Lindler, "The Universities and the People," *School and Society,* XXII (December 1925), 700. On public school reform, see also Marvin Lazerson, *Origins of the Urban School: Public Education in Massachusetts, 1870–1915* (Cambridge: Harvard University Press, 1971); Joel Spring, *Education and the Rise of the Corporate State* (Boston: Beacon Press, 1973); Raymond E. Callahan, *Education and the Cult of Efficiency* (Chicago: University of Chicago Press, 1962); David B. Tyack, "City Schools: Centralization of Control at the Turn of the Century," in Jerry Israel, ed., *Building the Organizational Society* (New York: Free Press, 1972), Chapter 4; David B. Tyack, *One Best System: A History of American Urban Education* (Cambridge: Harvard University Press, 1974); David B. Tyack, ed., *Turning Points in American Educational History* (New York: Wiley, 1967); Clarence J. Karier, ed., *Shaping the American Educational State, 1900 to the Present* (New York: Free Press, 1975); and Clarence J. Karier and Paul C. Violas, eds., *Roots of Crisis: Essays in Twentieth Century Education* (Chicago: Rand McNally, 1973).

70. Roman, quoted in *National Association of Corporation Schools Bulletin,* II (October 1915), 23.

71. Young, quoted in *ibid.,* p. 24.

72. Counts, *School and Society,* Chapter 8.

73. "Proceedings of the Organizational Meetings," *National Society for the Promotion of Industrial Education Bulletin,* No. 1 (1907), pp. 1–3, 15; "List of Members," *National Society for the Promotion of Industrial*

Education Bulletin, No. 7 (1908), p. 2. James Parton Haney, "The National Society for the Promotion of Industrial Education," *Manual Training Magazine,* XI (1910), 27–35.

74. Henry S. Pritchett, "The Arms of the National Society for the Promotion of Industrial Education," *National Society for the Promotion of Industrial Education Bulletin* No. 5 (1908), p. 22.

"National Society for the Promotion of Industrial Education Constitution," *National Society for the Promotion of Industrial Education Bulletin* No. 1, p. 10.

75. Wirth, *Education in the Technological Society,* pp. 160, 165. Grant Venn, *Man, Education, and Work* (Washington, D.C.: American Council on Education, 1964), p. 112.

76. William E. Wickenden, "Discipline or Discipleship?", Steinmetz Lectures, American Institute of Electrical Engineers, March 7, 1930, Wickenden Papers.

77. Cheeseman Herrick, "Commercial Education: Training of Business Men as a Branch of Technical Instruction," *Fifth Yearbook of the National Herbart Society,* 1899, p. 119.

78. O. H. Cheney, discussion in *The Teaching of Labor Relations in Engineering Schools: An Informal Conference of Engineering Educators at the Home of Sam A. Lewisohn* (privately printed, 1926), Harvard University Library.

79. Wickenden, "Discipline or Discipleship?"

80. Leon Alford, "Report of the Conference on Industrial Engineering, Schenectady, June 1925," *Journal of Engineering Education,* XVI (1925–6), 237.

Charles R. Mann, quoted by Fred H. Rindge, Jr., "Teaching Human Engineering in the College Curriculum," *American Machinist,* XLVIII (April 25, 1918), 714; Charles R. Mann, "Present-Day Conditions are Forcing the Engineer to Assume New Responsibilities," *Engineering News-Record,* LXXX (January 1918), 208.

Charles F. Scott, discussion in *Teaching of Labor Relations,* 1926.

Sam A. Lewisohn, *New Leadership in Industry* (New York: E. P. Dutton & Co., 1926), p. 93.

81. H. P. Hammond, discussion in *Teaching of Labor Relations,* 1926.

82. Leon P. Alford, "Ten Years' Progress in Management"; C. W. Beese, "Extension of Education in Engineering Management to Fields Other Than Factory Production," *Journal of Engineering Education,* XVI (1925–6), 240–52.

83. Erwin Schell, "Quiz," January 8, 1921, and "The Trend of Management," one of a series of lectures entitled "The Technique of Executive Control," *Course Record in Business Management, 1920,* MIT Archives.

84. Erwin Schell, "The Workmen—Their Impulses and Desires," *Course Record in Business Management, 1920,* MIT Archives.

85. Erwin Schell, "Executive Traits Which May Be Developed," *Course Record in Business Management, 1920,* MIT Archives.

86. Alford, "Report of the Conference on Industrial Engineering, 1925," pp. 239, 244. Lionel S. Marks, "The Five-Year Program in Engineering and Business Administration at Harvard University," *Society for the Promo-*

tion of Engineering Education Proceedings, XXXI (1923), 166–69. O. N. Leland, "Discussion," *Society for the Promotion of Engineering Education Proceedings,* XXXI (1923), 170. See also John B. Rae, "Engineering Education as Preparation for Management: A Study of MIT Alumni," *Business History Review,* XXIX (1955), 64–79; and Erwin H. Schell, "Trends in the Teaching of Management at MIT.," *Journal of Engineering Education,* XXVII (1938), 455–9.

87. William Wickenden, quoted by Ian Braley, "The Evolution of Humanistic-Social Courses for Undergraduate Engineers," unpublished Ph.D. dissertation, Stanford University School of Education, 1961, p. 115. For more on the evolution of the humanities-social stem of engineering education, see H. P. Hammond, "A Study of Evolutionary Trends in Engineering Curriculum," *Journal of Engineering Education,* XVIII (1927), 57–84; H. P. Hammond, "Report on the Humanistic-Social Studies in Engineering Education," *Journal of Engineering Education,* XXXVI (1946), 338–51; William Wickenden, "The Humanistic Band in the Engineering Curriculum (1939); "Memorandum of Humanistic and Cultural Interests in Technological Institutions" (1944); and "Memorandum on Background of Movements to Broaden Curricula on the Humanistic Side" (1946)—all in the Wickenden Papers, Case Western Reserve University Archives. See also Wickenden's "The Social Sciences and Engineering Education," *Mechanical Engineering,* February 1938, pp. 147–50. For a comprehensive bibliography on the subject, see Edwin S. Burdell *et al., The Humanistic Social Stem of Engineering Education* (New York: Cooper Union, 1955).

88. Robert I. Rees, discussion in *Teaching of Labor Relations in Engineering Schools,* 1926. See also the extensive surveys of practicing engineers in the Wickenden Investigation. When asked to indicate what subjects they felt should be included, or enlarged, in the curriculum, a significant majority cited training in management or the social sciences in general.

89. Don D. Lescohier, "The Place of the Social Sciences in the Training of Engineers," *Journal of Engineering Education,* XXIV (1933–4), 417.

90. Samuel Stratton, quoted in Henry C. Link, *Education and Industry* (New York: MacMillan Co., 1923), p. 168.

91. Willard E. Hotchkiss, "Social Sciences in Engineering Schools," *Journal of Engineering Education,* XXVI (1935), 94.

92. William E. Wickenden, "The Social Sciences and Engineering Education," address of the retiring chairman of the American Association for the Advancement of Science, Engineering Section, December 29, 1937, Wickenden Papers; "Humanistic Band in the Engineering Curriculum," n.d., Wickenden Papers. "Report of Committee No. 14, Business Training for Engineers," *Society for the Promotion of Engineering Education Proceedings,* XXXI (1923), 113. See also "The Engineer as a Leader in Business," *SPEE Proceedings,* XXXI (1923), 109.

93. Minutes of the Division of Engineering, January 14, 1924, NRC Archives.

94. Dugald C. Jackson, "Lighting in Industry," *Journal of the Franklin Institute,* CCV (March 1928), 289–302. See also his reports to the president on the activities of the Electrical Engineering Department, *Report*

of the President, 1925–1930, MIT Archives, and his "Reports of Work in Progress," Engineering Division Files, NRC Archives.
    Baritz, *Servants of Power,* Chapters 5 and 6.
95. Jackson, "Lighting in Industry," p. 302.
96. See, for example, Wickenden's *Professional Guide for Junior Engineers.* For a more recent example, see "The Crown Princes of Business," *Fortune,* XLVIII (October 1953), 152.
97. William Wickenden, "Final Report of the Director of Investigations," *Report of the Investigation of Engineering Education,* II, 1056, 1059, 1060.
98. *Ibid.,* II, 1059.

Epilogue

1. See Peter F. Drucker, "The First Technological Revolution," *Technology and Culture,* VII (1966), 143.
2. "The Crown Princes of Business," *Fortune,* XLVIII (October 1953), 264, 270.
3. A. A. Potter, Letter to President Roosevelt, *New York Times,* October 25, 1936, p. 1.
4. On the proletarianization of engineers, see André Gorz, "Technical Intelligence and the Capitalist Division of Labor," *Les Temps Modernes,* Summer 1971 (also in *Telos,* Vol. XII); "Technicians and the Capitalist Division of Labor," *Socialist Revolution,* Vol. II, No. 3 (May–June 1972); and Britta Fischer and Mary Lesser, "An Examination of Some Myths and Contradictions Concerning Engineers," *Science for the People,* II, 16–19. See also Fischer's Ph.D. dissertation, Department of Sociology, Washington University, 1976 (forthcoming).

Index

Rice, E. Wilbur, 113–14, 127, 191, 237–8, 278
Richards, C. Russ, 131–2
Richards, Charles R., 302, 309
Ricketts, Palmer, 25
Rochester, University of, 299
Rochester Mechanics Institute, 197 *n.*
Rockefeller, John D., 126
Rockefeller, John D., Jr., 254
Rockefeller Foundation, 126–7, 153, 155, 164, 212, 233, 255
Roe, J. W., 239, 277
Roman, Frederick, 308
Roosevelt, Franklin D., 64, 272
Root, Elihu, 110, 126, 151–3, 155–7
Rorty, Malcolm, 284–5, 291
ROTC (Reserve Officers Training Corps), 225–6
rubber industry, 117
Rumford, Count, 21

Sadtler, Samuel P., 38
SAE (Society of Automobile Engineers), 79
safety, industrial, 289–90, 305
SATC (Student Army Training Corps), 217–23, 245
Schell, Erwin, 288 *n.*, 294 *n.*, 298 *n.*, 302 *n.*, 313–15
Schneider, Herman, 183, 185–9, 212, 214, 215, 239, 302
School of Chemical Engineering Practice (MIT), 141, 193–4
schools, *see* colleges and universities; corporation schools; engineering education; engineering schools; technical education
Schumpeter, Joseph, 118
science: applied, 4, 24–6, 112, 114–15, 140–1, 147, 151, 152; capitalism and, 3; engineering and, integration of, 159; industrial monopolization and control of, 6, 43, 109, 110, 114, 118; pure, 112, 141, 151, 152; social, 170, 242, 274, 275, 322, 323; social, in engineering curriculum, 31, 316–18; useful arts and, union of, 3–4, 24; *see also* professional engineer; science-based industry; technical education; technology
science-based industry: definition of, 5; rise of, 3–19, 33; *see also specific industries*
Science Service, 128 and *n.*

scientific discovery and investigation, corporate control of, 110, 114, 118
scientific management (Taylorism), 40, 209, 264–78, 314; corporate liberal management reform and, 277–8; engineers and, 265–6, 273–4; government investigation of, 272; "human factor" in, 274–5; industrial psychology and, 297; industrial standardization and, 82–3; labor unions and, 264, 265, 267 *n.*, 271–3, 276; motivation of workers and, 266, 269, 274, 275; opposition and resistance to, 271–3; phases of, 267–8; revisionist movement within, 264, 265, 275–7; scientific pretensions of, 273–5; Taylor's contributions to, 268–71
Scott, Charles F., 46–8, 174–6, 189, 203 *n.*, 238, 243 *n.*, 312
Scott, Howard, 63
Scott, Walter Dill, 160, 183, 207–8, 215, 221, 230 *n.*, 239, 254, 275, 296–8
Seashore, Dean, 160, 233
selection and evaluation of college graduates, 171, 174, 176, 177, 187–8, 198–9, 201, 205; *see also* personnel and placement systems; recruitment of college graduates; tests
Sellers, William, 76, 82 *n.*
Sherman Act, 88
shop culture, 27, 40, 76, 77 *n.*
shop management, *see* scientific management
shop movement, 185
shop work in technical schools, 26, 28
Sibley College (Cornell University), 198, 276
simplification of production, 81–2
Skill, W. M., 181
skilled workers, demand for, 300–3
Skinner, Charles E., 78, 156
Sloan, Alfred P., 137, 282, 283, 313
Sloan, M. S., 182
Slosson, E. E., 127–8 and *n.*
Smith, Oberlin, 266
Smith-Hughes Act, 214, 310
Smithsonian Institution, 125
Smith-Towner Bill, 249
social engineering, 264, 265
socialism (socialist movement), 42 *n.*, 60–1
Socialist Party, 58, 60
socialization for management, 175–6
social production, *see* production